# Weed Control Methods for Public Health Applications

Editor

**Edward O. Gangstad**

Aquatic Plant Control Specialist
Army Corps of Engineers
Washington, D.C.

CRC Press, Inc.
Boca Raton, Florida

**Library of Congress Cataloging in Publication Data**

Gangstad, Edward O
Weed control methods for public health applications.

Bibliography: p.
Includes index.
1. Aquatic weeds——Control. 2. Aquatic weeds——
Biological control. 3. Vector control. I. Title.
SB614.G35 628.9 79-9836
ISBN 0-8493-5326-2



Direct all inquiries to CRC Press, Inc., 2000 N.W. 24th Street, Boca Raton, Florida 33431



International Standard Book Number 0-8493-5326-2

Library of Congress Card Number 79-9836
Printed in the United States

# PREFACE

This volume includes measures of control of aquatic vegetation that harms human health, since water-related diseases exist in this environment. Although malaria has receded internationally due to the combined chemotherapeutic-insecticidal programs, recently it has resisted both medicines and insecticide control. Active malaria cases in the U.S. were fewer than a dozen before the Vietnam War, but in 1973 the figure was about 700, almost all traceable to returning military personnel. The disease could again become prevalent. Other diseases exist whose transmission is indirectly affected by aquatic weed conditions including filariasis, and various trematodiases, especially from the schistosomes, Chinese liver fluke, cattle liver fluke, Guinea worm, giant intestinal fluke, Asiatic lung fluke, and broad tapeworm. Waterweeds also support disease-pest arthropods, i.e., snipe flies, tabanids (horse, gad, deer, and greenheads), Clear Lake gnats, May flies, black flies, sandflies, and sewage flies.

Ecosystem studies of impounded water research and development of herbivorous fish, and utilization of herbivorous fish in China, are also included in this volume.

**Edward O. Gangstad**

# THE EDITOR

**Dr. E. O. Gangstad,** Aquatic Plant Control Specialist, transferred to the Office of the Chief of Engineers in October of 1966 from the Texas Research Foundation, Dallas, where he was employed as Principal Agronomist of the Hoblitzelle Agricultural Laboratory. Dr. Gangstad was graduated from the University of Wisconsin with an M.A. in Biochemistry (1947) and from Rutgers University with a Ph.D. in Agronomy (1950) and minors in Plant Pathology and Plant Physiology.

During the past 25 years he has established an outstanding career in the environmental and biological fields and has over 80 scientific publications. He is a contributor to the *Journal of Biological Chemistry, Agronomy Journal, Crop Science, Journal of Heredity, Weed Science, Crops and Soils, Tropical Agriculture, Turrialba, Soil Science Society of Florida. Hoblitzelle Agricultural Bulletin, Wrightia, Hyacinth Control Journal, Environmental Quality Journal,* and the *Journals of the American Society of Civil Engineers.*

Dr. Gangstad's experience includes research and field studies on the physiology and pathology of plants at the Universities of Wisconsin and New Jersey; research and development studies on strategic long-vegetable fibers and related programs in Florida with the United States Department of Agriculture, Office of Naval Research, and Department of Defense; and agrobusiness studies to develop renewable natural resources of the State of Texas with the Texas Research Foundation and management of natural resources for the Department of the Army.

Dr. Gangstad is listed in *American Men of Science, Leaders in Science, American Men of Achievement* and the *International Who's Who of Intellectuals.*

# *IN MEMORIAM*

## Lyle W. Weldon

Dr. Lyle W. Weldon, one of the nation's foremost aquatic weed researchers, drowned February 1, 1970 while scuba diving in Lake Susannah on the Naval Training Base at Orlando, Florida. He was investigating suitable sites to conduct submersed weed research at the time of his death.

Dr. Weldon, native of Oregon, had worked for the USDA since 1965. After four years at Laramie, Wyoming, he transferred to Fort Lauderdale, Florida, in 1960 to work in aquatic weeds on cooperative studies of the U.S.D.A. and the Army Corps of Engineers. He earned a B.S. degree in Farm Crops from Oregon State University in 1955, the M.S. degree in Agronomy with minor in Botany from the University of Wyoming in 1965, and Ph.D. in Agronomy with minor in Plant Physiology and Statistics also from Wyoming in 1959. He took additional graduate work at North Carolina State College and at Oak Ridge Institute of Nuclear Studies.

Dr. Weldon authored or co-authored some 90 articles and a handbook. Many of these articles were published in *Weed Science* and *Hyacinth Control Journal.* Dr. Weldon was active in several regional and national organizations. He was chairman of the outstanding program sessions of the Aquatic and Marginal Weed Section at the Weed Science Society of America meeting in 1968 and was, for several years, a member of the Program Committee of the Southern Weed Conference. He helped organize the Hyacinth Control Society and, in 1969, received a citation for four years of outstanding service as editor of the Hyacinth Control Journal. Dr. Weldon had acted as an aquatic weed control consultant to the governor of the Panama Canal Zone and to the Suriname Aluminum Company on aquatic weed problems in South America. He was a Kiwanis Club member and a member of the Governor's Aquatic Weed and Research Committee in the State of Florida.

# TABLE OF CONTENTS

# *Part I*
# *Aquatic Plant Survey and Assessment*

Chapter 1

# AQUATIC WEEDS AND MAN'S WELL-BEING*

## INTRODUCTION

Aquatic vegetation harms human health in several ways since water-related diseases are part of our environment. Malaria has receded internationally due to chemo-therapeutic-insecticidal programs, but recently has resisted both medicines and insecticides. Lately, active malaria cases in the U.S. were kept to fewer than a dozen annually; in 1967, the figure was about 700, almost all traceable to returning military personnel. Other diseases exist whose transmission is indirectly affected by aquatic weed conditions: filariasis, and various trematodiases especially the schistosomes, Chinese liver fluke, cattle liver fluke, Guinea worm, giant intestinal fluke, Asiatic lung fluke, and the broad tapeworm.

Water weeds support the other disease pest arthropods: snipe flies, tabanids (horse, gad, deer, and greenheads), Clear Lake gnats, May flies, black flies, sandflies, and sewage flies.

Bacteria in the wrong place and in harmful population density degrade the health of urbanized man and as water pollutants require upgrading and maintenance of all water bodies. Lay understanding of conservation must include aquatic vegetation management. A pest plant should be removed mechanically, by chemical herbiciding, or by recently developed biocontrol techniques using several plant-eating snails, fishes, ducks, and manatee and aquatic insects. Weed removal allows the natural or introduced predators of larval stages of fluke infections to operate without mechanical hindrance near habitations in endemic areas. Seaweeds must be protected from coastal pollution to insure the future of a growing industry giving human and animal food, medicinals, mucilages, bacterial media, fertilizers, and various organic compounds. Marine Protista toxify edible shellfish and fish and seriously threaten consumers. Blooms (red tides) of the Protista somehow destroy fish and other marine life. Control of fresh water aquatic weeds strengthens the alliance between agriculture, navigation, and public health and will serve man's survival in the 21st Century.

What is the value of considering the relation of water weeds to man's physical and mental state? The subject forms no composite whole and will never form one. Yet the ancient problem can be viewed from angles usually overlooked. Current public acceptance of some unpleasant facts about increased human population, food and water supply, and air and water pollution spurs growing reverence for conservation of natural resources whether it be expressed as a birth control pill, development of fungus resistant wheat, conversion of salt to fresh water, regulated usage of exhaust reducers in vehicles, or the simple contemplation of the present condition of the stream in which one swam as a child. Within this melancholy framework, there is room to outline some ways natural plants living in or near water bodies affect man's health and equanimity. Such an art cuts across several disciplines of natural history, and the supporting data are hidden in many bibliographic niches and in the memories of biologists.[22]

Water then is our basic interest, and weeds (plants in the wrong place, wrong population density, and serving wrongly) are defined as pollutants. To the farmer whose

* By Frederick F. Ferguson, Chief, Tropical Disease Section, Ecological Investigations Program, National Communicable Disease Center, Health Services and Mental Health Administration, Public Health Service, U.S. Department of Health, Education and Welfare, San Juan, Puerto Rico. From Ferguson, F. F., *Hyacinth Control J.*, 7, 7, 1968. With permission.

pond use has been curtailed by an almost impenetrable cover of water lettuce (*Pistia stretioties*), each plant is a particulate pollutant. According to the lay concept, water pollution means sewage contamination firstly and industrial wastes secondly. According to the technical concept, the most important vital component of the debased aquatic ecosystem is the bacterial-mycotic complex.[27,32] Here we think of bacteria as aquatic plant life in the polluted maze. Although chemical control is a necessary part of current programs, the current list of known organisms useful as biocontrol agents of aquatic weeds[18] includes the snails, *Marisa* and two *Pomacea*; ducks—for example, black muscovies;[5] several fishes (species of *Tilapia* and the Chinese grass carp, *Ctenopharyngodon idella*); the beetle, *Agasicles,* consumer of alligator weed;[36] the aquatic wingless grasshopper, *Paulinia*, eater of water fern;[8] and the manatee which ingests both floating and rooted plants.

## SPECIFIC PROBLEMS

### Water Supply and Pollution

By the year 2000, we will need 35% more unpolluted water than will be available. The present supply must be maintained for use and re-use. The aquatic budget is limited. All we can do is locate water, decontaminate it, transport it, and be prepared to exert eternal vigilance for its proper maintenance.[2] Of the total annual precipitation in the U.S. in 1966, only 8% was used by man, 71% evaporated, and 21% was returned to the ocean.[32]

Although the term ''bacterial flora'' is glibly used to refer to microbiotic residents of the human gut, the public does not connote these key organisms as the plant life they are. The current pollution problem in American water bodies is an appalling example of the significance of bacteria whether the bacteria are pathogenic or not. The natural biological cycle in aquatic environments is still a fact. The key element in the normal cycle is oxygen; the key aquatic plant is the bacterium that converts soluble organic matter into other bacterial cells and inorganic elements. The inorganics are absorbed by algae and metabolized into other algal cells. Both cell types become food for protozoans, rotifers, and crustaceans. Some bacteria, algae, and animal life serve as basic food for minnows and young fish; small fish are devoured by large fish. Man discharges his wastes into the water body where bacteria metabolize organics. The cycle is then completed.[2,4,15,27,32] Thus, a certain amount of waste discharged into water bodies is normal and biologically required.

However, if organic waste concentrations increase over normal, bacterial populations explode with increased oxygen (BOD). As this life-giving element decreases, the higher animal forms perish — game fish first, then crustaceans, rotifers, and the higher protozoa.[32] Bacteria remain dominant and in the absence of dissolved oxygen (DO) undergo anaerobic metabolism producing vile odors and black water. Man thus fouls his aquatic ecosystem. That it required 100 years of varied pollution to bring Lake Erie to its current status as a massive cesspool is a monument to the strength of the elemental aquatic plant, water-fouling bacterium.[12] In contrast, management of fishing ponds and lakes is now an advanced art providing both beauty and recreation.[25]

### Taste-Odor Problem in Water Supply

Public acceptance of drinking water supplies, whether purified or not, hinges to a great degree on absence of detrimental tastes and odors. If poor taste and odor are of chemical origin (saline or sulphurous) the public is resigned to its fate; but if these are of microbiotic origin, great clamor is raised and the sanitary engineer is expected to defy nature and quickly bring the supply back to levels to which the user is accustomed.[15] Remarkably enough, publication of *Standard Methods* has stimulated codification and description of common odors that may emanate from the domestic tap.[4] These are spicy, balsamic, geranium, nasturtium, sweetish, violet, chlorinous, hydrocarbon, medicinal, sulferetted, earthy, peaty, grassy, musty, moldy, vegetable, septic, fish and pigpen! Abatement of taste-odors is not easy and at times impossible for a particular season.

Much accumulated information on the kinds of algae implicated exists, and recent studies show that the actinomycetes are potential troublemakers. Blooms of microorganisms die and release certain oils that cause undesirable taste-odor problems. Good algae and actinomycetes result in agricultural areas where run-off water carries fertilizer residues into impoundments increasing the phosphorous content.[7] In impoundments that receive treated or untreated sewage, algal problems may be increased by use of polyphosphates in household detergents. At certain population levels, algacides must be applied.[5] The most common one is copper sulfate applied at 0.1 to 1.0 ppm in only the upper 10 ft of the reservoir.[27] Since the algacide kills the protozoan predators of bacteria there is a temporary surge of these organisms. In treatment plants, activated carbon, chlorination, and aeration are indicated treatments to alleviate the problem should reservoir source treatment be insufficient.[15]

## Toxic Shellfish and Fish

Seaweed and certain types of plankton (termed Protista) cause a serious form of human intoxication — shellfish poisoning — and produce disasters in marine biology.[33] Shellfish poison is one of the most lethal known; the minimal lethal oral dose for man is thought to be between 1 and 4 mg/kg of body weight. Clinical effects are not completely understood, but the heart is definitely harmed. Death is usually attributed to respiratory paralysis. *Protista* are single-celled, animal-plant organisms in that their energy is derived by photosynthesis and their motion by flagellation. Widely distributed in marine waters, they reach public attention as blooms called "red tide" and "red water." Massive fish destruction may accompany these blooms produced by a variety of environmental changes. The cause of mass mortality of marine life during and after the blooms is not clear, yet the best evidence points to physical rather than chemical factors.

Paralytic shellfish intoxication is caused by certain molluscs and a few echinoderms and arthropods that eat toxic *Protista* (dinoflagellates) and are then ingested by man.[33] Some 21 dinoflagellates have been implicated; the best known is *Gonyaulax catenella* of the Pacific coast of North America. How the poison accumulates in molluscs is unknown. At least 28 species are involved including commonly known forms — clams, mussels, and oysters. Mostly the toxin is concentrated in the digestive glands of the mollusc. In Japan, nonparalytic form of poisoning may be contracted by ingestion of certain clams, oysters, and gastropods that consume toxic plankton.

Ichthyotoxism is a very complex subject [24,33] viewed either scientifically or as a basis for folklore. At least 16 factors have been studied as probable causes of concentration of very potent toxins in tissues of various marine fishes. The most important form of fish poisoning implicating over 300 species is ciguatera. It seems certain that it is associated with the food chain relationship of fishes. Probably the toxin originates in a benthonic organism eaten by herbivorous fishes and in turn consumed by carnivorous ones. The latter fishes accumulate poison without harm. Herbivores feed on toxic algae, fungi, or Protista, and current evidence favors the blue-green algae as the most probable causative factor. Clinical symptoms resemble those of shellfish poisoning.

## Health-Related Arthropods

### *Malaria*

This significant disease is gradually being brought under international control.[3] Indicative of its recent seriousness, of 36 great diseases, it ranked fourth in morbidity in 1957 with 4,153,250 cases reported in a population of 1.25 billion. It ranked third as a cause of mortality with 310,842 fatalities attributed to it.[14]

Water plants are favorable to anopheline mosquito production. In Puerto Rico, this includes coontail, fanwort, widgeon grass, waterweed, pondweed, stonewort, bladderwort, and filamentous green algae, especially *Spirogyra*. Also, the most important wetland or semiaquatic plants are Para grass, Carib grass, tropical carpet grass, sour paspalum, and small trumpet grass. Malaria mosquito larvae are inhibited by the duckweeds, waterfern, water hyacinth, water lettuce, pond lilies, and water snowflake, certainly by total cover, and probably also by associated water quality changes.[30,31] Iron bacteria, productive of rust colored flocculent masses and an oily surface in stagnant water, almost totally repels anophelines. Research by the Tennessee Valley Authority (TVA) provided leads on beneficial reduction of anopheline-supporting vegetation by seasonal manipulation of reservoir water levels.[1,22]

In recent decades, great advances have been made in development of pest mosquito control districts designed to improve public health. Control efforts are directed at abatement of these pests in urban areas although Culicoides, or "no-see-ums", is also an object of attack in places. As an important part of these local, tax-supported programs, much is done to remove ditch band vegetation by herbicides and cutting equipment during the off-season.

Recreation facilities imply healthful recreation. In the U.S., 41% of the population prefers water-based activities to any other. This is clearly shown in the booming development of family recreational facilities at large, public hydroelectric reservoirs. Here it is mandatory that aquatic weeds do not hinder fishing, swimming, boating, or esthetic interests, and as such are a key facet in the mosaic of a specialized, healthy environment.

### Arboviruses

Sleeping sickness or virus encephalitis has been a problem in parts of Latin America and the U.S. for decades. From 1926 to 1950, 27,749 human cases were reported in the U.S. — the lowest number, 702, in 1932; the highest, 3516, in 1941.[17] Only a few hundred cases have been reported annually during recent years. The decrease is most probably due to health education, better screening on houses, and promptness and efficiency of mosquito control incident to epidemics. Larvae of *Culex tarsalis*, the feared vector of St. Louis and Western encephalitis in midwestern and western U.S., thrive in vegetation-clogged irrigation and drainage ditches, seeps, and roadside impoundments.[11,21,22] Two other species, *Culex pipiens* and *C. quinquifasciatus*, are not vegetation associated, but prosper in domestic containers, catch basins, cesspools, and filthy ditches. All mosquito larvae except *Mansonia* must breathe at the water surface. *Mansonia*, host of Eastern encephalitis, is unique in that both larvae and pupae derive their oxygen by puncture from air trapped in hollow plant stems.[21]

Dengue fever, normally of low endemicity, exhibits occasional outbreaks — Panama in 1941, Yemen in 1954, and Puerto Rico in 1963.[24] In the 1963 Puerto Rican epidemic, infections reached 33,000 in a population of 2.5 million. In a carefully monitored urban population, the prevalence was 35%. Dengue does not kill, but is a miserable illness.[19] Yellow fever fortunately remains at a low ebb in the Old World and has been reported in man but a few times in recent decades in the Americas.[14] It remains endemic as a sylvan animal form in limited parts of Latin America.[19] *Aedes aegypti*, vector of both dengue fever and yellow fever against which there is a Western Hemisphere eradication program, lives well with inside water plants, but is otherwise confined to breeding in outside artificial containers.[11]

Some 50 mosquitos representing *Culex, Aedes, Mansoni*, and *Anopheles* are reported natural or artificial vectors of filariasis or elephantiasis in the tropics and subtropics. Other diseases borne by mosquitos are tulaemia (Swedish), fowl pox, and Rift Valley Fever.[16,19]

Mosquito species are attracted to particular aquatic habitats for egg laying and development of larval stages. In the water quality intergrades from well balanced to oxygen-poor, all species fit in.[11] As water quality changes, so does the mosquito fauna capable of living in that ecosystem. Of possible interest here is the gradual dominance of culicines over anophelines as impounded or sluggishly flowing water becomes degraded by total plant coverings or by massive plant decomposition. Conceivably, this could provide a new habitat for important disease vectors of encephalitis instead of commonly resident pest mosquitoes.

Other arthropods of public health interest linked permanently with waterweeds include snipe flies, tabanids (horse flies, gadflies, deer flies, and greenheads), Clear Lake gnats, May flies, sewage flies, black flies, sand flies, and tsetse flies.[21,22]

The tabanids and sandflies or "no-see-ums" are fierce biting pests which seriously affect tourist use of recreation areas.[22] Larval development occurs in weedy roadside pools and rice paddies. However, semiaquatic habitats are utilized by some species seasonally. In the southern U.S., cattle may suffer greatly from bites and this must adversely affect meat and milk production. In the U.S., tabanid species carry tularemia, and in African countries *Chrysops* transfers the eye worm, *Loa loa*.[28] The New England black fly pest problem is well known to the tourist industry. Control of stream-side vegetation has become an effective means of reducing tsetse fly vectors of African sleeping sickness (trypanosomiasis) near endemic villages.[21]

Breeding in vegetated margins of both fresh water and brackish water swamps, the "no-see-ums" occupy extensive environments on the Atlantic and Gulf Coasts and in the Caribbean.[24] The principal offender, *Culicoides furens*, is well named. Its bite can immobilize highly reactionary victims, and it has been incriminated as the vector of one minor filarial worm, *Mansonella*, in the Caribbean.

In the Great Lakes area, asthmatic attacks occur in recreation groups due to contact with wing dust from massive shoreline production of May flies.[21] The nonbiting Clear Lake gnat of the Northwest U.S. poses significant nuisance problems in the summer simply by density of adult insect population.

Another most undesirable urban pest is the Psychodid sewage fly borne massively at times in sewage treatment facilities and in oxygenation ponds.[15,24] Its potential for mechanically carrying pathogens to contacted persons seems unlimited.

**Worm Diseases**

Aquatic plants are definitely associated with worms affecting man. The key point is that snail vectors and larval stages of the parasitic life cycles are naturally abated by animal predators: (1) by duck, turtle, and fish consumption of snails; (2) by fish ingestion of crustaceans (crayfish, crabs, copepods); and (3) by fish predation of both miracidial larvae infective for snail intermediate host and cercarial larvae infective for the final host, man. Numerous flukes of fish, amphibians, reptiles, and other insectivorous animals use aquatic arthropods as hosts such as dragonflies, caseworms, May flies, and stone flies.[16,24,28] Considerable research data are available on the ability of guppies to control cercaria of the human blood fluke life cycle[29] which involves some 200 millions of cases internationally. Significantly, for predators to operate efficiently, they must be impeded by masses of aquatic vegetation. In this sense, a water body relatively clean of water weeds is judged a safer place than one clogged with vegetation and located in endemic foci of certain helminthic diseases.[18,22]

Correlations between presence of snail hosts of various trematodiases (especially the schistosomes) and aquatic plant preferences have been worked out in ecological surveys fundamental to control planning in many countries. In Puerto Rico, for example, of the 38 aquatic plants,[30] the *Biomphalaria glabrata* association is good in 14 instances and of these, it is generally consistent in arrowhead, elephant's ear, stonewort, and water hyacinth; it is fair in nine instances, and poor in 14 instances.[18] *Caladium* (or elephant's ear) which grows in and marginally to small sluggish streams, is the best snail indicator plant. It provides shelter, semishade, and a rich source of snail food in disintegrating leaves, stems, and roots. In fact the roots are sliced, dried, and ground as an important ingredient in a special diet for mass production of all laboratory snails.

Some parasitoses representing flukes, tapeworms, and roundworms[16,24,28] having stages vulnerable to predation may be listed with respective host and larval stages, shown parenthetically: filariasis which attacks some 250 million persons (Mosquito larvae); Chinese liver fluke, *Clonorchis* (Buliminid snails, miracidia, and cercaria); cattle and sheep liver flukes, *Fasciola hepatica* and *F. gigantica* (Lymnaeid snails, miracidia, and floating cercaria); broad tapeworm *Diphyllobothrium* (crustacea); guinea worm, *Dracunculus* (copepods); giant intestinal fluke, *Fasciolopsis* (Planorbid snails, miracidia, and cercaria); Asiatic lung fluke, *Paragonimus* (Thiarid snails, crabs, crayfish, miracidia, and cercaria); blood flukes, *Schistosoma* (Planorbid snails, miracidia, and cercaria); and avian schistosomes responsible for swimmer's itch (several snail genera, miracidia, and cercaria).

Although the larvae of the blackfly *Simulium* (vector of onchocercosis) are aquatic, these insects are not always plant associated, the larvae cases being attached to rocks in clean, rapidly flowing streams in the mountains. They breed also in low velocity drains. Other kinds of water-based flies affect health of economic animals as vectors of poultry disease or as important blood-sucking pests of domestic cattle—the Southern Buffalo Gnat.[21]

Significantly, the giant intestinal fluke infection is obtained by the Asian custom of eating raw aquatic plants—water chestnuts, lotus parts, water bamboo, and water caltrop—on which cercariae are encysted.[19] Normally, liver fluke infection occurs when encysted cercariae are ingested with grass. Thus, herbicide application might provide some control, but would not affect floating encysted cercariae normally drunk by cattle. *Fasciola* can be a human health problem as in Arequipa, Peru, where cercariae encysted on watercress are ingested with salads.[1]

Keeping water bodies comparatively free of weeds near human habitation in areas endemic for diseases noted above could become a reality because of availability of efficient chemical herbicides[1,2,5,7,20,23,26,31] and biological control techniques.[1,8-10,13,17,18,26,34-36] The combination of chemical herbiciding and biocontrol of water plant pests has been strikingly illustrated in a film recently produced by the Central and Southern Flood Control District, Florida, entitled *Marisa* and the Mermaid. As shown, chemicals designed for specific control of surface matting water weeds are applied with the use of the manatee and the large snail, *Marisa cornuarietis*, for control of rooted submersed plants such as *Elodea*. *Marisa* has a double attack since it reduces the aquatic weed problem and is an efficient predator of the snail intermediate hosts, *Schistosoma mansoni* and *S. haematobium*, schistosomes common to Africa, the Mediterranean, and parts of the New World. In addition, it can destroy the lymnaeid vectors of cattle liver fluke, but since these snails are amphibious, herbicidal vegetation removal is probably of no import.[2] Although liver fluke infestation in a particular pond or stream may be possibly eradicated by fish predation of floating cercariae and by *Marisa* eradication of the vector snails within the water body and at the margins, the snail continues to thrive in the adjacent humid soil-plant complex. Regional usage of the technique instead of focal application is the indication. In this connection, in laboratory observations another fairly large water snail, *Pomacea australis*, discovered (in Belo Horizonte, Brazil) to be a predator of the Brazilian blood fluke snail, *Biomphalaria*, is an avid eater of certain kinds of water weeds.[6]

## BENEFICIAL SEAWEEDS

All aquatic weeds are not detrimental to man's welfare. For many decades, seaweeds have formed the basis of a useful maritime industry from which human and animal food, medicinals, mucilages, bacterial medias, fertilizers, and various inorganic and organic extractives are taken.[33] Although the bulk of the industry (mostly confined to coastal reaches of the temperate zone) is concerned with extracting organic compounds, there is continuing research on utilization of seaweed as animal food. Six species are currently eaten as fodder by cattle, and various ground meals are used as dietary supplements. Meals having a vitamin-mineral content higher than dried grasses are useful components of sheep food, but objectionable in poultry food. Use as organic fertilizer is mostly limited to coastal area farming and is supplemental in nature. For man, seaweeds have a relatively low protein and fat content. Although there is carbohydrate available, digestion of algal polysaccharides is poor. Efforts to make the polysaccharide acceptable by enzymic predigestion are under investigation.

Seaweed stands are harmed by both fungal and bacterial disease. Low sewage contamination favors growth; high concentration is destructive. Fuel oil contamination is decidedly harmful. Seaweeds absorb various types of contaminating radioactive materials. Interestingly, seaweed production is adversely affected by several browsing animals—limpets, barnacles, sea urchins, sea hares, and some kinds of fishes.

## IMPORTANCE OF AQUATIC SNAIL CONTROL

In many areas where Americans are now operating or are likely to be in the next few decades, various aquatic snails are alternate hosts of many trematode parasites of man (see Figures 1.1 and 1.2). Snails of the genus *Onchomelania*, the intermediate host of *Schistosoma japonicum* in Japan, Taiwan, the Philippines, and China, and *Australorbis glabratus*, the intermediate host of *Schistosoma mansoni*, in the West Indies (particularly Puerto Rico) and tropical South America, are of real significance. These snails are most numerous near the water's edge and control of marginal vegetation is very important in a program of disease control.[37]

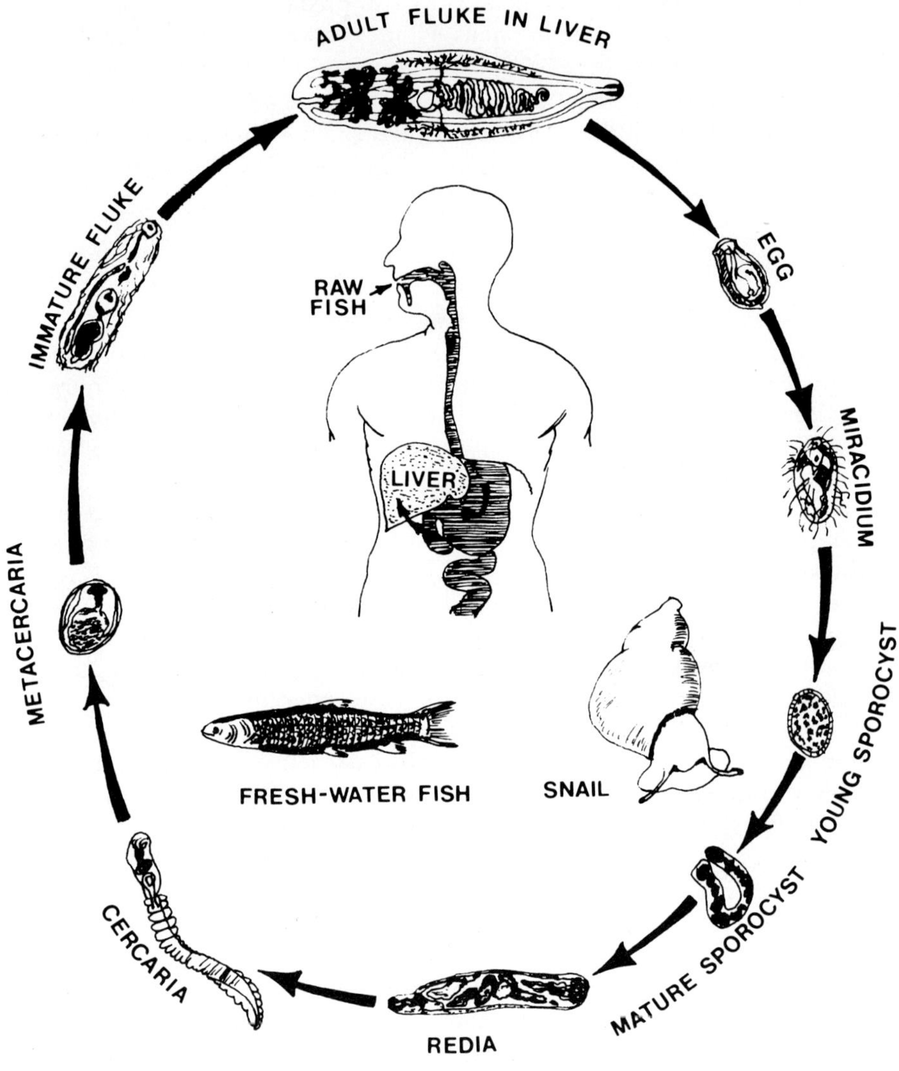


FIGURE 1.1. Life cycle of *Clonorchis senensis*, a liver fluke disease. (Adapted from *Military Entomology Operational Handbook*, Departments of the Navy, the Army, and the Air Force, 1971.)

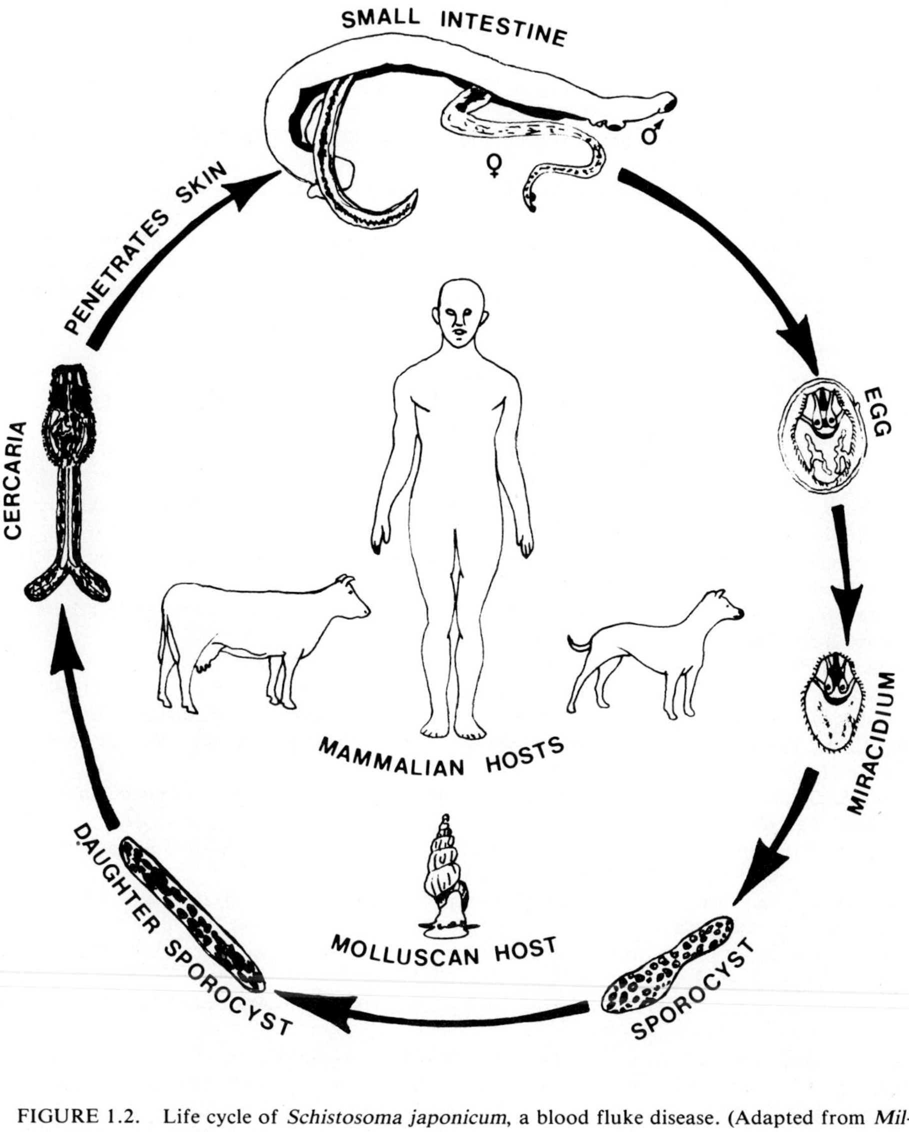


FIGURE 1.2. Life cycle of *Schistosoma japonicum*, a blood fluke disease. (Adapted from *Military Entomology Operational Handbook,* Departments of the Navy, the Army, and the Air Force, 1971.)

## REFERENCES


1. **Anon.**, Malaria control on impounded water, Tennessee Valley Authority, No. 697591, U.S. Government Printing Office, Washington, D.C., 1947.
2. **Anon.**, Water, in The Yearbook of Agriculture, U.S. Department of Agriculture, Washington, D.C., 1955.
3. **Anon.**, Epidemiological and Vital Satistic Report: Malaria, World Health Organization, Geneva, 1964, 89.
4. **Anon.**, Standard methods for the examination of water and wastewater, *Am. Public Health Assoc. Yearb.*, 1965.
5. **Anon.**, Waterweed control of farms and ranches, *U.S. Dept. Agriculture Farmers' Bull.*, Vol. 2181, 1965.
6. **Anon.**, Biological control of aquatic weeds, *Agric. Res.*, 8, 9, 1968.
7. **Bennett, G. W.**, *Management of Artificial Lakes and Ponds*, Reinhold, London, 1962.
8. **Bennett, F. D.**, Investigations on insects attacking the aquatic ferns, Salvinia spp in Trinidad and Northern South American, *Proc. South. Weed Conf.*, 19, 497, 1966.
9. **Blackburn, R. D.**, A freshwater snail as a weed control agent, *Proc. South. Weed Conf.*, 18, 589, 1965.
10. **Blackburn, R. D.**, Weed control in fish ponds in the United States, in Proc. Food and Agricultural Organization World Symp., Rome, 1966.
11. **Carpenter, S. J., Middledauff, W. W., and Chamberlain, R. W.**, The Mosquitoes of the Southern United States, University Press, Notre Dame, Ind., 1946.
12. **Commoner, B.**, The Killing of a Great Lake, World Book (Year Book) Enterprise Corporation, Chicago, 1968, 127.
13. **DeBach, P.**, *Biological Control of Insect Pests and Weeds*, Reinhold, New York, 1964, 844.
14. **Donaldson, A. W.**, The Status of World Health, Senate Commission on Government Operations, U.S. Senate, 86th Congress, 1st Session, U.S. Government Printing Office, Washington, 1959.
15. **Fair, G. M. and Geyer, J. C.**, *Water Supply and Wastewater Disposal*, John Wiley & Sons, New York, 1956.
16. **Faust, E. C., Beaver, P. C., and Jung, R. C.**, *Animal Agents and Vectors of Human Disease*, 3rd ed., Lea & Febiger, Philadelphia, 1968.
17. **Ferguson, F. F.**, Biological factors in the transmission of arthropod-borne virus encephalitides, *US Public Health Serv. Public Health Monogr.*, 23, 37, 1954.
18. **Ferguson, F. F. and Butler, J. M., Jr.**, Significance of animal control of aquatic weeds, *Proc. South. Weed Conf.*, 21, 304, 1968.
19. **Gordon, J. E.**, *Control of Communicable Diseases in Man*, 12th ed., American Public Health Association, New York, 1965, 282.
20. **Harrison, D. S.**, Aquatic Weed Control, Circular No. 219A, University of Florida, Agricultural Extension Service, Gainesville, 1965.
21. **Herms, W. B.**, *Medical Entomology*, MacMillan, New York, 1950, 582.
22. **Hess, A. D., Hall, T. F., and Iseley, D.**, Weeds injurious to the health of man and animals, in Monograph on Weeds, Agricultural Board, National Academy of Science—National Research Council, Washington, D.C., 1968.
23. **Hotchkiss, N.**, Underwater and floating-leaved plants of the United States and Canada, Research Publication No. 44, U.S. Bureau of Sports Fisheries and Wildlife, Washington, D. C., 1967, 124.
24. **Hunter, G. W., III., Frye, W. W., and Swartzwelder, J. C.**, *A Manual of Tropical Medicine*, W. B. Saunders, Philadelphia, 1967, 931.
25. **Lagler, K. F.**, *Freshwater Fishery Biology*, W. C. Brown Co., Dubuque, Iowa, 1961, 421.
26. **van der Lingen, I., Phipps, J. B., and Templeton, A. A.**, Weed control in dams and irrigation channels, *Rhod. J. Agric. Res.*, 57, 353, 1962.
27. **McKinney, R. E.**, *Microbiology for Sanitary Engineers*, McGraw-Hill, New York, 1962, 293.
28. **Noble, E. A. and Noble, G. A.**, *Parasitology—The Biology of Animal Parasites*, Lea & Febiger, Philadelphia, 1964, 724.
29. **Oliver-Gonzalez, J.**, The possible role of the guppy, *Lebistes reticulatus*, on the control of *Schistosoma masoni*, *Science*, 104(2712), 605, 1946.
30. **Pratt, H. D.**, Key to Aquatic and Semiaquatic Plants Found in Puerto Rico, U.S. Public Health Service, Washington, D.C., 1946.
31. **Pratt, H. D.**, Relation of Plants to Malaria Control in Puerto Rico, *Public Health Rep. Suppl.*, 200, 1, 1947.
32. **Reid, G. K.**, *Pond Life — A Guide to Common Plants and Animals of North American Ponds and Lakes*, Golden Press, New York, 1967, 160.

33. **Russell, F. S.,** *Advances in Marine Biology,* Vol. 3, Academic Press, London, 1965, 402.
34. **Seaman, D. E. and Porterfield, W. A.,** Control of aquatic weeds by the snail, *Marisa cornuarietis, Weeds,* 12, 87, 1961.
35. **Wilson, F.,** The biological control of weeds, *Annu. Rev. Entomol.,* 9, 225, 1964.
36. **Zeiger, C. F.,** Biological control of alligatorweed with *Agasicles* n. sp. in Florida, *Proc. South. Weed Conf.,* 20, 299, 1967.
37. **Anon.,** *Military Entomology Operational Handbook,* Departments of the Navy, the Army, and the Air Force, U.S. Government Printing Office, Washington, D.C., 1971.

# Chapter 2

# CONTROL OF SNAIL-BORNE PARASITIC DISEASES*

## INTRODUCTION

Perhaps as many as 300 million human beings are victims of snail-borne parasitic diseases: schistosomiasis (which is of primary concern), fasciolopsis, angiostrongyliasis, clonorchiasis, and paragonimiasis. Fascioliasis, a major ailment of cattle and sheep, can also infect humans, and its control may thus serve a dual purpose. Various control methodologies, including chemical, which can be applied to the mollusc hosts of these diseases frequently are related to aquatic plant control.

Parasitic-disease control can emphasize prevention of infection or chemotherapy after infection. Obviously, the former is preferred, although the best approach in many situations may be a dual strategy. The basic prophylactic methods practiced to date have relied upon chemical treatment of snail habitats, destruction of such habitats through engineering technology, or prevention of human contact with the larval form. Prevention of human contact, which involves sanitary facilities and/or the supplying of safe water, requires a tremendous capital outlay in most endemic regions. Similarly, the engineering approach to widespread snail control is not economically feasible at this particular time in history — especially in view of the depressed economies of many of the afflicted nations. For although irrigation and other water supply systems can be designed in such a manner as to minimize movement of snails from infected habitats to human or domestic animal contact points, in practice this is not done. Mechanical snail barriers, such as screening, have been utilized with minimal success.

Biological control concepts, using snail predators, parasites, and pathogens, have been the subject of a number of studies and a great deal of discussion. However, as yet no widely useful techniques have been developed.

In practice, public health authorities have relied almost exclusively on chemical control strategies. Molluscicides are applied through various techniques to the water course as granules, wettable powders, solutions, or emulsions. Methodology varies from site to site as regards application, dosage, treatment regimen, and so on.

Nonetheless, it is well recognized that schistosomiasis is spreading, especially through the introduction of harnassed water supplies in irrigation schemes, man-made lakes, and reservoirs. The boundaries of the endemic areas enlarge year by year in most of the afflicted nations. Consequently, the number of human victims and those at risk increases.

Failure in many instances of snail-control efforts to halt the transmission and spread of schistosomiasis arises from a variety of causes. The usual lack of proper funding and adequately trained personnel leads to incomplete programs. Habitats are overlooked during molluscicide application and serve as foci for future snail repopulation. Because most molluscicides in conventionally used concentrations are extreme irritants to the target snail, they induce various forms of often successful avoidance behavior. Not only economic but also environmental considerations govern dosages used and points of application. The destruction of food fish, for instance, cannot be tolerated in many locales. In any case, it is necessary to retreat the same areas at intervals varying

* By Cardarelli, N. F., Engineering and Science Division, University of Akron, Akron, Ohio, abridged and updated from an unpublished report, Controlled Release Molluscicides, University of Akron, Akron, Ohio, 1977.

from 2 months to 1 year. The lack of long-term control may be the crucial negative factor in halting disease transmission.

It is believed that controlled-release molluscicides offer a practical and economic method of achieving long-term snail control with the possibility of permanently interrupting the transmission cycle. Furthermore, this emerging technology may provide a method of intervention through destruction of the larval forms of the trematode parasite — a total impracticality from the viewpoint of the *conventional* methodology. On the other hand, controlled release does not represent a panacea. A number of serious problems are involved in the use of sustained-release molluscicide formulations. These include our lack of knowledge concerning environmental chemodynamics and environmental impact, the possibility of inducing resistance, questions concerning phytotoxicity, effects on the food chain, dosage requirements under varying water quality conditions, methodology for treating flowing water systems, cost/effectiveness, and the like.

## DEVELOPMENT OF CONTROLLED-RELEASE FORMULATIONS

In order to meet the need for long-term antifouling surfaces, beginning in 1964 various agents known to be toxic or repellent to barnacles, marine algae, bryozoans, and other sea-dwelling organisms that in the adult form are sessile and collectively termed "fouling" were incorporated in molded and sheet rubber goods. Proper compounding of such formulations led to the development and commercialization of antifouling elastomers used in naval sonar domes, buoy coverings, ship hulls, and various other marine apparatus.[1-5] The material presently used consists of 6% by weight of *bis* (tri-n-butyltin) oxide (TBTO) in a chloroprene base. Release life of this material appears to be in excess of 8 years for the initial commercial application, and to date no known biological failures have occurred on the numerous ship hulls and other objects covered, some of which have been in service for 5 or more years.[6,7] Recently it has been demonstrated that natural rubber, various nitrile polymers, polyisobutylene, and other elastomeric matrices will retain and slowly release organotins with considerable antifouling efficacy.[8,9]

The development of controlled-release molluscicides was a logical extension of the antifouling rubber program of the B. F. Goodrich Company, Akron, Ohio. In 1965, the University of Akron was asked to evaluate the molluscicidal potency of the several antifouling elastomeric formulations. Original formulations contained TBTO, tributyltin fluoride (TBTF), and tributyltin acetate (TBTA). Lack of data concerning the environmental impact and mammalian toxicity of organotins encouraged the development or attempted development of controlled-release niclosamide, trifenmorph, and other conventional materials. In recent years, copper sulfate has been successfully incorporated in ethylene-propylene-diene type elastomers.

Organotin/elastomer formulation development at the B. F. Goodrich Company, the University of Akron, and the Creative Biology Laboratory (Barberton, Ohio) was extended to encompass bactericides and algicides, insecticides, and cercariacides.[1] Work with controlled-release aquatic herbicides, initially utilizing various esters of 2,4-dichlorophenoxyacetic acid in elastomeric matrices, was initiated in 1969, and later successful field trials demonstrated efficacy.[10,11]

Development of plastic-based carriers containing insecticides commenced in the 1960s, with the Shell Chemical Company's No Pest® strip being the first to reach the marketplace. Currently, a number of controlled-release products have been approved by the Environmental Protection Agency. These include various flea and tick collars for domestic pets, cockroach control devices,[12,13] Dow Chemical Company's CR-10 controlled-release mosquito larvicide (based upon the extensive effort of the U.S. Army Environmental Hygiene Agency),[14] and Zoecon Corporation's Altosid materials, which slowly release a morphogenetic agent preventing maturation of mosquito larvae.[15] In the agricultural area, Pennwalt Corporation's Pencap® encapsulated methyl parathion has been registered and is in use. In all instances, increased efficacy, economy, and safety have been demonstrated over conventional usage methodology.

**Controlled-Release Principles**

Mechanisms for the sustained release of a pesticidal material are varied. The simplest is encapsulation based upon coacervation techniques wherein the control agent is surrounded by a regulating membrane. The agent may pass through the membrane by diffusion or leaching processes, or the polymeric envelope may rupture under environmental stresses, spewing forth its contents. At least 284 U.S. patents exist describing various encapsulating methods and usages. To date, this technique has not been used with molluscicides. More pertinent to this treatise is the integral incorporation of the agent in a polymeric matrix. The matrix element may be elastomeric or plastic in its thermodynamic and kinetic aspects. Agent release is effected by one or more of the following mechanisms: diffusion-dissolution, leaching, and exfoliation.

Diffusion-dissolution systems are dependent upon the solubility of the agent in the polymeric matrix. For instance, trialkyl organotins are soluble in natural rubber, polychloroprene, styrene-butadiene copolymers, and other elastomers. If the agent is added to some level below the solution limit, a typical liquid-in-solid or solid-in-solid solution results. If the agent is slightly soluble in water — as is true of the organotins, organoleads, and niclosamide — upon immersion molecules on or near the elastomer surface pass, through dissolution processes, into the water/elastomer interface. This, in turn, leads to a solution disequilibrium, and internal solute molecules migrate under solution pressure to the depleting surface. Since the process is continuous as long as dissolution continues, a long-term loss process is established. The NoFoul® antifouling sheet rubber acts in this manner as well as BioMet-SRM®, CBL-9B, and other controlled-release molluscicides.

Certain factors are characteristic of this mechanism. As a rule, water velocity over the exposed surface does not materially affect loss rate. Hydrostatic pressure likewise shows little effect; rubber panels immersed at depths of over 6000 feet in the Atlantic Ocean showed an organotin loss rate similar to that of panels held at 6-ft depth.[16] No pore structure is established as a function of agent loss. Water permeation of the matrix is unimportant in process dynamics. Examination of the depleting matrix by microtoming and X-ray spectrophotometry disclosed that no interior-to-surface toxicant gradient exists.[1]

In diffusion-dissolution systems, either diffusion or dissolution may be the rate-controlling step. Harris[17] has developed equations, based upon the Higachi expression,[18] that describe systems wherein diffusion is the release-rate controlling factor. Biological lifetime can be computed for such systems based upon diffusion constants, matrix factors, and surface-to-volume ratio of the prepared granule, pellet, or sheet.

Agent loss through exfoliation, the mechanism for earlier antifouling paints, has not (to the author's knowledge) been used with controlled-release molluscicides. In such systems, the binding matrix is gradually lost through mechanical erosion or solvation in water. The process has been mathematically described and systems designed wherein toxicant loss is at a predictable rate.[19,20]

In controlled-release systems lacking agent solubility in the polymeric matrix, as appears true of most plastic-based devices, loss is generally dependent upon a leaching mechanism. Unlike diffusion-dissolution, wherein the agent moves towards the watery interface, in this case the external water must penetrate the matrix and contact the immobile agent molecule. Solution occurs and the agent passes via a water path outward following a concentration gradient. Excluding the use of a third-phase carrier material, agent movement is dependent upon the porosity of the binding element. In elucidating the mechanism involved, a quantitative consideration of the number of pores or total pathways through the matrix is not in itself adequate. The tortuosity of the pore structure plays a key role in water transport. Approximate mathematical expressions have been developed to allow loss-rate prediction.[17,21,22,30]

## Formulation

Controlled-release molluscicides, whether elastomer or plastic based, are formulated and processed by techniques well known to the state-of-the-art using conventional equipment. Formulations are composed basically of four elements: (1) a binding matrix; (2) additives necessary for vulcanization ("curing"), enhanced durability, and ease of processing; (3) additives essential to the regulation of the agent loss rate; and (4) the molluscicidal agent. Selection of the binding matrix is dependent upon the release system desired, i.e., diffusion-dissolution or leaching. In diffusion-dissolution systems, solubility of the molluscicide in the binder is essential. In leaching systems, a binder is sought that, in general, will retain a fairly high agent loading that is conducive to pore growth. Elastomer systems are almost always vulcanized in order to achieve a processible compound and, as a rule, to retard toxicant loss and thus create long-term release. Additives, such as carbon black or fine structured clays, can be used in elastomeric formulations to retard loss rate while plasticizers are sometimes utilized in plastic-based compounds to increase loss. Compounding techniques and associated problems have been summarized elsewhere for both loss mechanisms,[1,23-25] and the fundamental relationship between molluscicide loss rate and matrix properties has been studied and a computerized model established by Kanakkanatt.[26]

Controlled-release formulations based upon the use of elastomers or plastics as the binding matrix can be processed in geometries most conducive to pest control in a specific habitat. Materials can take the form of discs, granules, powders, sheets, or tubes in any shape or size normal to plastic and rubber goods. In contrast, conventional molluscicides are limited to granules, powders, and emulsifiable concentrates.

The versatility of controlled-release design is evidenced in the creation of sinking and floating herbicidal formulation *and* suspended forms that hover at specific depths in the treated water course. This allows phytozone type treatments, which in turn reduce chemical costs and environmental impact.[11,27]

## Laboratory and Small-Scale Field Evaluations

Laboratory and small-scale field evaluations have demonstrated that candidate controlled-release molluscicides are efficacious under varying environmental conditions. It is not in doubt that, properly formulated, such materials release the toxic agent at a slow, continuous rate over an extended period of time. Furthermore, snails exposed to organotin, or ganolead, and copper sulfate at ultralow concentrations will succumb through a chronic intoxication mechanism. Since concentrations in the low part-per-billion range will destroy snails — provided the exposure period is over several weeks and the dosage level is maintained through the use of controlled-release formulations — it is reasonable to hypothesize that the environmental impact may be substantially less than that observed with more conventional chemical control methods.

Usage of controlled-release molluscicide will necessitate the development and evaluation of a methodology substantially different from the conventional.[28,29] General methods of evaluating molluscicidal potency based upon lethal concentration (LC) and lethal dose (LD) values do not appear to be especially pertinent. The elaborate system designed over the years to determine cost/effectiveness ratios, dosages, and the like will need extensive revision if not abandonment.[30,31] It has long been recognized that the basic problems in adequate snail control arise from the resurgence of snail populations when conventional molluscicide treatment is halted and from the detoxifying effects of little-known origin that recur in an unpredictable fashion. Likewise, treatment of large though major transmission sites — such as rivers and natural and man-made lakes — provides no control or only very transient control.[28,30]

Schistosomiasis intervention can rely upon techniques other than chemical control. Provision of noninfected water for drinking and general purposes would probably offer the greatest benefit, as evidenced by a considerable number of positive reports.[32] Ths approach has been meritoriously applied by Jordan et al. on St. Lucia, West Indies.[33] However, other than in densely populated areas, the cost of supplying pure water and of modern wastewater disposal systems is exhorbitant and far beyond the means of the underdeveloped nations. If the human condition continues to improve, this will undoubtedly be done — but hardly in the foreseeable future.

Medical therapy as it stands today can only be of permanent value if the infected person is isolated from sources of infection after treatment. To date, a vaccine has not been developed even though considerable effort and resources have been expended in this area over the last 30 years. Certainly no one can predict when such a vaccine will be available.

The immediate and future necessity for snail control can hardly be overemphasized. Nonchemical control methods would certainly have less environmental impact than present techniques. Methods advanced as means of biologically controlling snails include the use of competing refractory species and predatory organisms including parasites, genetic manipulation via radiation or chemically induced sterility, habitat alteration, and the like. Hairston et al. have summarized the extensive literature in this area.[34] Biological control schemes might well be the most acceptable manner of snail control. However, there have been few field techniques developed pertinent to the schistosomiasis problem, and those have only been applicable to rather isolated environmental situations. For instance, irrigation reservoirs can be subjected to periodic water fluctuation in such a manner as to retard development of snail populations.[35]

Engineering techniques coupled with the use of chemical agents have brought schistosomiasis under control in most formerly endemic areas of Japan, and actual eradication may have been achieved in some instances.[36] Increase in water flow rate will dislodge snails; thus, appropriate hydrodynamic design of irrigation canals might be a suitable technique for reducing the desirability of that infested environment. Concrete lining of canals tends to deprive snails of sufficient vegetation and thus renders the habitat unsuitable. Likewise, periodic drying of irrigation canals and drainage ditches tends to retard snail population growth — although some snail hosts can withstand dessication and extremes in temperature to a remarkable degree.[37]

Materials that have low biological impact on major segments of the biota but that will suppress cercarial production and/or shedding (or otherwise interfere with sporocyst development) might be used as control agents. Substances such as choramphenicol do have this effect, and their usage essentially results in "curing" the snail through medical therapy.[38] However, various evaluations of such substances indicate that, while they do confer protection, their removal leads to further shedding of cercariae or future infection of the snail if miracidia are present.

A very large body of relevant literature reveals that the majority of those responsible for schistosomiasis control feel that, in all save very isolated situations, chemical control of the snail host coupled with medical therapy produces the most favorable cost/effectiveness ratio. Difficulties arise in determining the allocation of scarce resources between chemical control and medical therapy in order to achieve the greatest benefit.[39-41]

## Conventional Chemical Control

Conventional chemical control need not focus on the snail host. Destruction of the molluscan food supply can obviously retard or eliminate snail populations. Also, some herbicides have adjunct properties useful in snail control.[42] As well as profoundly affecting the snail, soluble copper salts, whether used in the conventional manner or through controlled-release techniques, will destroy aquatic weeds and algae vital to snail nutrition.[1]

Molluscicides as conventionally used suffer a number of disadvantages:

1. Nonpersistence of the molluscicidal effect
2. High cost of application arising from nonpersistence
3. High cost of chemical agents due to the need for relatively massive dosages to overcome detoxifying effects of the environment
4. Limitation on available formulation types — materials being confined to emulsions, granules, and wettable powders
5. Require volume treatment to achieve practical control dosages
6. Undesirably high levels of environmental impact manifested in the contamination of bottom soils and destruction of nontarget biota, including fish and other elements of the food chain
7. Health hazards to consumers of treated waters
8. Health hazards to application personnel
9. Difficulties involved in the treatment of large water bodies such as lakes, or in efficacious applications to water courses having moderate to high flow rates

The use of controlled-release molluscicides would appear to overcome several of the above disadvantages while ameliorating the rest. Laboratory and field trials to date suggest that the formulations discussed in this monograph provide longer lasting results than those normally achieved by conventional usage. Periodicity of treatment is a major economic factor as well as being the underlying cause of the general inability to achieve permanent interruption of transmission. Thus, extending the between-treatment interval can effect substantial labor and chemical savings. Also snail population resurgence, stemming from targets that have successfully evaded intoxication through avoidance of conventional molluscicides, ought to be prevented if the water habitat is permanently or semipermanently toxic.

Those individuals committed to snail control through conventional methodologies seek persistent molluscicides as a means of extending the treatment interval; on the other hand, controlled-release techniques permit the use of substances that are *persistent* at the point of application — the controlled-release pellet — and essentially nonpersistent at the point of action — the snail.

The generally realized low to moderate persistence of conventional molluscicides, such as copper sulfate and niclosamide, also indicates that the chemical must be applied in relatively massive dosages to insure target contact at a lethal level. In other words, dosages must be large enough to overcome natural detoxifying effects arising from dilution, insolation, adsorption, chemical interactions with dissolved minerals, and the like over that period of time necessary for the target snail to absorb or ingest a lethal quantity. Although field treatment levels vary with the toxicant used and the quality of the water treated, concentrations usually vary from 1 ppm upwards. Acute intoxication requires a relatively high concentration × time exposure.

In contrast, controlled-release molluscicides rely upon the inducing of a terminal, chronic intoxication. *All* evidence amassed to date indicates that target snails can be destroyed at levels of 0.01 ppm or less if the exposure duration is several weeks or longer. In terms of environmental impact, it is reasonable to expect that long-term intoxication of a water course at 0.01 ppm would have substantially less environmental impact than repeated treatments at 1 ppm. Nontarget toleration among fish and various plants examined in this regard is far greater for a continuous exposure to ultralow concentrations of a toxicant than for *one* application at the higher level. Since the underlying mechanism for chronic intoxication remains essentially unknown, it is useless to speculate upon subtle damage to nontarget biota. On the other hand, it is quite obvious that large doses of molluscicides lead to trauma even where death does not occur. Fish exposed to 1 ppm of trifenmorph may well recover since this toxicant is rapidly detoxified. However, activity during recovery is drastically slowed, and in a natural environment, the ability to resist predation and disease, find food, and reproduce would certainly be lessened.

**Economic Considerations**

The economic aspects involved in the use of controlled-release materials are not known since cost/effectiveness evaluations against conventional molluscicides under field conditions have not been performed. Chemical costs based upon active-agent content are higher for controlled-release formulations, while costs on a formulation per pound basis are considerably lower or will be lower once some level of use is realized. A comparison of ingredient cost and processing expense does allow prediction of future selling price. A conservative present price and reasonable future price based upon economies of mass production are depicted below in U.S. dollars.

| Controlled-release molluscicide | Present cost (per pound) | In-production cost[a] (per pound) |
|---|---|---|
| BioMet® SRM (6% a.i.[b]) | $10.00 | $6.50 |
| ECOPRO® 1320 (20% a.i.) | 6.00 | 3.50 |
| Incracide® E-51 (50% a.i.) | 6.50 | 2.00 |
| ECOPRO® 1330 (30% a.i.) | 6.50 | 4.25 |
| ECOPRO® 1000 (43% a.i.) | 2.25 | 1.80 |

[a] Author's estimate.
[b] a.i. = active ingredient.

Controlled-release formulations can be processed in any form taken by rubber or plastic goods. Whereas the conventional materials are limited to emulsions, solutions, and granules, controlled-release molluscicides can be prepared as paints — applicable, for instance, to the cement walls of irrigation canals — sheet rubber, or floating, sinking, and suspending strands, strips, pellets, discs, hollow cylinders, and so on. Biozone treatment, localizing the molluscicide in the area of infestation, becomes possible. Sheet rubber, adhered to concrete structures or simply attached to stakes and driven into the subsoil of the water course, will release at that point without being lost through water movement. Bottom-browsing snails can be attacked by use of sinking pellets or bottom-suspending strands that release toxicant at the water/subsoil interface and, at least with the organotin materials, saturate the bottom soil with little upward displacement. Top-dwelling snails can be similarly reached by the use of floating pellets or discs or top-suspending strands. In this case, the toxicant, under stillwater conditions, tends to disseminate along the surface with relatively little downward transport.

The problems involved in treating flowing-water systems with conventional formulations are well known. But with controlled-release formulations, computation of dosage and treatment of an infested habitat based upon *area* rather than water volume appears feasible in both static and flowing-water systems, since snails which dwell in running water systems cling to a surface, usually the bottom soil but sometimes rooted vegetation. Although the actual effectiveness of an area treatment based on the dispersal of pellets along the water bottom is unknown, the development of theoretical models is underway which, to date, indicates that efficacious use of the hydrophobic organotins can be expected.

Since the toxic element is bound within a polymeric structure, controlled-release molluscicides would appear to offer far less of a health hazard than conventional systems to human handlers. At any given time the number of toxic molecules on the elastomer or plastic surface — and, thus, available for transfer to an external organism—is small compared to the number of unavailable molecules. It is not readily conceivable that a person handling controlled-release materials (excepting paints) in the geometries made to date could inhale pellets, have them lodge in the eye, or "spill" on the unprotected dermis. Since the matrix elements are not digestible by humans, accidental ingestion would likely lead to passage through the gastrointestinal tract and excretion with relatively little degradation or loss of toxicant. This has not been tested, of course, but past experience with forced bovine, avian, and piscine ingestion has resulted in whole pellet recovery in the feces and no grossly observable damage to the animal.

Treatment of an infested habitat with controlled-release molluscicides has an accumulative negative effect on the total schistosome transmission mechanism. As Ritchie et al. have so succinctly said, "It is possible that the toxicant could be leveled against the most susceptible stage of the snail or parasite, yet if a multi-level control concentration is used, transmission might be interrupted through the cumulative effect of several parameters at less than 100% efficiency."[43] It has been noted that the organotins are toxic to snails, snail ova, cercariae, and miracidia and that they affect sporocyst development. At levels sublethal to snails, certain effects can be pronounced — including suspension of feeding and reproductive activities. Intervention need not require absolute destruction of a given segment of the transmission cycle but merely a reduction in population of each segment. However, laboratory tests indicate that the schistolarvae can be destroyed at concentrations that do not affect adult snails, although ova mortality is increased.

The basic controlled-release arsenal available for field evaluation contains formulations of TBTO, TBTF, and copper. Organolead substances have been developed, but environmental questions of paramount importance remain to be addressed. Trifenmorph and niclosamide systems are presently under laboratory evaluation with negative results to date. Copper-based controlled-release elastomers have an immersion longevity of 2 or 3 months (ECOPRO® 1000) to 6 or 7 months (Incracide® E-51). In contrast, the organotin-based formulations will provide two or more years of active agent release. Based upon collateral effort in the antifouling area, biologically efficacious release for over 10 years is conceivable.

Questions regarding the environmental effects of copper sulfate in controlled-release systems have not engendered the controversy noted with the advocated use of the organotins. The toxicology and environmental impact of copper ion are far better understood, and copper has long been used in snail control. Niclosamide and trifenmorph were in use prior to the general realization that man's environment cannot forever tolerate various toxic burdens thrust upon it. In this period of enlightenment concerning ecological disturbances, questions have been raised regarding usage of organotins as molluscicides and schistolarvicides that were never addressed to copper, the pentachlorophenates, niclosamide, and other conventional in-use materials. An impartial examination of the literature indicates that (1) much more is known about the toxicological and environmental effects of organotin than of niclosamide or trifenmorph and (2) not enough is known about either the conventional molluscicides or the organotins to insure that their benefits outweigh their damage to the environment.

Basic unresolved issues concerning the potential uses of the organotins, specifically TBTO and TBTF, as molluscicides and schistolarvicides can be summarized as follows: (1) toxic effects on nontarget biota, (2) ultimate fate in the environment, and (3) the possibility of snails developing tolerance.

One basic problem in determining the impact of ultralow but continuous exposure of the biota to organotins is the present impossibility of detecting tin, much less its speciation, in the parts-per-billion range. Standard techniques are inadequate. Newly developed concepts for speciation are under evaluation, but much remains to be done prior to the ability to actually analyze for TBTO or TBTF in parts-per-billion or parts-per-thousand range under field conditions.[44] Present work at the Environmental Management Laboratory and elsewhere is keyed to determining the degradation pathways of the specific organotins, the toxicity of the intermediates, and their ultimate environmental fate. One aspect involves the processing and use of $C^{14}$-labeled TBTO in elastomeric pellets in an attempt to determine accumulation in plants, animals, and soil.

A controlled-release material consists not only of a bound toxic element but also of an elastomeric or plastic matrix with suitable vulcanizing, processing, and protective additives. It is believed that the natural rubber and ethylene-propylene matrix elements degrade through microbial activity, hydrolysis, and cleavage arising from electromagnetic radiation through a series of steps to carbon, carbon dioxide, and water. In fact, rubber goods need various pesticidal additives and protective agents if use-life in a natural environment is to exceed a few months. The presence of an organotin will, of course, substantially retard degradation through microbial attack. BioMet® SRM pellets returned after 14 months of immersion at field sites show considerable surface degradation but retain overall structural integrity.

## SUMMARY AND CONCLUSIONS

It is believed that, at this time, the chemical control of snails remains an integral element in the holistic approach to containing schistosomiasis. Conventional methods of applying chemicals have often led to haphazard control and, in many instances, have at best slowed the spread of schistosomiasis. It is possible that chemical treatment through the use of newly developed controlled-release materials will provide long-term snail control with cost/effectiveness and risk/benefit ratios far more advantageous than those presently achieved with conventional strategies. It is contended that controlled-release formulations can be efficaciously used against schistolarvae, although application and monitoring techniques are still to be developed. The greatest needs at this time are (1) the development of application methods and regimens under field conditions and (2) a better understanding of the environmental chemodynamics of such systems.

# REFERENCES


1. **Cardarelli, N. F.**, *Controlled-Release Pesticides Formulations*, CRC Press, Cleveland, 1978, 7.
2. **Wuerzer, D. R., Sanderling, R. L., and Cardarelli, N. F.**, Protection buoys with antifouling rubber coverings, *Rubber World*, 157, 177, 1967.
3. **Cardarelli, N. F. and Neff, H. F.**, Composition de Preservations Controles Incrustations Sons-Marines, French Patent 1,506,704, 1964.
4. **Cardarelli, N. F. and Caprette, S. J.**, Antifouling Coverings, U.S. Patent 3,426,473, 1969.
5. **Cardarelli, N. F. and Neff, H. F.**, Biocidal Elastomeric Compositions, U.S. Patent 3,639,538, 1972.
6. **Bollinger, E. H.**, Controlled-Release Antifouling Rubber Coating, Rep. No. 19, in Proc. Controlled-Release Pesticide Symp., University of Akron, Akron, Ohio, September 16 to 18, 1974.
7. **James, C. A.**, Polymeric antifouling coatings on pleasure craft, in Proc. Controlled-Release Pesticide Symp., Wright State University, Dayton, Ohio, September 8 to 10, 1975, 293.
8. **Woodford, J. M. D.**, Underwater Marine Biocidal Rubbers Containing Organotin Toxics, Rep. 496, Defense Standards Lab, Maribyrnong, Victoria, Australia, March 1972.
9. **Dunn, P. and Oldfield, D.**, A novel concept in marine antifouling elastomers, paper presented at 3rd Australian Rubber Technology Convention, Terrigal, New South Waters, Australia, September 18 to 27, 1974.
10. **Bille, S., Mansdorf, S. L., and Cardarelli, N. F.**, Development of slow release herbicidal materials for controlling aquatic plants, final rep. DACW 73-70-C-0039, Office of the Chief of Engineers, U.S. Department of the Army, Washington, D.C., July 1971.
11. **Thompson, W. E.**, Field Tests of Slow Release Herbicide, Rep. No. 15, in Proc. Controlled-Release Pesticide Symp., University of Akron, Akron, Ohio, September 16 to 18, 1974.
12. **Kysonieus, A. F. and Smith, I. K.**, Efficacy of Hurcon® Polymeric Controlled-Release Dispensers (PCRD) in Cockroach Control, Herculite Protective Fabric Corporation Rep., New York, 1974.
13. **Olton, G. S. and Killmaster, A.**, Controlled-release insecticide for cockroach control, in Proc. Controlled-Release Pesticide Symp., Wright State University, Dayton, Ohio, September 8 to 10, 1975, 258.
14. **Nelson, J. H., Evans, E. S., Bennington, N. E., and Young, W. W.**, The U.S. Army Environmental Hygiene Activities in the Area of Slow Release Insecticides, Rep. No. 30, in Proc. Controlled-Release Pesticides Symp., University of Akron, Akron, Ohio, September 16 to 18,1974.
15. **Anon.**, EPA grants limited use registrations to new mosquito control pesticide, *Weeds, Trees and Turf*, 14(5) 30, 1975.
16. **Cardarelli, N. F.**, Biologically active rubber, Elastomer and Plastics Group New England Section, American Chemical Society, Cambridge, Mass., January 14, 1969.
17. **Harris, F. W.**, Controlled-release herbicides — polymer formulations for aquatic weed control, *Hyacinth Control J.*, 11, 61, 1975.
18. **Higachi, T.**, Mechanisms of sustained action medication, *J. Pharm. Sci.*, 52, 1145, 1963.
19. **Beasley, M. L. and Collins, R. L.**, Water degradable polymers; controlled-release herbicides and other agents, *Science*, 169, 769, 1970.
20. **Collins, R. L.**, A Theoretical Foundation for Controlled-Release, Rep. No. 9, in Proc. Controlled-Release Pesticides Symp., University of Akron, Akron, Ohio, September 16 to 18, 1974.
21. **Harris, F. W., Norris, S. V., and Post, L. K.**, Factors influencing release of fenas from polyethylene matrices, *Weed Sci.*, 21, 318, 1973.
22. **Desai, S. J., Simmonelli, A. P., and Higachi, W. I.**, Influencing release of solid drug dispersed in insert matrices, *J. Pharm. Sci.*, 54, 1459, 1965.
23. **Harris, F.**, Controlled-release from plastic matrices, Planary Lecture, in Proc. Controlled-Release Pesticide Symp., University of Akron, Akron, Ohio, September 13 to 15, 1976.
24. **Cardarelli, N. F.**, Compounding methods for controlled-release elastomers, Planary Lecture, in Proc. Controlled-Release Pesticide Symp., University of Akron, Akron, Ohio, September 13 to 15, 1975.
25. **Neogi, S. A. N.**, Polymer Selection for Controlled-Release Pesticides, Ph.D. thesis, University of Washington, Seattle, 1970.
26. **Kanakkanatt, S. V. and Cardarelli, N. F.**, Matrix factors affecting the controlled-release of pesticides, in *Controlled-Release Pesticide*, Scher, H. B., Ed., Am. Chem. Soc. Symp. No. 53, Washington, D.C., 1977.
27. **James, G. A. and Cardarelli, N. F.**, Aquatic Herbicides Chronicity Study, Final Rep., DACW 73-73-C0042, Office of the Chief of Engineers, Washington, D.C., February 14, 1974.
28. **Anon.**, Snail control in the prevention of bilharziasis, *WHO Monogr. Ser.*, No. 50, 1965.
29. **Doncan, J.**, Development and Application of Molluscicide in Schistosomiasis Control, Rep. Overseas Pool Research, London, September 1972.

30. **Crossland, N. O.**, Molluscicide screening and evaluation, *Bull. WHO*, 33, 567, 1975.
31. Scientific Group on Research in Bilharziasis (Molluscicides), Rep. to Director General, World Health Organization, Geneva, Switzerland, February 9 to 13, 1959.
32. **Deom, J.**, Water resources development and health: a selected bibliography, *WHO Tech. Rep. Ser.*, MPD/76.6, 1976.
33. **Jordan, P., Unrau, G. O., and Dalton, P.**, Preliminary results of a scheme to control *Schistosoma mansoni* through provision of water, *WHO Tech. Rep. Ser.*, WHO/SCHISTO/73.28, 1973.
34. **Hairston, N. G., Wurzinger, K. H., and Busch, J. B.**, Nonchemical methods of snail control, *WHO Tech. Rep. Ser.*, WHO/SCHISTO/75.40, 1975.
35. **Jobin, W. R. and Michelson, E. H.**, Operation of irrigation reservoirs for the control of snails, *Am. J. Trop. Med. Hyg.*, 18, 297, 1969.
36. **Yakagawa, M.**, The Decline of *Schistosoma japonicum* infections in Japan, *WHO Tech. Rep. Ser.*, WHO/SCHISTO/74.34, 1974.
37. **Chernin, E. and Adler, V. L.**, Effects of dessication on eggs of *Australorbis glabratus*, *Ann. Trop. Med. Parasitol.*, 61, 11, 1967.
38. **Warren, K. S. and Weisberger, A. S.**, Suppression of schistosomiasis in snails by chloramphemicol, *Nature (London)*, 209, 422, 1966.
39. **Ansari, N.**, Meeting of directors of collaborating laboratories on molluscicide testing and evaluation, *WHO Tech. Rep. Ser.*, WHO/SCHISTO/71.6, 1971.
40. **Jobin, J. R.**, Cost/benefit analysis for application of molluscicides in the prevention of schistosomiasis, *WHO Tech. Rep. Ser.*, WHO/SCHISTO/73.24, 1973.
41. **Paulini, E.**, On the problem of allocating funds for molluscicides and drugs in schistosomiasis control, *WHO Tech. Rep. Ser.*, WHO/SCHISTO/73.24, 1974.
42. **Dawood, I. R., Clu, K. Y., and Dazy, B. C.**, Herbicide trials in the snail habitat of the Egypt 49 project, *Bull. WHO*, 32, 269, 1965.
43. **Ritchie, L. S., Berrios-Duram, L. A., and Wessel, H. B.**, Molluscicidal time-concentration relationships of organotin compounds, *Bull. WHO*, 31, 147, 1964.
44. **Blair, W. R.**, Advances in specification of trace organotin compounds in aqueous environments, in Proc. Controlled-Release Pesticide Symp., 8.41, University of Akron, Akron, Ohio, September 13 to 15, 1976.

Chapter 3

# AQUATIC WEED PROBLEMS IN THE PANAMA CANAL*

## INTRODUCTION

Mechanical removal of floating aquatic weeds from the Chagres River, main tributary of the Panama Canal, has been a successful operation. After heavy rains, huge islands of water hyacinth and other obnoxious aquatic plants float down the Chagres River and are trapped by a log boom. When the boom is loaded, one end is disconnected from the bank and powerful launches pull the free end upstream and across the river. Weeds caught on the boom are pushed into a prepared basin in which a permanently installed Sauerman Slackline operates. The specifically designed bucket picks up as much as 2 tons of weeds and dumps them high on the bank where they are dried and burned. Under current situations, open burning may eventually be prohibited.

Mechanical removal of submersed aquatic weeds and disposal at sea by dump barges have not been successful. The physical tearing apart of the plant tended to have a cultivating effect which stimulated regrowth, and the barges did not dump well in the water. Attempts to control aquatic vegetation by grazing with an aquatic mammal, the manatee *Trichechus* (spp.) or sea cow, has been unsuccessful in the Panama Canal.

The Panama Canal joins the Atlantic and the Pacific Oceans across the Isthmus of Panama in an area designated as the Panama Canal Zone. It is administered by the Panama Canal Zone Government by a governor appointed by the President of the U.S. with the advice and consent of the Senate. The Canal and its related businesses are operated by the Panama Canal Company. The Zone itself has a total area of 1680 $km^2$: 970 $km^2$ of land and 710 $km^2$ of water including the 4.8 km limit of the ocean coasts. The Canal has an overall length of approximately 80 km including the dredged approaches.

From the Atlantic Ocean, ships pass through 16.5 km of dredged channel in Limon Bay. The bottom of the channel is 12.4 m below mean low water level and 150 m wide. The Canal proceeds for a distance of 10.7 m at sea level to Gatun Locks. Ships are lifted 25.7 m through a series of three locks to Gatun Lake which was formed by damming of the Chagres River. The Canal passes through Gatun Lake and through Gaillard Cut, an excavated channel 13 km to Pedro Miguel Lock. Ships are then lowered 9 m to Miraflores Lake and after passing through the lake for a distance of 2 km to the Miraflores Locks, they are lowered to the Pacific tidewater level. They then pass on through a dredged channel in the Bay of Panama for approximately 8 km and out to sea.

## AQUATIC PLANT SURVEY

The Canal operates on the runoff water from the surrounding area. Most of the water comes from the catchment of the Chagres River Basin and enters Gatun Lake about midway between the two oceans. The Chagres River is typical of most tropical rivers in rain forest areas. Waters of the Chagres River are normally clear with a minimal current, but after tropical storms, there is considerable flooding, and a deluge of water moves on down toward Gatun Lake.

Climatic conditions are ideal for growth and spread of floating and submersed aquatic plants. Flooding causes water hyacinth and other floating aquatic plants in the

* This chapter was prepared prior to the implementation of the Panama Canal treaty, and many situations will change during the ensuing years.

TABLE 3.1

**Commonly Observed Aquatic Plants in the Canal Zone**

| Vegetation type | Common name | Scientific name |
|---|---|---|
| Floating | Water hyacinth | *Eichhornia crassipes* (Mart) Solms. |
| | Water lettuce | *Pistia stratioties* L. |
| | Waterprimrose | *Jussiaes repens* L. |
| | Smartweed | *Polygonium* spp. |
| | Waterfern | *Salvinia* spp. |
| | Duckweed | *Spirodela oligorhiza* (Kurtz) Hegelm. |
| Emersed | Arrowhead | *Sagittaria* spp. |
| | Pickerelweed | *Pontederia Lanceolata* Nutt. |
| | Water paspalm | *Paspalm fluitans* (Ell Kunth.) |
| Submersed | Hydrilla | *Hydrilla verticillata* Royale |
| | Egeria | *Egeria densa* Planchon |
| | Bladderwort | *Utricularia* spp. |
| | Marine naiad | *Najas marina* L. |
| | Coontail | *Ceratophyllum demersum* L. |
| Bank | Narrow cattail | *Typha angustifolia* L. |
| | Common cattail | *T. latifolia* L. |

tributaries and lagoons to move out into the mainstream and into Gatun Lake. Huge islands of water hyacinths are collected in the process and carry with them a threat of interference with canal operation. Sometimes they clog the intakes of cooling systems of large ships and render them helpless, unable to move on their own power.

This chapter is limited to discussion of mechanical contrivances that were tested in the never-ending battle against aquatic vegetation under tropical conditions and studies on biological control of aquatic vegetation with the sea cow or manatee. Vegetation types, common names and scientific names of aquatic plants in this study are given in Table 3.1.

## FLOATING AQUATIC PLANTS

Several mechanical methods have been tried for removal of obnoxious aquatic plants in the Panama Canal over the past 25 years. The most successful of these has been the use of a floating boom to collect vegetation, moving the vegetation to a holding lagoon, and removal from the lagoon with a Sauerman Slackline to a disposal area and then burning the residue after drying. For this method, a floating boom (Figure 3.1) was stretched across the Chagres River as an interceptor and overlapped in the center to allow passage of small boats. On one bank just above the boom, a lagoon was prepared, capable of holding several acres of floating vegetation. When the boom is full, it is disconnected from the other bank. The free end is towed upstream and across the river by launch and thus it is unloaded in the lagoon. When the lagoon is sufficiently loaded, the permanently installed Sauerman slackline is used to lift the vegetation from the water and dump it high upon the bank (Figure 3.2). The specially designed bucket is capable of removing 2 to 3 tons of vegetation in a single pass.

November is the month of highest rainfall in the Canal Zone and the vegetation pile increases day by day. By Christmas it becomes mountainous. Shortly thereafter, the rains stop and there are few water hyacinths entering the lake. During the next few months, there is little or no rainfall. The trade winds blow from the north and dry out the vegetation mass. When it is sufficiently dry, it is set afire and allowed to burn until completely destroyed. Although this operation has been very successful, occasionally

FIGURE 3.1. Floating boom across the Chargres River to hold floating aquatic plants.

FIGURE 3.2. Sauerman Slackline used to lift water hyacinths from the lagoon and dump them high on the bank. (Courtesy of the Panama Canal Company.)

the boom has become overloaded during tropical storms and vegetation has escaped downriver. The result is a pileup of water hyacinths against the Gamboa Bridge. The water hyacinths are so tightly wedged that workmen can walk over them, dry-foot. This jam is removed by surrounding the mass with cables and pulling it under the bridge and into Gatun Lake with winches and tugboats. The mop-up operation employs launches with two pangas fanned out, one on portside and one on starboard. Booms on either side of a launch would probably work equally well, but the Panama Canal has many pangas. They are available whenever required (Figure 3.3).

Use of the winch barge is another mechanical method of removing water weeds in remote areas of Gatun Lake and feeder streams. The winch barge hauls in on a wire rope through a block on a tree on the shore. By this method, debris is sledged upon the bank and kept from moving into the Canal and interfering with traffic. Although this method is practical for cleanup operations and removal of miscellaneous logs and heavy floating objects, it is not very efficient because equipment must be kept at a small scale to be mobile for field operation. Final cleanup is done by hand with crews in small rowboats (Figure 3.4).

## SUBMERSED AQUATIC PLANTS

To remove submersed aquatic plants from Gatun Lake, a unique experiment was undertaken to use a derrick barge, an 8 m rake and 1000 cubic yard dump scows. The derrick barge and rake were used (Figure 3.5) to tear out submersed aquatic vegetation, load it into a dump scow, and tow it out to sea for disposal (Figure 3.6). After some effort, it was possible to remove the vegetation from the water. However, when the scows were towed out to sea, the vegetation matted and would not dump when the scow doors were opened. Even with small dynamite charges, the offensive vegetation still would not drop out, and the scows were eventually unloaded with a clam shell. This method was not investigated further because raking seemed to have a cultivating effect on vegetative growth.

## MOSQUITO CONTROL PROBLEMS

Study of aquatic vegetation related to the *Mansonia* mosquito was initiated in April 1965 by the Health Bureau of the Canal Zone government. The study consisted of trapping adult mosquitoes emerging from aerated roots of aquatic plants by use of floating bamboo traps. The traps were placed in new locations at the first of each month over different plants in shallow and deep water along the banks of the Chagres River for daily collections of any adult mosquitoes caught as they emerged from the pupa stage. Larvae collections were also made by removing individual plants from the river banks, washing larvae into pans and counting those that were removed by the process. Plants that have provided the largest numbers of adult *Mansonia* mosquitoes in the floating collection traps include arrowhead, pickerelweed, and sedge. The largest larval collection was from water lettuce, arrowhead, pickerelweed, and sedge. These are mostly plants growing along the shore, but water lettuce also is found floating in deep water. No evidence was found that submersed aquatic vegetation such as coontail, waterweed, bladderwort, and related plants supported *Mansonia* populations. The pest mosquito has larval and pupa development states occurring under water where they attach their breathing tubes to aerated roots of aquatic plants. Submersed vegetation supports anopheline mosquitoes (the malaria vectors) by protecting them from underwater fish predators.

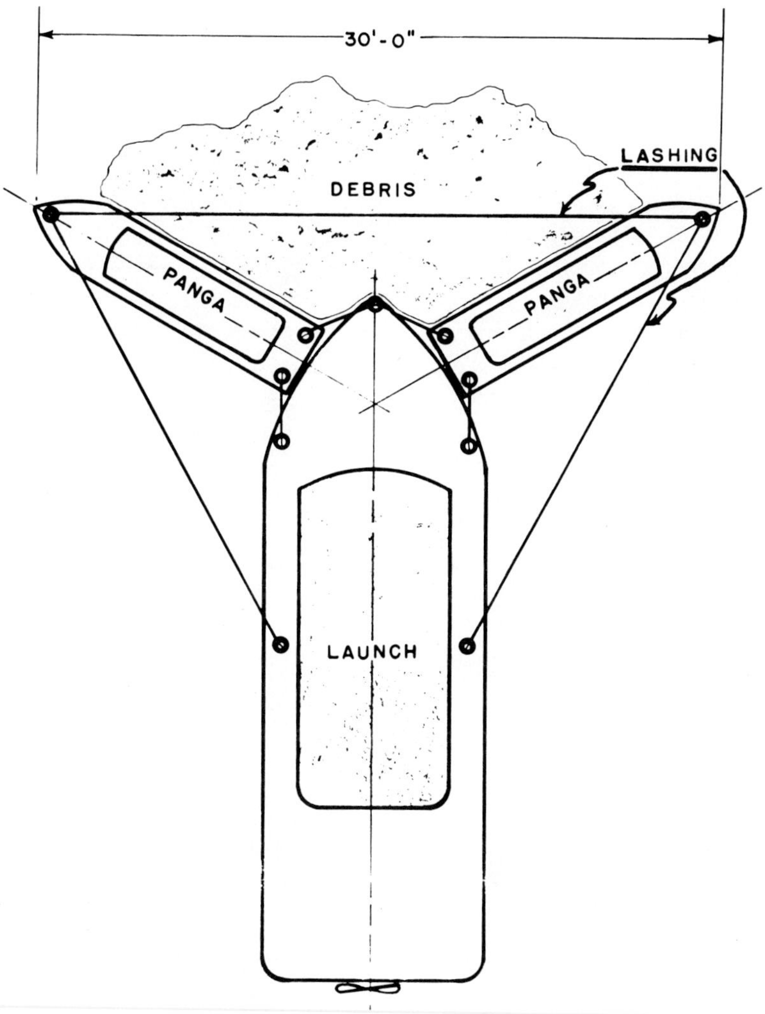


FIGURE 3.3. Mop-up operations are done with a launch with two pangas fanned out on the portside and starboard. (Courtesy of the Panama Canal Company.)

## MANATEE RESEARCH

The Division of Sanitation, Health Bureau, Canal Zone government, became interested in manatees (*Trichechus* spp.) because the manatees had been reported from British Guiana as voracious eaters of aquatic vegetation. Two male manatees weighing approximately 70 kg were transported from Iguiloes, Peru, to a fenced lagoon on the Chagres River. Three females (360 kg each) from Changuinola, Republic of Panama were added later. Although only five grown manatees were in the lagoon during the 2-year study period, general conclusions can be made that manatees are selective in their eating habits where there is abundant vegetation, and they will not eat enough of offending vegetation to reduce mosquito breeding. Cut tops of water hyacinths were the only plants showing results of what the manatees were eating. Since water temperatures in the Canal Zone vary around 27°C throughout the year, manatee deaths from pneumonia, as reported from Florida,[2] were not a problem. Although efforts to reestablish the manatee in the Panama Canal Zone have been successful, there has been

FIGURE 3.4. Final cleanup operations are done by hand with crews in small rowboats. (Courtesy of the Panama Canal Company.)

FIGURE 3.5. Derrick barge and rake were used to tear out submersed vegetation. (Courtesy of the Panama Canal Company.)

FIGURE 3.6. Derrick barge and rake used to remove submersed vegetation and load it into a dump scow. This method was not successful. Matted vegetation would not dump. (Courtesy of the Panama Canal Company.)

no noticeable evidence that the manatees are effective as an aquatic biocontrol agent. It is indeed definite that the manatees have not removed (eaten) the aquatic vegetation that supports *Mansonia* mosquito breeding or the submersed plants that support the anopheles mosquito breeding.

## CHEMICAL CONTROL

Plants observed in 1966 have continued to be a problem.[3] During the following 5 years extensive studies were made to investigate chemical control.[8] Hydrilla were effectively controlled for periods of 12 to 14 weeks by applications of copper sulfate crystals, diquat, diquat-copper sulfate combinations, and Hydrothol 191 (liquid). Treatments of diquat and diquat-copper sulfate were observed to have no toxic effect on fish. Copper sulfate and Hydrothol 191 (liquid) at the level required for weed control were toxic to fish.

## NEW CONSTRUCTION

The Panama Canal is now obsolete for many ocean-going vessels, and there have been a number of different proposals to change or develop new canals in the area. One of the most interesting and plausible is a sea-level canal designated as Route 10.[7] In this study,[7] various options were investigated, including operation of the canal with and without locks and with and without a bypass.

**Accuracy of estimates** — Adequate data are available for assessing the engineering feasibility of constructing a canal along this route. Climatic, hydrologic, and geologic records for the Canal Zone area have been kept for nearly 100 years. The climatic and hydrologic records generally can be extrapolated to Route 10, and on-site geologic investigations were conducted to answer specific questions relating to this route. In general, the accuracy of Route 10 estimates is considered good (see Figure 3.7).

**Route 10 trace** — The Pacific terminus of Route 10 is at the town of Puerto Caimito at the mouth of the Caimito River. The trace follows the river northwestward for 5 mi, crossing the Pan American Highway about 3 mi northeast of La Chorrera. It continues to the north through generally open, rolling terrain; crosses the Continental Divide through the Chorrera Gap; and parallels the Pescado River until it reaches an arm of Gatun Lake at La Laguna. Turning in a more westerly direction, the trace continues over relatively flat terrain and crosses the Trinidad Arm of Gatun Lake to a point about 3 mi southwest of the town of Escobal. From there it runs northwesterly through low ridges which become more open toward the coast.

**Construction** — Most of the excavation along Route 10 would employ open-pit mining techniques, using rail haul for soil disposal. Truck haul would be used at higher elevations, while dredges would excavate the approach channels and the layer of muck at the bottom of Gatun Lake. Barrier dams would maintain Gatun Lake at levels needed for operating the Panama Canal during construction, at the same time permitting excavating at controlled water levels or in the dry. Muck underlying these dam sites would be removed by hydraulic dredging, after which soil from dry excavation would be brought in to construct the embankments.

Diversion of streams on Route 10 would be relatively simple because their drainage basins are small. Most of them would be diverted into the Caribbean Sea; the Caimito River would be the only stream of consequence to discharge into the canal.

Because construction and operation of Route 10 could be supported largely from existing facilities in the Canal Zone and the metropolitan area of Panama, supporting construction requirements would be minimal. Required items would include a trans-isthmian highway, crossing Gatun Lake over the barrier dams, breakwaters on the Caribbean coast, a jetty on the Pacific, and a high-level bridge over the canal.

**Problem areas** — The most critical engineering problems involve the geology of the divide area which consists of a hard basalt cap overlying much weaker materials. Soft rocks, similar in strength to those along the Panama Canal, are found at the depth of the navigation channel. These highly altered volcanic rocks have the undesirable properties of clay shales. Slope angles required for stability in the divide area would have to be verified through detailed investigations or modified by observation during construction.

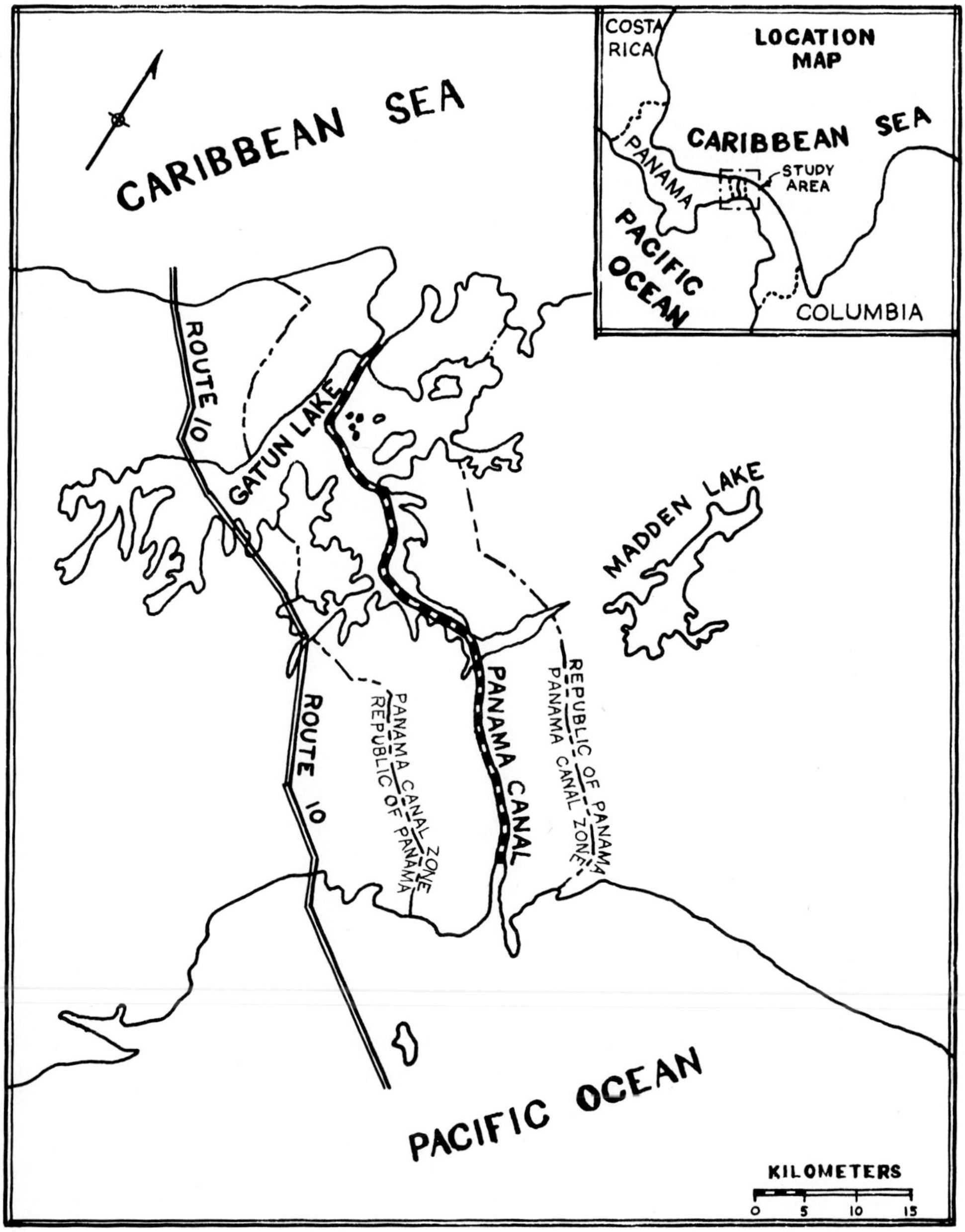


FIGURE 3.7. Route 10 Canal Zone and vicinity.

In the design of the barrier dams, particular attention must be given to the poor foundation material in Gatun Lake and to the stability of fill material in embankments. Throughout the construction period, the lake would have to be maintained at its present level. Although failure of these dams is highly unlikely, it would have disastrous effects not only on construction of the sea-level canal, but on the operation and safety of the Panama Canal as well.

Aquatic weed problems are likely to be about the same as those already present in the Panama Canal, and the same control measures would be required.

## REFERENCES

1. **Allsopp, W. H. L.**, The manatee-ecology and use for weed control, *Nature (London)*, 188, 762, 1960.
2. **Blackburn, R. D., Sutton, D. L., and Taylor, T.**, Biological control of aquatic weeds, *Proc. Am. Soc. Civ. Eng.*, 97, 421, 1971.
3. **Hearne, J. S.**, The Panama Canal's aquatic weed problem, *Hyacinth Control J.*, 5, 1, 1966.
4. **Moore, J. C.**, Distribution of marine mammals in Florida waters, *Am. Midl. Nat.*, 49(1), 117, 1953.
5. **Squros, P. L., Monkus, T., and Phillips, C.**, Observations and techniques in the study of the Florida manatee-reticent, but superb weed control agent, *Proc. South. Weed Conf.*, 18, 5188, 1965.
6. Threatened Wildlife of the United States, Resource Publication No. 114, U.S. Department of the Interior, Washington, D.C., 1973.
7. Interoceanic Canal Studies, Atlantic-Pacific Interoceanic Study Commission, Washington, D.C., 1970.
8. **Hearne, J. S.**, Aquatic weed control trials, Gatun Lake, Panama Canal, *Hyacinth Control J.*, 10, 33, 1972.

Chapter 4

# AQUATIC WEED PROBLEMS OF PUERTO RICO*

## INTRODUCTION

In Puerto Rico, the main aquatic plant problem areas are along the north side of the island from Rio Grande to Arecibo. Lesser infestations exist in the river systems along the southern part of the island from Naguabo to Ponce and in the southwestern part from Mayaguez through Valle de Lajas to Yauco. The major reservoirs infested are Lago de Cidra, Lago de Loiza, Lago de la Plata, Lago Dos Bocas, Lago Coamo, and Laguna Cartagena. There is relatively little infestation in the streams and lakes of the high mountainous interior. Water hyacinth (*Eichhornia crassipes* (Mart.) Solms-Lauback) is the major aquatic plant problem, but alligator weed (*Alternanthera philoxeroides* (Mart.) Griseb.) is likely to become a problem when water hyacinth is brought under control. Chemical and mechanical methods of control have been used with some success. Efforts are underway to introduce biological methods of control that will be more efficient and economical.

A control program is needed for aquatic weed problems in Puerto Rico. The desire of local interests to participate in a program is such as to warrant establishment of a priority for expenditure of available funds. It seemed advisable to explore the situation with detailed coordination and joint appraisals of other agencies before undertaking a general program. A meeting to discuss the program was held in the U.S. Army Corps of Engineers District Office in San Juan on June 5, 1974 with representatives of the Puerto Rico Department of Natural Resources, Puerto Rico Aqueduct and Sewer Authority, Puerto Rico Water Resources Authority, and the Puerto Rico Environmental Quality Board.

During a preliminary reconnaissance of aquatic plants in the watersheds of Puerto Rico, representatives of the Corps of Engineers, Jacksonville District (June 1968) observed localized growths of alligator weed and recommended introduction of the alligator weed flea beetle (*Agasicles hygrophila* Selman-Vogt) as a measure of biological control. Research personnel of the U.S. Department of Agriculture, Fort Lauderdale, Florida, visited the Commonwealth of Puerto Rico to observe problems of aquatic weeds in large reservoirs, drainage and irrigation canals, and small farm ponds with the assistance of Dr. F. F. Ferguson,[3] Chief, Puerto Rico Field Station, Communicable Disease Center and Mr. Rafael Reyes, Assistant General Manager of Irrigation Services of the Puerto Rico Water Resources Authority. The largest and probably most pressing problem appeared to be the Lajas Valley Irrigation and Drainage District of the Puerto Rico Water Resources Authority.

The primary aquatic weed problem in the drainage laterals of the Lajas Valley area appeared to be paragrass (*Panicum purpurascens* Raddi.) and water smartweed (*Polygonum natans* Eaton). The use of 2,2-dichloropropionic acid (dalapon) and the dimethylamine salt of 2,4-dichlorophenoxyacetic acid (2,4-D) is recommended. A combination of these herbicides can also be used to control the cattails (*Typha* sp.) growing in the canals.[6-8]

Probably the easiest and most efficient way of spraying canals of 10 m in width would be from the canal bank. Since roads have been constructed along canal banks, it would be very easy to drive a truck along these canals and spray the aquatic vegeta-

* From Gangstad, E. O., *J. Aquatic Plant Management*, 15, 3, 1977. With permission.

tion. It would be very easy to have 30 m of hose connected to the tank pump with a hand-type quick release gun connected to the hose end. A man walking along the canal bank could apply the spray solution over the aquatic weeds. A boom that would hang out over the canal surface could also be mounted on the truck or trailer.

Human health aspects of aquatic weed control are well documented. The problem is particularly acute in tropical areas such as Puerto Rico.[5] Water weeds support many parasite and disease organisms including malaria, encephalitis, filariasis, schistosomes, liver flukes, and disease-pest arthropods. In the interest of national health, these problems must be dealt with throughout Central America and the Gulf Coast of the U.S.

## BIOLOGICAL CONTROL

During the past two decades, intensive efforts have been made to research biological methods of control as a means of more effectively controlling aquatic weed species such as alligator weed and water hyacinth. The U.S. Army Corps of Engineers has supported varied and wide-spread efforts to achieve this end. Natural enemies of these plants have been studied throughout South America, Central America, Southern United States, Thailand, and India.[1-3,4,8-11]

Preliminary results on the use of the Brazilian snail (*Pomacea australia* d'Orbigny) by the U.S. Army Waterways Experiment Station in Puerto Rico and results of the use of white amur (*Ctenopharyngodon idella* Val.) by the U.S. Department of Agriculture Research Center at Fort Lauderdale, Florida to control water hyacinth are not considered satisfactory for field control.

Insects have been used successfully as a form of biological control to suppress alligator weed in Florida and other states of the Southeast under the Corps of Engineers Aquatic Plant Control Program in cooperation with the Division of Entomology Research of the United States Department of Agriculture. The alligator weed flea beetle was the first host-specific insect to be introduced. Other insects, alligator weed thrips (*Amynothrips andersoni* O'Neill), and a stem-boring moth (*Vogtia malloi* Pastrana) are also host-specific and have been introduced for alligator weed control. Infestations of alligator weed are reduced to a maintenance status in most situations in the southeastern states where these insect controls have been released. These insect enemies could be released in Puerto Rico for the control of alligator weed. Insects and diseases for control of water hyacinth are not fully researched for field application, but current results indicate that biological control of water hyacinth is a very real possibility.

## WATERSHED AREAS

### Rio Grande de Loiza

The Rio Grande de Loiza is the most prominent river in Puerto Rico. Its drainage area, the largest in Puerto Rico, is about 18 $km.^2$ Major tributaries are the Rio Canovanas, Rio Canovanillas, Rio Caguitas, Rio Turabo, and Rio Gurabo. The basin is situated in the eastern central part of Puerto Rico.

A broad valley extending from Aguas Buenas in the west to Pena Pobre in the east comprises a large part of the upper basin. It is in the valley near Caguas where the Rios Turabo, Caguitas, and Gurabo join the Rio Grande de Loiza to flow to the coast. Flowing north from this valley, the Rio Grande de Loiza enters the Loiza or Carraizo reservoir, the water supply for the metropolitan San Juan area. A power installation at the dam is capable of generating 2600 kW of electrical power. The water hyacinth infestation starts just upstream of the Loiza reservoir, and heavy infestations occur in the reservoir causing problems in the generation of electrical power and water supply.

Water hyacinths are flushed from the reservoir and flow into the coastal plain east of San Juan.

**Rio de Bayamon**

The drainage area of the Rio de Bayamon originates in the central mountain interior of Puerto Rico where small tributaries collect at Cidra Reservoir before emerging into a narrow valley. The stream flows 30 km through this valley and is joined by the Rio Guaynabo before reaching the coastal plain. This river drains an area of 16 $km^2$ and has caused serious flooding until the past few years. Lago Cidra storage is used to regulate runoff in the upper basin and serves as a source of municipal water. The lake is the major source of water hyacinth infestation in the Rio de Bayamon watershed.

**Rio de La Plata**

The Rio de La Plata is a major river in the central part of Puerto Rico. The river drains a basin of 150 $km^2$ in area. This river, unlike the other large rivers to the west, has a long main stem. Significant tributaries are the Rio Hondo, Rio Arroyata, Rio Usabon, and Rio Guavate. The terrain of this basin is basically of two types: the flat coastal plain and the interior mountains. The majority of the water hyacinth infestations are in the coastal plain area and La Plata Lake.

There are three existing water resource developments in the Rio de La Plata. Lago Carite in the upper part of the basin collects and diverts all the normal runoff to the Rio Guamani on the south coast. The Puerto Rico Water Resources Authority has two small dams where turbines and generators are installed.

Lago La Plata located above Toa Alta has been recently constructed by the Aqueduct and Sewer Authority to provide water supply to the western part of the San Juan metropolitan area. Water hyacinth infestations apparently start above this lake. During the filling of the lake, large amounts of plants flowed into the lake from upstream and accumulated at the dam. One of the heaviest infestations of water hyacinths on the island is in the Rio de La Plata from Lago La Plata by Toa Alta to Dorado. In June 1974, the river was solidly covered with water hyacinths approximately 1 m high.

Coastal flooding is a major problem in this basin. During major floods, the town of Toa Baja is completely inundated. Dorado, Toa Alta, Ingenio, and Companillas are also affected by flooding. Agricultural interests are also adversely affected by floods.

**Rio Grande de Manati**

The Rio Grande de Manati basin lies in the central part of northern Puerto Rico. Encompassing some 100 $km^2$ of drainage area, this basin is one of the largest river basins in Puerto Rico. In the areas checked the river had a fairly rapid flow and steep banks. Water hyacinths were noted in small pockets along the banks; however, free-floating plants apparently flowed downstream into the Atlantic Ocean and were a minor problem.

**Rio Grande de Arecibo**

The Rio Grande de Arecibo is one of the large river basins of Puerto Rico. The basin is composed of three distinct types of terrain. A coastal plain 5 to 8 km wide lies adjacent to the coast and is heavily infested. A 12 km strip of limestone formation separates the coastal plain from the upper definable drainage basin.

There are several reservoirs in the basin: Dos Bocas, on the main stem at a point that controls flows from all the basin streams, and a chain of small reservoirs, Lakes Adjuntas, Pellejas, and Vivi that tunnels into Lake Caonillas. All of these have been

TABLE 4.1

**Problem Areas, Classes of Aquatic Plants, Surface Area Infested, and Methods of Control**

| Problem areas | Aquatic plants | Surface area (ha) | Control methods | Costs (monthly) |
|---|---|---|---|---|
| Lake Loiza (Carraizo) | Water hyacinths<br>Water lettuce<br>Alligator weed<br>Paragrass | 700 | 2,4-D<br>Mechanical | $2500 |
| Lake Cidra | Water hyacinths<br>Paragrass<br>Alligator weed<br>Cattail | 200 | 2,4-D<br>Mechanical | $1500 |
| Lake La Plata | Water hyacinths<br>Water lettuce<br>Coontail<br>Alligator weed<br>Paragrass | 500 | 2,4-D<br>Mechanical | $1500 |
| Botijas River | Water hyacinths<br>Water lettuce<br>Paragrass | 1 | Mechanical | — |
| Usabon River | Water hyacinths<br>Paragrass | 1 | Mechanical | — |
| Caguitas River | Water hyacinths | 1 | Mechanical | — |

built and operated for power generation. Lago Dos Bocas is the only lake reported to have a water hyacinth problem.

The principal community, Arecibo, is located on the west bank of the river along the coast. Industrial and commercial activity predominate along the coastal plain along with sugar cane production. The municipality has suffered flooding which has affected the area's established development.

The major water hyacinth infestation is in the numerous channels in the delta area. Branches and side channels are also highly infested with water hyacinth; however, the main river generally maintains itself free from the plants. Serious problems develop during high water when floating plants accumulate at the bridges and impede the water flow.

## CHEMICAL AND MECHANICAL CONTROL

Information contained in Table 4.1 includes operations in infested areas of lakes and rivers used in Puerto Rico for the production of potable water. Problem areas, obnoxious plants, surface area infested, and methods used for control are given. Physical and economic damages caused by the presence of these massive aquatic growths are indeed extensive, but the costs listed are limited to those involved in the production of potable water.

It is important to recognize that bad odors and tastes arising from the growth of these plants affect the services rendered. To cope with this problem, mechanical aeration has become necessary in Puerto Rican water treatment plants. Extensive use of activated carbon has been particularly necessary in the Sergio Cuevas filtration plant. This plant produces about 300,000 $k\ell$/day from raw water withdrawn from Lake Loiza, a heavily infested resource.

Total or average costs are not available, but during a recent 10-month period[12] the Puerto Rico Aqueduct and Sewer Authority at San Juan, Puerto Rico spent over $130,000 to mechanically harvest water hyacinths from Lake de Cidra to eliminate the aquatic plant problem (see Figures 4.1 to 4.4).

FIGURE 4.1. Collecting water hyacinth plants for mechanical harvesting in Lago de Cidra. The total cost of harvesting is approximately $1000.00/ha.

FIGURE 4.2. Water hyacinths removed from lake by dragline hoist.

FIGURE 4.3. Loading water hyacinths with clamshell buckets onto a standard dumptruck for removal to disposal area.

FIGURE 4.4. Unloading water hyacinth from the field trucks at the disposal site by mechanical hoist and buckets.

## REFERENCES


1. **Bennett, F. D. and Twolfer, H.**, Exploration for natural enemies of the water hyacinth in northern South America and Trinidad, *Hyacinth Control J.*, 7, 4, 1968.
2. **Blackburn, R. D., Sutton, D. L., and Taylor, T. M.**, Biological control of aquatic weeds, *ASCE J. Irrig. Drain Div.*, 97, 421, 1971.
3. **Charudattan, R.**, Pathogenicity of fungi and bacteria from India to hydrilla and water hyacinth, *Hyacinth Control J.*, 11, 44, 1973.
4. **Coulson, J. R.**, Prognosis for control of water hyacinth by arthropods, *Hyacinth Control J.*, 9, 31, 1971.
5. **Ferguson, F. F.**, Aquatic weeds and man's well-being, *Hyacinth Control J.*, 7, 7, 1968.
6. **Gangstad, E. O.**, Aquatic plant control program, *Hyacinth Control J.*, 9, 7, 1971.
7. **Gangstad, E. O.**, Herbicidal control of aquatic plants, *ASCE J. Sanit. Eng. Div.*, 98, 397, 1972.
8. **Gangstad, E. O., Seaman, D. E., and Nelson, M. L.**, Potential growth of aquatic plants of the Lower Mekong River Basin, *Hyacinth Control J.*, 10, 4, 1972.
9. **Rintz, R. E.**, A zonal leaf spot of water hyacinth caused by *Cephalosporium zonatum*, *Hyacinth Control J.*, 11, 41, 1973.
10. **Rushing, W. N.**, Water hyacinth research in Puerto Rico, *Hyacinth Control J.*, 12, 48, 1974.
11. **Zeiger, C. F.**, Biological control of alligator weed with *Agasicles n.* sp. in Florida, *Hyacinth Control J.*, 6, 31, 1967.
12. Preliminary Reconnaissance Report for Puerto Rico, Aquatic Plant Control Program, U.S. Army Corps of Engineers, Jacksonville, Fla., 1974.

Chapter 5

# REMOTE SENSING OF AQUATIC PLANTS*

## INTRODUCTION

Native aquatic plants have existed for approximately as long as there have been aquatic environments, usually in reasonable equilibrium with the lakes and streams supporting them. Nonnative plants are frequently introduced into U.S. waters. Lacking the biological and climatological constraints that may have existed in their native habitats, and nourished by the high nutrient effluents currently encountered in most local waters, these noxious plants often spread prolifically. Their unchecked growth causes not only displacement of native plant species, but substantial takeover of previously open water in reservoirs and waterways. This diminishes the free flow of water, reduces recreational potential and property values, increases transpiration losses, restricts navigation, increases water treatment costs, and provides ideal breeding grounds for disease vectors.

Newly formed reservoirs, popular spots for pleasure boating, are particularly vulnerable to infestation from these exotics. A trailered boat can move from lake to lake and from state to state with viable plant material clinging to trailer, hull and outboard motor. On relaunching of the boat, these plant fragments drop to the bottom and soon take root in the new habitat.

Texas typically exemplifies a state with new and growing aquatic weed problems. Over the past decade, lakes in the state have become infested with several aggressively growing nonnative aquatic plant species. Chief among these is water hyacinth (*Eichhornia crassipes* (Martius) Solms-Laubach). There are also spreading populations of Eurasian watermilfoil (*Myriophyllum spicatum* L.), hydrilla (*Hydrilla verticillata* Royle) and alligator weed (*Alternanthera philoxeroides* (Martius) Grisebach).

Information on infested areas comes from incidental boat reconnaissance; from reports of nearby property owners, recreational boaters, and marina operators; and occasionally, from chance aerial observation. Thus, there is no specific program for determining how effective ongoing measures are in controlling the total problem in Texas. Lacking current synoptic data on spread and condition of aquatic plants, the control program manager is forced to react to localized complaints of outbreaks rather than act on a coherent body of information.

### The Need for Monitoring

Existence of aquatic plant infestations constitutes an environmental problem relating to public health, to water resources, to water recreation, and simply to aesthetics. For whatever reason, a management methodology is required. A successful plant control program needs means of detection and assessment of infested areas and of monitoring treated areas to determine the value of the utilized control method. Operational realities of the aquatic plant control problem often make detection and monitoring by ordinary means rather difficult. New infestations frequently occur well off the beaten path. Inaccessibility of a large part of these plant-clogged areas makes it almost impossible to measure the efficiency of a chosen control method.

The situation is ideal for application of remote sensing. This concept, discussed at

* Abridged and edited from Remote Sensing of Aquatic Plants, by Arthur R. Benton, Jr., Associate Research Engineer, Texas A & M University, College Station, Texas. With permission.

the 15th annual meeting* of the Aquatic Plant Management Society, San Antonio, Texas, found near unanimous agreement — remote sensing is the best monitor of plant infestation. Nevertheless, no one could justify the assumed "high cost of remote sensing," let along pioneer its application. The Texas Parks and Wildlife Department, responsible for most of the state's aquatic plant control, would like to use remote-sensing monitoring for aquatic plant inventory work but lacks funds and expertise.[1]

This chapter describes a study[2] whose purpose was to demonstrate the need for remote sensing monitoring of aquatic plants and to examine the economic feasibility of its use in moderately well-funded control programs.

**Water Loss from Aquatic Plant Infestation**

The study's primary objective was to develop and refine an economical remote sensing procedure. It is inappropriate to consider a remote sensing procedure unrelated to the aquatic plant overgrowth problem, as though mere existence of plants was sufficient reason for eradication. The study therefore investigated the economics of aquatic plant infestations and the impact of water loss through evapotranspiration, a process seldom considered.

Terrestrial plant transpiration is well known. Potential transpiration from floating aquatic plants in reservoirs is even more significant from the standpoint of water loss per unit area. Water hyacinth is the floating plant of importance in the U.S. today. Several researchers have investigated its transpiration rates. Their results were expressed as the ratio of plant transpiration to open water evaporation per unit area. The highest figure was 5.3, obtained by Rogers and Davis in Florida.[3] Timmer and Weldon[4] also in Florida found a ratio of 3.7. In Louisiana, Penfound and Earle obtained 3.2.[5] The Rogers and Davis experiment utilized periodic harvesting to reduce crowding. The latter two groups left plants in their crowded natural state. Timmer and Weldon also took data on assumed causative factors such as temperature, relative humidity, wind speed, and solar radiation. They found little direct correlation of transpiration with temperature, humidity, or air movement. They did, however, discover a strong correlation between solar radiation and transpiration, a relationship that held almost constant over the time frame of their experiment (late April to early September). The mean average weekly transpiration rate was found to be 10.06 cm (3.96 in.), and the mean cumulative solar radiation per week was 3505 lys. This equates to 2.870 cm (1.130 in.) of transpiration per 1000 lys.[4]

The same experiment showed that the relationship between transpiration and evaporation is not constant. Although the Timmer and Weldon mean value was 3.7, the ratio varied between 13.4:1 and 2.2:1. For this reason, it seemed inappropriate to use the 3.7 figure as a factor to determine transpiration at a different place or time. However, it did seem reasonable to use their derived relationship between solar radiation and transpiration as a predictive factor for Texas.

Using this relationship and climatological records for the state of Texas, a projection of the anticipated ratio of transpiration to evaporation was obtained. The average was slightly over 3.0.[2] Interestingly enough, the watershed adjacent to the Louisiana border was found to have a projected ratio of 3.22, almost the same figure as that found by Penfound and Earle during their Louisiana research.[2,5] The final result of the substudy on evapotranspiration effects in Texas was the indication that the net losses in terms of percent of planned reservoir yield from water hyacinth transpiration would be numerically equal to the mid-season percentage of water hyacinth coverage on the average reservoir. A mean water hyacinth coverage of 20% would result in a 20% loss of the

* July 1975.

planned yield of reservoirs involved. This unexpected loss would be roughly half of the anticipated loss through evaporation.[2]

## REMOTE SENSING TEST PROGRAM

The foregoing postulates the feasibility of using aerial photography to effectively and economically monitor aquatic plants. A 3-year test of low cost aerial photographic procedures will be discussed in detail.

At the beginning of the test, it was assumed that a few overflights would sufficiently establish the typical appearance of a variety of aquatic plants in Texas lakes on color infrared film. Earlier research by Gustafson and Adams in Wisconsin indicated that aquatic plant species could be uniquely defined on color infrared imagery by texture and Munsell color.[6] Seher and Tueller, working in Nevada, also described fresh water marsh plant species in terms of absolute color on color infrared film.[7] Several researchers apparently obtained similar results using color infrared photography for differentiation of coastal wetland vegetation.[8-10]

However, after three flights (beginning in April of 1974) it was discovered that many factors dictated color appearance of a given plant on color infrared film, only one of which was the species' spectral reflectance characteristics, i.e., absolute color. Further, the spectral characteristics themselves varied with seasonal maturity, relative biomass, and stress conditions. Study procedures were modified to compensate for as many as possible extraneous factors governing color variability. Also, to document seasonal changes, it became necessary to photograph the various species throughout the annual sequence of youth, maturity, and senescence or dormancy.

The 3-year study resulted in the definition of an effective, low cost remote sensing methodology. The theory and test procedures are discussed first.

### Theory of Remote Sensing in the Visible and Near-Infrared Ranges

Basic tools of this research program were color and color infrared film. Color film simulates the color response of the human eye, detecting and portraying objects in terms of their spectral reflectance within the visible range — the relative energy within the 400- to 700-nm range reflected from those objects and detected.

#### *Spectral Response of Photographic Films*

Perceived by the human eye, the spectral reflectance of a tree would begin at approximately 400 nm, show a low level of reflected energy in the violet to blue range (400 to 500 nm), rise to a peak in the green to yellow range (500 to 600 nm), then decrease through the orange to red range (600 to 700 nm). In other words, a tree looks green to the eye. Considering only the 400 to 700 nm range, a plot of the spectral signatures for a large variety of plants would show a great deal of similarity since they would appear (with rather subtle differences) essentially green. Color photography, designed to reproduce the colors seen by the eye, faithfully records these plant-to-plant similarities.

Given a sensor that can detect reflected energy in the near infrared (from 700 to 850 nm), a whole new world of vegetation detection and differentiation becomes possible. Figure 5.1 shows spectral signatures for several plant classifications.[11] Note the similarities of these signatures in the visible range and the significant differences in the near infrared range. Herein lies the advantage of color infrared photography.

Color film is made with three emulsion layers that respond to blue, green, and red light so that the final film positive exhibits colors quite similar to that seen by the human eye. On the other hand, color infrared film is designed to respond to green, red, and near-infrared light, but recorded as blue, green, and red, respectively.

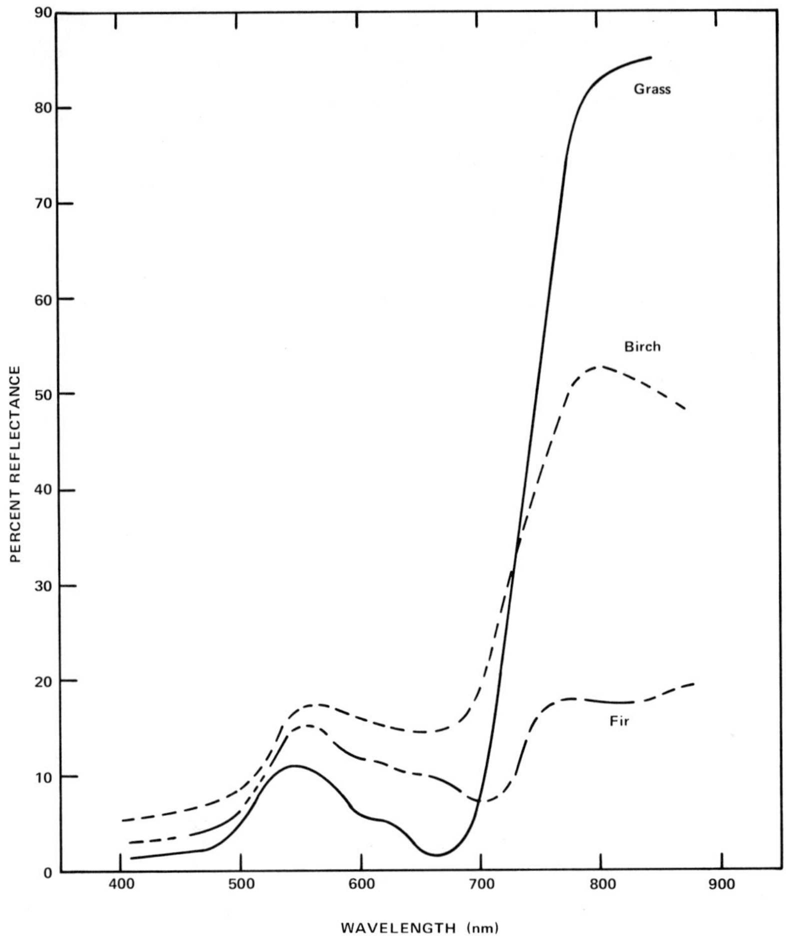


FIGURE 5.1. Typical spectral signatures for vegetation. (From Fritz, N. L., *Photogramm. Eng.*, 34, 1128, 1968. With permission.)

Actually, all three emulsion layers respond to blue light as well. For this reason, color infrared film should be used with a filter that eliminates all light of wavelengths shorter than approximately 500 nm. Figure 5.2 shows response of the various emulsion layers in color infrared film with a 500 nm filter curve superimposed.[12] Note that the infrared-sensitive cyan layer, which provides the red color in the final image, is substantially less sensitive than the other two. This is done purposely to compensate for the extremely strong, near infrared reflectance normally associated with healthy vegetation.

The result is a photograph in which the colors are shifted so that blue is unseen, green appears blue, red appears green, and the near-infrared appears red. Thus, on color infrared imagery, green vegetation assumes colors ranging from pale lavenders to bright reds, often mixed with greens and browns, depending on species, season, and stress conditions.

Another characteristic of color infrared film is the clarity with which these false colors are seen in aerial photography. This is a result of the haze effect reduction due to use of the 500-nm filter. The blurring effect of haze is primarily due to scattering

PLATE 1. Color infrared photo of the east tip of Lake Livingston "Jungle", April 24, 1974.

PLATE 2. Color infrared photo of the east tip of Lake Livingston "Jungle", June 26, 1974.

PLATE 3. Color infrared photo of the Skaines Lake area of Lake Livingston, August 21, 1974.

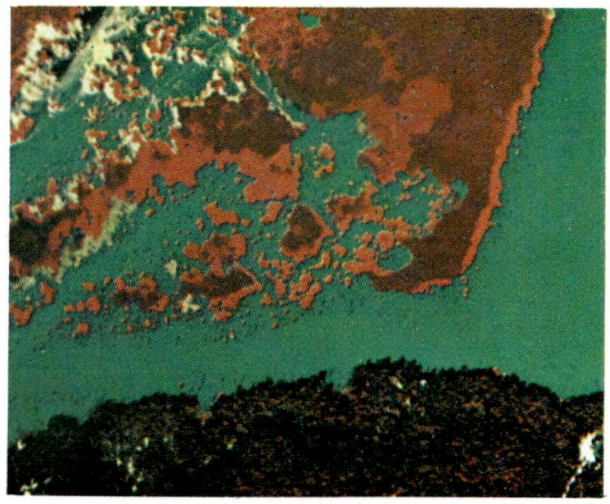

PLATE 4. Color infrared photo of the east tip of Lake Livingston "Jungle", October 2, 1974.

PLATE 5. Color infrared photo of the Sebastopol area of Lake Livingston, September 23, 1976.

PLATE 6. Color infrared closeup of the Sebastopol area of Lake Livingston, September 23, 1976

PLATE 7. Color infrared photo of the Penwaugh Slough area of Lake Livingston, September 23, 1976.

PLATE 8. Color infrared photo of the lower end of Lake Austin, Texas, October 18, 1974.

PLATE 9. Color infrared photo of the upper end of Lake Austin, Texas, October 18, 1974.

PLATE 10. Color infrared photo of the southeast tip of Lake Livingston "Jungle", November 14, 1975.

PLATE 11. Color infrared photo of Brushy Creek inlet, Lake Livingston, October 17, 1975.

PLATE 12. Filtered color photo of Brushy Creek inlet, Lake Livingston, October 17, 1975.

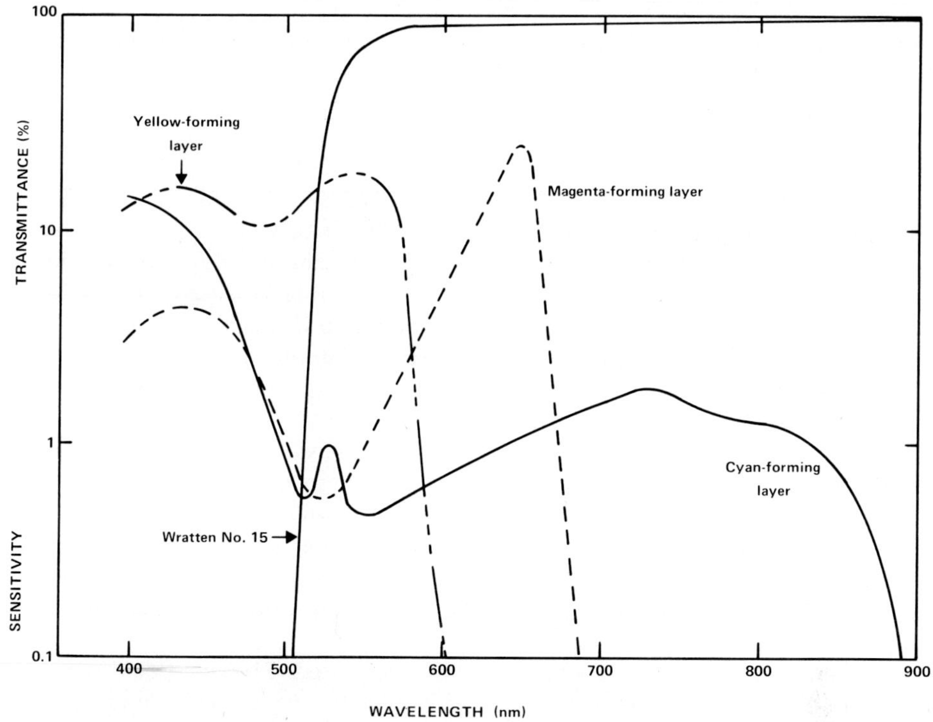


FIGURE 5.2. Spectral sensitivity for color infrared film. Also shown is spectral transmittance for Wratten filter No. 15. (From Fritz, N. L., *Proc. 3rd Biennial Workshop on Color Aerial Photography in the Plant Sciences*, Gainesville, Fla. 1971, 34.)

of blue light by particles composing the haze layer. This fuzziness is detected by the eye and by color film as well. This is not the case with color infrared film imagery on which the blue light has been entirely filtered out.

### *Film Qualities other than Spectral Sensitivity*

Photographic remote sensing is particularly effective since it faithfully records object characteristics other than color — shape, texture, and context. Shape refers to the outline of an image; texture refers to its roughness or smoothness. The two go hand in hand and are somewhat dependent on scale. A close-up of a tree shows the intricate branch pattern, leaf shape and clusters, perhaps roughness of the bark, and a shiny or dull leaf surface. At a distance, these characteristics would vanish completely or assume mass characteristics. Thus, a hardwood forest looks pebbly from a considerable height, and an evergreen forest appears somewhat spiky. "Coarse" and "smooth" describe different types of vegetation masses seen in small scale. Similarly, forests may have sharp or diffuse borders depending on natural environmental conditions and those imposed by man.

Context is an important part of remote sensing interpretation. We associate certain types of vegetation with certain environments. Given a desert environment, the expected vegetation classes would be limited. A tree growing in wet or swampy ground is much more likely to be a willow than a spruce. A eutrophic lake would be expected to contain one or more species of aquatic vegetation growing around its periphery. Conversely, the presence of such plants would indicate that eutrophication had set in.

Considered separately, spectral signature, shape, texture, and context are seldom conclusive. Taken together (particularly when spot correlated by filed verification), a very high confidence level can be assumed for plant identification. Color infrared film, since it combines enhanced spectral range with a faithful portrayal of shape, texture, and context, is an excellent remote sensing tool with particularly significant application in aquatic vegetation analysis.

**Test Methodology**

The research field work began in the spring of 1974, roughly coincident with youthful emergence of noxious species then known to inhabit east Texas. The intention was to document the seasonal appearance of the various species on color and color infrared film preliminary to further investigations into the remote sensing of eutrophication factors. The complexity and eventual broad application of aerial photographic detection and monitoring of aquatic plants were not realized at the time.

*Equipment*

Two 70-mm electrically driven Hasselblad® cameras were mounted with their optical axes parallel. They were triggered simultaneously to provide dual coverage of a single scene with two separate film and filter combinations. One camera held Kodak® Aerial Infrared Film 2443 and used a Wratten® 12 filter to eliminate blue light; the other used Kodak Ektachrome® MS Aerographic Film 2448, a standard aerial color film, without filter. Use of two different films hopefully would enhance the potential for accurate species differentiation.

Two different De Havilland® Beaver aircraft were used. The first, lacking a vertical camera mount, was flown with the cameras bolted to an oblique mount and aimed out the open door. The second, used after August 1974, was equipped with a camera hatch and mount for vertical photography. Film imagery was analyzed on a Richards® binocular microscope viewing table, a first-rate instrument whose bright light source closely approximates daylight color temperature.

*Test Area*

The 1974 test program was carried out initially on several different lakes within an hour's flight radius of College Station, Texas. The purpose was to document the photographic appearance of as many different aquatic plant species as possible. Because of the logistics problem involved in multilake operations, and because Lake Livingston (see Figures 5.3 and 5.4) was found to contain all but one of the plant species in abundance, the project concentrated on Lake Livingston after October. The only other major lakes photographed in 1974 were Lake Austin and Camp Creek Lake.

The 1975 and 1976 work concentrated almost exclusively on Lake Livingston. The exception was some late fall target-of-opportunity photography of Lake Conroe, a large, new lake lying along the flight route between College Station and Lake Livingston.

Lake Austin is a 730-ha (1820 acre) reservoir lying along the west side of Austin, Texas. The lake has a large watermilfoil infestation which (though seemingly limited to the shallow areas) covered approximately 37% of the water surface in late 1974.

Camp Creek Lake is a 380-ha (750-acre) private lake lying 40 km (25 mi) north of College Station. The dam was completed in 1948. For the past decade, the lake has had a continuing problem with coontail. Recently it has also suffered watermilfoil infestation.

Lake Conroe is an 8500-ha (21,000 acre) reservoir lying 70 km (45 mi) north-northwest of Houston, Texas. The dam was completed in January 1973, and the reservoir filled by early 1974. The lake contained both coontail and watermilfoil from the start. By late 1975, it had become infested with hydrilla and water hyacinth as well.

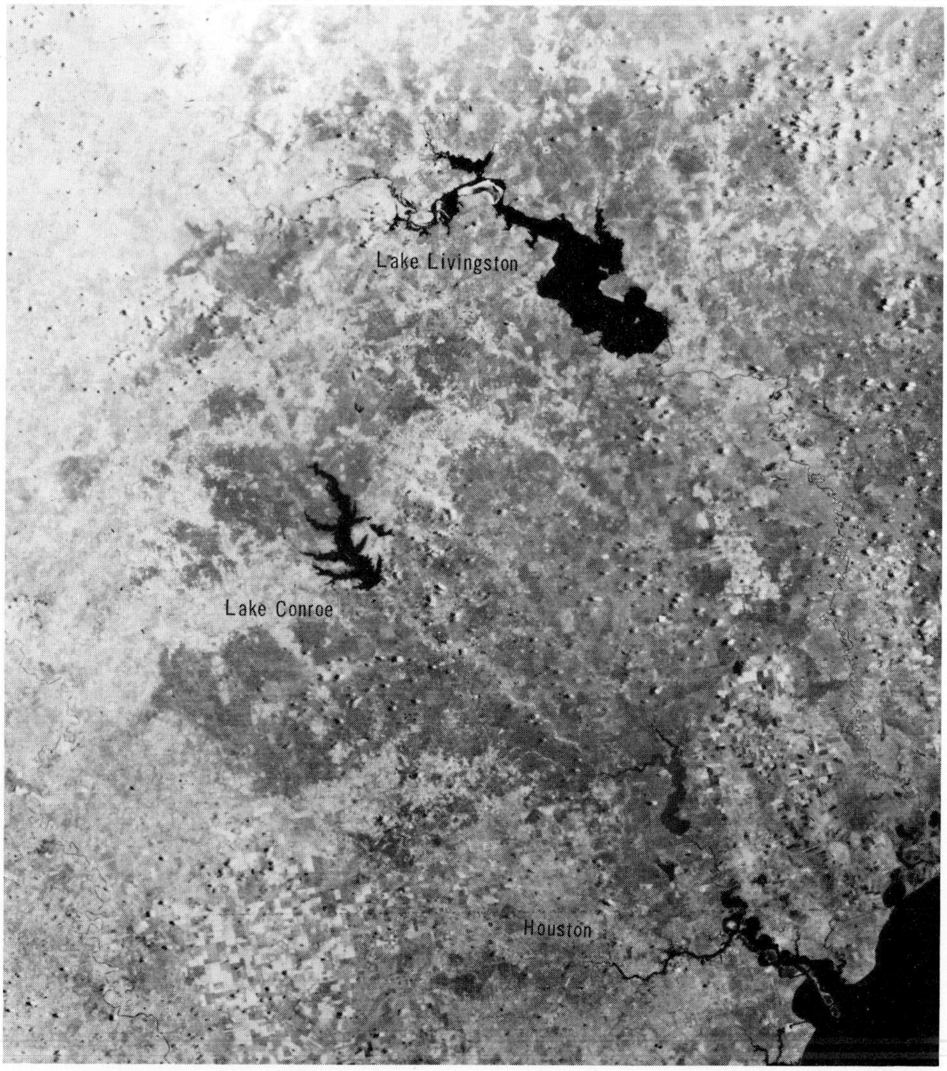


FIGURE 5.3. Landsat Band 6 image of Houston, Texas and vicinity.

Lake Livingston is a 36,000-ha (90,000 acre) reservoir lying 100 km (62 mi) north of Houston, Texas. The dam was completed in early fall of 1969 and the reservoir filled by the fall of 1971. The aquatic plant population in 1971 consisted of 20 ha (50 acres) of water hyacinth. By the fall of 1974, it had grown to nearly 800 ha (2000 acres) of water hyacinth, duckweed, hydrilla, and coontail. The Texas Parks and Wildlife Department conducts a continuing aquatic plant control program on the lake. This program primarily uses 2,4-D applied by spray boat to water hyacinth.

Figure 5.3, a Landsat Satellite Band 6 image of Houston and vicinity, shows the locations of Lake Livingston and Lake Conroe. Figure 5.4, a detail of Figure 5.3, shows the major areas of Lake Livingston photographed during the 3-year test.

*Procedures*

Throughout the 1974, 1975, and 1976 growing seasons, aerial photography was taken at approximately 1-month intervals, depending on clear weather. For compari-

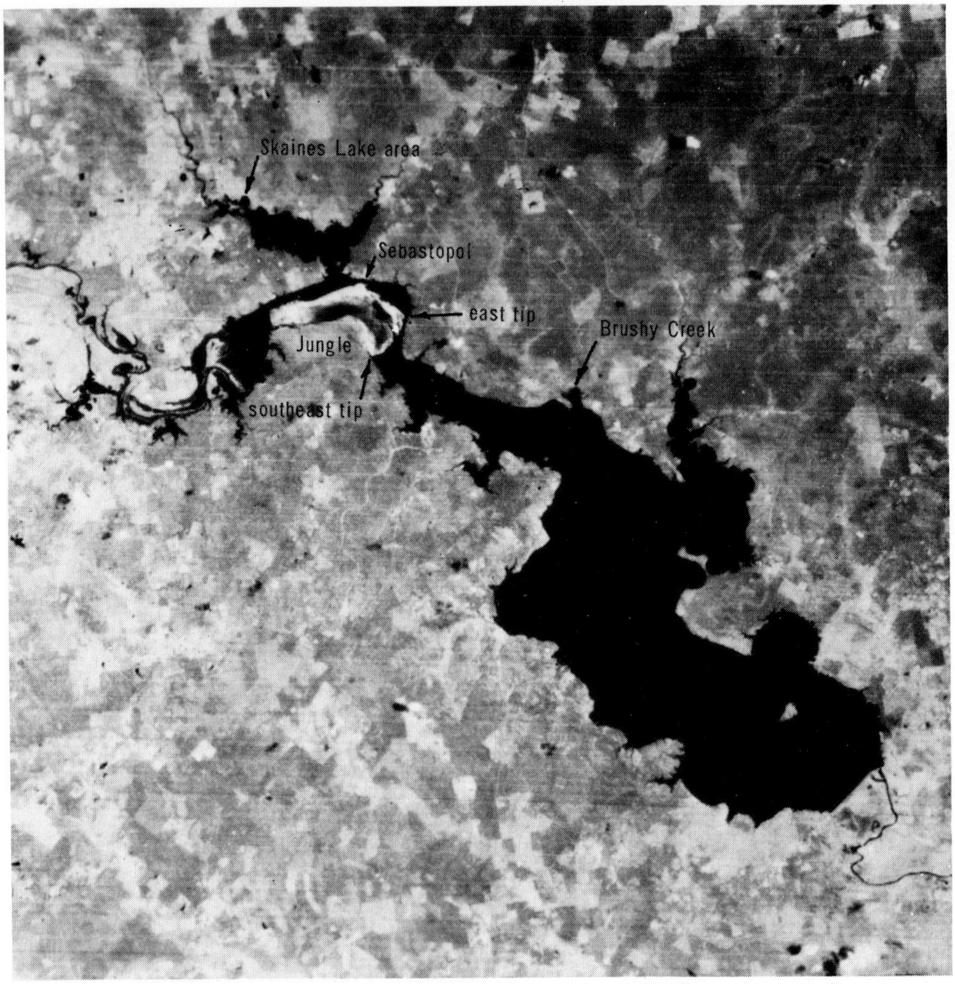


FIGURE 5.4. Lake Livingston, Texas, enlarged from Landsat Band 6 image.

son, photos were taken along the same ground path each flight, concentrating on the major problem species of each lake photographed.

As previously stated, photographs were taken with color and color infrared film in tandem cameras. The film was developed immediately after each flight with Kodak E-4 chemicals according to manufacturer's instructions. The short processing turnaround permitted preliminary photoanalysis the day following photography and ground truth the day after that. Since conditions on the ground were little changed in the 2-day period between photography and ground truth, this procedure permitted use of the imagery for planning the ground truth itinerary. This is a timesaving advantage that does not accrue when concurrent ground truth is attempted.

### Evolution of Technique

The 3-year test produced about 1000 separate images of aquatic plant concentrations in various lakes. The majority of the photographs taken were of Lake Livingston. There are many more photos illustrative of the test results than could possibly be included in this chapter. The images selected illustrate the films' capabilities for plant detection and differentiation as well as some important information about characteristics of the plants themselves discussed in detail later in this chapter.

*Initial Image Quality*

Plate 1* illustrates initial color infrared photography taken at the east end of Lake Livingston's "Jungle" on April 24, 1974. The Jungle (as it is locally known) lies at the lake's upper end. It is an area of drowned forest with dead tree trunks protruding starkly out of the usually murky water. Water hyacinth has found the Jungle a viable habitat, providing good anchorage, strong sunlight, and a steady supply of nutrients from the upper Trinity River watershed. The photo shows the extreme eastern tip of the Jungle where the old river channel made a sharp bend.

The quality of the color infrared imagery was surprisingly good for a first effort. Though slightly underexposed and the film apparently slightly overage, the colors are fairly typical of juvenile water hyacinth — generally a pink lavender. The light-colored area within the youthful hyacinth mat indicates plant stress resulting from recent herbicide application.

*Reticulation*

Seen under magnification, Plate 1 exhibits a condition known as reticulation or "alligatoring," in which the emulsion takes on the cracked appearance of sunbaked mud flats. This condition is due to poor temperature control during processing. The April imagery was processed in a borrowed, substandard darkroom. Shortly thereafter, a complete new darkroom facility was installed for project use; a commercial temperature control valve was included.

The May color infrared imagery again exhibited a slight tendency toward reticulation caused by the dark room air conditioning vent being too close to the film drying area. The cold air input was then deflected toward a different part of the darkroom. This permanently solved the reticulation problem.

*Film Procurement*

Shortly after the second flight over Lake Livingston, the supply of fresh color infrared film ran out. It was then discovered that this film, being quite perishable, is not stocked regularly by retail outlets or even by regional distributors, nor is a supply kept on hand at the manufacturer's plant. Depending on the extent of the manufacturer's stock at a given time, procurement of color infrared film can take 2 weeks to 2 months. It became apparent that color infrared film purchases should be planned far enough ahead to allow up to 90 days for delivery.

*Film Storage*

A supply of year-old color infrared film, apparently kept at room temperature, was found and used during a reflight on May 28, 1974. The results were poor, with almost total loss of the red forming layer due to long exposure to ambient temperatures. The scene which should have shown a preponderance of reds both on land and in the aquatic vegetation was instead a gloomy setting of deep, dark purples. Film subsequently purchased was kept in the freezer at −18°C in accordance with manufacturer's instructions. There was no further problem encountered on the May 28 flight, even though color infrared film was later used as long as 24 months after the purchase date.

*Value of Color Film*

The tandem color photography obtained in 1974 generally provided relatively little contrast in the vegetation and was somewhat affected by haze. The chief value of the color film was found to be its slightly improved delineation of submersed vegetation, a result of its somewhat better water penetration.

* All plates appear following page 48.

Color film was used without a filter in 1974. However, an earlier study by Lockwood et al. indicated that excellent water penetration could be achieved by using a 500-nm filter with color film to eliminate the effect of blue light scattering within the water column.[13] The report recommended using Kodak SO-397 high speed color film to compensate for the one-stop reduction in light caused by filter use.

Following this advice, SO-397 film with a Wratten 12 filter was used in tandem with the Kodak 2443 color infrared film during all 1975 test photography. Comparison of the two films is made in a later section of this chapter. Results are definitely superior to those obtained from the use of unfiltered color film.

Kodak Ektachrome 64 film, introduced in 1976, dispelled the need to procure expensive and inaccessible SO-397 for tandem use. Kodak 64 film has essentially the same speed and color characteristics as SO-397 and is readily available from most suppliers in format sizes up to 70 mm.

*Improper Exposure Settings*

The next flight over Lake Livingston after May 28, 1974 was scheduled for mid-June, but no fresh film was yet available. On June 26, when a Landsat satellite overpass was scheduled, the weather was clear, bright, and dry. Rather than miss the opportunity for coincident aerial photography, a flight was contracted with another aircraft carrying a vertically mounted Zeiss® RMK mapping camera loaded with Kodak 2443 color infrared film. Plate 2 is an enlargement of a portion of a single 9 by 9 in. format photograph showing the same general area as that covered by the 70-mm format.

The Zeiss RMK camera is equipped with a light-actuated, automatic exposure control which may be used if desired. By mistake, the exposure system was left on "Automatic" rather than "Manual". This caused the light sensor to react to sun reflection off the water, reducing the aperture about one stop under its optimum setting.

The underexposure effect was compounded by lack of an antivignetting filter. This is a filter with a heavy neutral density coating at the center, decreasing toward the edges to compensate for normal cosine-to-the-fourth-power ($\cos^4$) light falloff, an unavoidable consequence of the laws of optical physics. Plate 2 coverage was obtained from near the original format center; thus, the apparent falloff is negligible. Viewing the entire original, with its uncompensated $\cos^4$ light falloff, one clearly sees the particular sensitivity of color infrared film to changes in light value. Color film allows a fairly substantial latitude in aperture setting before significant color shifts occur. On the other hand, color infrared film evidences quite noticeable color shift with aperture differences on the order of half a stop. A color infrared photograph taken through a wide-angle lens without benefit of light falloff compensation amply demonstrates the effect.

The Zeiss RMK camera, having large format and relatively short focal length, is subject to light falloff in the corners equal to approximately two stops, meaning that only about a quarter as much light reaches the corners as reaches the photograph center. Examination of Zeiss RMK photography taken on June 26, 1974 clearly showed that color shifts due to half-stop light reduction or more are readily detectable. The implication is obvious. Color shifts are readily introduced by improper and inconsistent camera settings.

*Exposure Determination*

One problem with obtaining camera settings using Kodak's Aerial Exposure Computer is that the device does not account for scene brightness. Another difficulty is its reliance on subjective evaluation of haze. On the other hand, since Eastman Kodak does not provide American Standards Association* (ASA) equivalent ratings for aerial color film, the user must rely on the Aerial Exposure Computer.

However, Harris[14] suggests an alternative. It requires that test photographs be taken with a given film at a number of aperture settings. The optimum camera setting is selected by subjective comparison of the developed imagery, and a dummy ASA rating is obtained for the film by reversing the procedure for using the analogue calculator on the exposure meter. This method was used for all photography from mid-1974 on with good, consistent results.

Because emulsion characteristics of color infrared film vary significantly from one film batch to the next, use of film from different batches on the same project requires making new test exposures each time.[15] Thus, it is better to use film with the same batch number for all flights during a given season's photography.

When using the exposure meter method during operational flights, care must be taken to obtain light readings from representative scenes. Forests tend to be darker than mats of emersed aquatic vegetation. Open water is light or dark depending on turbidity and sun angle. The best target for light readings, other than a truly enormous mat of aquatic vegetation, seems to be a grassy field.

*Choice of Lens Focal Length*

It would be appropriate to examine the falloff aspect of lenses used in this research. As previously stated, the cameras were 70-mm Hasselblads. The format is 56 by 56 mm, or 40 mm from center to corner. Early on, the lenses used were 150 mm focal length. Therefore, the maximum angle off center was $\tan^{-1}$ (40/150) or 15°. Light level in the corner of the photograph is $\cos^4$ (15°), equal to 0.87 or 87% of the light at photo center. This 13% light loss is less than a quarter of a stop and therefore negligible.

From August 1974 on, 80-mm lenses were used in order to obtain greater ground coverage at moderate altitude. With this shorter focal length lens, the maximum angle off center is $\tan^{-1}$ (40/80), or 26.6°. $\cos^4$ (26.6°) is 0.64, indicative of a 36% maximum light falloff in the corners, slightly greater than half a stop equivalence. Center to edge equivalent loss is 21%, somewhat less than half a stop. The effect in terms of color shift is difficult to detect. Thus the 80-mm lens affords fairly good ground coverage with minimal shift across the photo frame.

*Spectral Response of Different Films*

A new batch of film was acquired prior to the August flight. The initial flight with fresh film was made on August 21, 1974. Plate 3 is a color enlargement taken during that flight. It shows a different area within Lake Livingston. The scene included water hyacinth and coontail (to be discussed later).

The camera settings for the April, May, and August photography were the same — 1/500 sec at f 5.6. Sun elevation was close to 45° in each case. The films were all from different batches with significant differences in film age. As previously stated, film used in May was overage and improperly stored. The August film was fresh and kept frozen until just prior to use.

Examination of the three sets of imagery showed (besides the obviously poor May imagery) a fair agreement in color tone between the April and August photos. Closer examination of all the imagery acquired on April 24 and August 21 (not shown here) indicates that April's is "cool" or slightly more blue, and August's is "warm" or more red and yellow. The colors in the June imagery (Plate 2) are rather different from

* Now reorganized under the name United States of America Standards Institute.

those encountered in either April or August. However, this is probably the result of the difference, not only in film batch, but in exposure and filter as well.

Plate 4 shows the color infrared imagery taken on October 2, the second last flight of 1974. Note that the picture was taken vertically, rather than as an oblique since a new mount was fabricated after the August flight. Color quality is reasonably consistent with the August imagery. The November imagery (not shown here) was also comparable in coloration to the October and August photography. Use of the same film with consistent photographic and processing procedures proved effective.

*Beneficial Effects of Overexposure*

The major emphasis on technique development during 1974 and 1975 concerned optimization of imagery quality and consistency. As will be discussed in greater detail, the achievement of color brilliance and contrast over the entire aquatic scene provided outstanding imagery of the emersed species but a somewhat poorer rendition of the submersed vegetation.

Except when the plants lie on the surface, the submersed species appear underexposed in a scene where the emersed species show up well. Since discrimination between submersed species is a significantly greater problem than discrimination of emersed species, it was decided during the latter part of the 1975 test to experiment with overexposure to enhance the appearance of submersed vegetation, particularly against dark water backgrounds.

The results expected for submersed plant discrimination did not materialize. Resultant enhancement in color quality and clarity was only moderate. However, with this minor improvement in discrimination capability, there emerged a substantial enhancement of detection capability. This improvement made detection of submersed beds a much simpler procedure, and detection of infestations is a key purpose of remote sensing monitoring.

The optimum exposure for color infrared film (a compromise between best exposure for emersed plants and submersed plants) seems to be about half a stop overexposed with respect to maximum brilliance of the emersed plants. This is a subjective evaluation based on analysis of many hundreds of submersed vegetation photos.

Water penetration film is of relatively little help in species discrimination. Its chief use is detection and delineation of submersed plant communities. One-stop nominal overexposure provides better contrast and a significantly sharper line of demarcation between submersed plants and open water. Overexposure much beyond one stop tends to lighten the image too much with corresponding contrast loss.

**Discrimination of Species**

The early expectation was to discriminate all, or nearly all, noxious species solely on the basis of color differences on color infrared film. Differentiation of water hyacinth, potamogeton, and duckweed, the major emersed plants at the test sites, was almost as easy as anticipated. The problem with submersed species is more complex, but it does yield to analysis.

*Water Hyacinth*

Young water hyacinth appears on color infrared film as a characteristic lavender as seen in Plate 1 taken April 24, 1974 at the eastern tip of Lake Livingston's "Jungle". By June 26, 1974, most hyacinth in the same area has assumed the magenta color of early maturity (Plate 2). By observation made on October 2, 1974 (Plate 4) and November 15, 1974 (not shown here), most of the plants had assumed the deeper, richer, slightly brownish red of late maturity.

Note that all scenes in that part of Lake Livingston show effects of ongoing herbicide application. The interesting aftereffect is the regrowth of juvenile water hyacinth from the spray-stressed mat. As a result, late summer imagery shows youthful, early-mature and late-mature hyacinth beds adjacent to one another, clearly differentiable due to their color infrared appearance.

Close examination of the April imagery in Plate 1 shows that the main body of the hyacinth mat is essentially the same color except for white stressed areas and the slightly bluer appearance of the newest growth in open areas outlined by the spray patterns. Note also the typically smooth hyacinth texture, indicating uniform vegetative density. Although poorly defined in this oblique view, the boundaries of the rapidly expanding areas in the upper left are cumuloform in appearance; that is, like puffy clouds. The equilibrium boundary area (foreground and far right) is more linear. The June 26 photo (Plate 2) clearly shows this effect.

Conformity of a plant boundary to a limiting condition is fairly typical of water hyacinth. In this case, it seems to result from the proximity of deeper water in the adjacent river channel, from the presence of shear currents, or from a combination of the two. In any event, the hyacinth mat tended to cease its spread along this line. When this boundary was sprayed some time between the June 26 and August 21 flights, there occurred a regrowth in two distinct stages toward the original boundary. In the October 2 imagery (Plate 4), a single regrowth band of youthful to early mature growth extends out toward the channel from the darker mass of mid-to late-mature growth. In the November 15 imagery (not shown), yet another band was seen to spread out toward the channel.

In each of the photos for August, October, and November 1974, the youthful lavender was found typical for new growth and for regrowth areas. Close comparison of October and November imagery with that of August and November with that of August and October showed consistent change in color from youth through the two maturity stages.

Although the above examples were photographs taken of a single location over a single year, the color infrared appearance of water hyacinth has been very consistent from place to place and year to year. Some rather interesting photos exemplifying this were taken in 1976 after the basic test program was complete. Plate 5, taken on September 23, 1976, shows another part of Lake Livingston's "Jungle". Of significance here is the green coloration within the hyacinth mat. It denotes very recent herbicide spray application. Note also that only magenta and lavender are apparent within the healthy part of the mat — evidence that all growth in the scene is relatively young. Indeed, imagery taken of the same area 6 weeks earlier showed only bare beginnings of new growth. Thus, over 100 ha (250 acres) of hyacinth had sprung up in that particular area in less than 2 months.

Plate 6 shows a low altitude photograph of a portion of the same scene in Plate 5. There is a significant amount of detail here, including the dead tree trunks of the drowned forest. The recently sprayed area (top left) appears as a sharply delineated lime green against the lavender and magenta background of the untouched mat portion. Contrast that photo portion with the whitish tan areas sprayed several weeks earlier. Note in the older spray area how the spray borders diffuse due to replacement of the dying mat by new growth.

Of particular interest in Plate 6 is the fine detail along the large mat borders. Small clumps of hyacinth are shown breaking off and drifting out to open water. Note that older portions of each mat are turning the magenta color of maturity, the outer fringes show the lavender of youthful expansion. The most youthful of all shows palest lavender.

Plate 7 provides additional insight into the color infrared indications of youthful growth. It was taken during the same flight as Plates 5 and 6 and shows Penwaugh Slough, a campground area on Lake Livingston some distance from the Jungle. In the shallows (upper left) extensive mats of mature and late mature water hyacinth can be seen. These mats began their youthful growth in early June 1976, well before the mat seen in Plates 5 and 6 emerged. A physical barrier held growth in check some 20 m removed from the campground proper. In early September, the inlet area just right of the barrier had been draglined in order to construct the embankment seen at the top. The draglining apparently stirred up much viable hyacinth seed on the bottom, resulting in an explosion of growth over a short period. A ground truth trip soon after taking this photo disclosed that the new hyacinth was all in the preflowering growth stage. Hyacinth fringing the shoreline further down the inlet originated earlier, as may be verified by the color of the mat nearest the shore.

All in all, the color infrared image of water hyacinth is readily recognized by its smooth texture and the hue and pattern of its colors. The mat has a cumuloform boundary when its expansion is not constrained. The major variations in hyacinth appearance are its seasonal colors on the imagery but these too are quite typical.

### *Potamogeton*

Not a truly troublesome plant, and hardly competitive with more noxious species, *Potamogeton illinoensis* Morong has sufficient ubiquity in Texas to warrant recognition of its occurrence. Potamogeton matures quickly, briefly establishes a limited domain, and senesces earlier than most other water plants. In early maturity its color infrared image is deep reddish orange. It usually develops a circular boundary, after which the central growth becomes increasingly sparse, giving plant clumps a doughnut appearance. Senescing potamogeton acquired a light greenish brown color. Like hyacinth, potamogeton is quickly recognized on color infrared imagery. The combination of color and shape gives the clump its unique appearance.

### *Duckweed*

Mature duckweed is quite easy to identify. On a color infrared photo it is seen as bright white at its first appearance in spring or early summer. It also appears white later in the year. White streaks in the upper left of Plate 4 indicate duckweed. Duckweed has been found and photographed in many other lakes and small ponds. Except for a brief brownish maroon appearance at the time of spring flowering, duckweed appears in consistent color on color infrared film. There seems to be no other noxious plant remotely similar to it.

### *The Submersed Species*

Submersed species images on color infrared film were less differentiable than emersed species. During 1974, extensive coontail beds were photographed several times in Lake Livingston and Camp Creek Lake. Watermilfoil was photographed several times in Lake Austin. Finally, in late 1974, hydrilla was found and photographed in Lake Livingston. The images of all three plants tend to appear nearly black at 1-m depth. Subtle differences show up when the plants have grown to within a few centimeters of the surface. This is due to elimination of infrared reflectance in the deep water and its considerable reduction near the surface. Although the color infrared images of milfoil and hydrilla change from black to dark maroon as they approach the water surface, coontail becomes dark brown instead. The plant's arrival at the surface causes a few more significant differences to appear on the imagery.

Submersed vegetation presents an additional problem. After reaching the water surface they often support an overlying layer of algae. Thus their appearance can be camouflaged by an algae cover of varying density and species.

These factors make differentiation of submersed species the greatest problem encountered while monitoring aquatic plants with color photography. The 1975 effort was designed to enhance the appearance of submersed plants and to document color infrared appearance of various species in pure stands and in association with "typical" overlying algae species.

Since the primary target in 1975 was submersed vegetation, it was decided to experiment with various overexposure levels. This tactic, even though it could flatten and degrade images of emersed and floating vegetation, was expected to enhance submersed vegetation even against a dark water background.

### *Appearance of Eurasian Watermilfoil*

The best photographic watermilfoil documentation was obtained in 1974 at Lake Austin. Fortunately, here the stands were relatively pure, and enough photographic sequences were obtained to show that the plant's color infrared appearance went from black to reddish brown in passing from deep water to the water surface (Plates 8 and 9). Plate 8 (October 1974) shows a dense upthrust mat of surfaced watermilfoil along the shoreline at Lake Austin's lower end. Plate 9 (taken the same day) shows an equally dense stand near the upper end of Lake Austin. The mat was held just at or below the surface by current force of the hydroelectric outflow from Lake Travis Dam, 1 or 2 km upstream (at left in the photo).

The downstream watermilfoil stand is more mature than that at the upstream end. Note in Plate 8 (downstream) that the plant has surfaced on both sides of the lake, and there are areas of overlying algae mats (lower center at delta-shaped inlet mouth and upper right in area crisscrossed by outboards). On the other hand (Plate 9), swiftly moving currents may well have inhibited the algae attachment, or perhaps there was no algae available in that area. The watermilfoil seen in Plate 9 is reasonably pure and probably quite typical of the mature plant color on color infrared film. A small watermilfoil stand was also photographed at Lake Conroe in 1975. The plant was fairly well covered with algae at that time.

### *Appearance of Coontail*

The 1974 coontail photography over Lake Livingston (Plate 3) and Camp Creek Lake (not shown here) demonstrated that pure plant stands, unadorned by algae, are characterized on color infrared film by a dark brown color when just below the surface and a tan to cocoa color at the surface. In the Lake Livingston photo, part of the coontail is covered by the alga *Spirogyra*. This frequently seen combination gives a whipped cream on cocoa appearance easily identified on color infrared films.

### *Appearance of Hydrilla*

The 1975 season produced about 300 frames of color infrared imagery over Lake Livingston, slightly more than that of water penetration color imagery. Hydrilla was photographed in perhaps a third of the pictures — in red hues of color infrared imagery and the bronze tones of filtered SO-397 film. The color infrared is remarkable for variety of hydrilla appearance. The limited number of pictures used here illustrate that the appearance of hydrilla on color infrared film is dependent on age, density, and stress conditions.

The first illustration (Plate 10) is a closeup of the Jungle area of Lake Livingston (southeast tip) taken in November 1975. Along the channel at the photo bottom, nearest to the shore, the magenta of pure water hyacinth appears partly turning to the rusty red of late maturity. Next to that is a remnant of what had earlier been extensive duckweed mats. Offshore from the duckweed are clumps of surfaced hydrilla, partially covered by either duckweed or algae. Further along the shoreline to the right, beyond the water hyacinth, is a stretch of hydrilla overlain with an algae species that gives it a medium purplish color. Across the channel, extending up and to the right side of the photo, near-circular hydrilla clumps are seen with considerably less algae. The least affected are at the upper right corner of the photograph. Note the color change from purplish brown to a more typical, cinnamon brown of bare, surfaced hydrilla. Note also that the centers of some clumps show the dark maroon of newly submerged hydrilla. The edges show the black of deeper hydrilla.

This photo illustrates that the color infrared hydrilla appearance indicates its depth, maturity, density, and class of algae cover. The imagery of the following herbicide test indicates that its appearance also represents the degree of stress encountered by the plant.

### *Texas Parks and Wildlife Department Herbicide Test*

In late 1975, the east side of Lake Livingston's Brushy Creek Inlet was the site of a test of several different herbicides on an extensive hydrilla mat. A large number of test plots were staked out around an open water area ringed with dense hydrilla stands. Herbicides were applied to the test plots during a 2-day period in mid-September. Aerial photography was taken on September 3 (before any herbicide had been applied) and thereafter on September 24, September 29, October 17, November 14, and December 1. The September 3 imagery showed dense, surfaced hydrilla mats with their fairly typical light cinnamon brown color indicating that they were actually airborne due to a low water condition. The two sets of September photography subsequent to herbicide application showed slightly varying turbidity in the test plots as herbicides began to take effect.

By October 17 (Plate 11), the test plots appeared more quiescent as the initial, intense herbicide impact decreased and some longer-term effects appeared. Note that several less dense areas appear around the open water periphery. These are points of application where chemical action excised away much of the overstory and some of the bottom was visible as a result. This impact appears somewhat better in Plate 12 which shows the tandem SO-397 image of the same scene. Note the sharp demarcation of submersed plant boundaries obtained with the latter film and filter combination. This is fairly typical of filtered SO-397 imagery. Though it is of little help in differentiating species, it is often significantly valuable in defining submersed plant boundaries and densities which are often obscure on color infrared film.

Comparison of Plates 11 and 12 clearly shows that there are numerous clumps of water hyacinth imbedded in the hydrilla at various points within the scene on the color infrared photo. On water penetration color film, however, there is much less contrast between the various species, mostly because of the overall bronze tone. Clarity differences of the submersed vegetation boundaries (comparing the two photos) illustrate the chief advantage of the water penetration film.

The November and December herbicide test photography showed continued decrease in density of the test plots, as opposed to the continued health of the buffer zones beween herbicide application areas. However, the December imagery showed that normal senescence was starting to occur in the buffer zones. This somewhat reduced the contrast between the test plots and other areas where no herbicides had been put down.

The photo series demonstrated utility of color infrared film for test documentation. It also illustrated the wealth of easily understandable, valuable information on this medium for the program manager involved in hydrilla control activities.

*Photography over Lake Conroe*

Most foregoing illustrations concentrated on Lake Livingston; all were spot verified by ground truth. In the fall of 1975, opportunity arose to photograph infestations in a different lake where no ground truth work was intended.

Two flights were made over Lake Conroe, one on return from the October 17, 1975 flight over Lake Livingston, and one on return from the November 14 flight. A flight map furnished by officials of the San Jacinto River Authority and indicating names and locations of all marinas and boat ramps aided navigation. The imagery was analyzed concurrently with Lake Livingston photographs on the same roll. Aquatic plant species predictions were made by comparison with concurrent and earlier imagery of other lakes.

Analysis of the first flight's imagery indicated presence of potamogeton, coontail, and watermilfoil. After the second flight, on being advised that hydrilla presence was suspected in at least one marina, a still submerged patch less that 1 ha in extent was found in the imagery of a marina near the dam. Also found in the second flight's imagery was a garage-sized patch, apparently an initial hyacinth outbreak, located in a small marina at the lake's extreme upper end.

A chance opportunity at a later date permitted verification of the presence of potamogeton, coontail, and milfoil beds adjacent to a main road running alongside the lake at the predicted locations. Suspicions concerning locations of hydrilla and hyacinth were passed along to San Jacinto River Authority personnel at Lake Conroe who later verified those infestations. Although there was some confusion between pure coontail, pure watermilfoil, and a mixture of the two, hydrilla, hyacinth, and potamogeton were uniquely discriminated.

Accuracy of these determinations could have been luck. They are difficult to explain in terms of pattern recognition since species discrimination on color infrared film is a rather subjective pursuit. Nevertheless, there is something about the appearance of these various species on color infrared film that makes precise discrimination possible, increasingly so as the analyst gains experience.

*Overall Effectiveness of the Method*

An ideal aquatic plant monitoring system would provide for infestation detection, positive discrimination of all species of interest, and metric capabilities for mapping if desired.

The system that evolved over the 3-year test period approximates that ideal. Use of tandem color infrared photography and water penetration color photography provides excellent detection capability clearly showing the presence and extent of aquatic plants. Emersed and floating species examined to date are readily discriminated from one another and from submersed species.

Maps showing species and extent were made quickly and easily from the 1975 imagery and used for local update of quadrangle maps.

Uncertainty in differentiating between the submersed plants is a minor problem particularly when the uncertainty is quickly dispelled by ground verification. Furthermore, seasonal plant succession, to the extent that it occurs at all, does so slowly. Therefore, there is little need for repeat ground verification in the same area during a given year. Once identity of a submersed species or species mix has been established, the plant community may expand or contract but will not suddently vanish. It is thus seen that the system works well.

**Analysis of Aquatic Plant Characteristics**

An important by-product of the 3-year study is insight into behavior of the plants themselves. A great deal about plants' life cycles, spread characteristics and spread rates, algal associations, cohabitation characteristics, and reaction to herbicide stress can be learned by observation of sequential imagery.

*Water Hyacinth*

Water hyacinth has a life cycle whose length depends on seasonal temperatures and degrees of stress. Hyacinth emerged on Lake Livingston in mid-April of 1974, following a mild winter during which many plants went dormant and did not senesce. General die-off began with the first hard freeze of early December. By mid-January of 1975, the entire Jungle hyacinth mat (some 650 ha) had senesced from the impact of several successive freezes. Desiccation of the dead mat inhibited new growth well into late spring of 1975. Mid-summer growth was slow. Peak coverage in the Jungle occurred somewhat under that reached by early May of 1974. In 1976, hyacinth coverage in the Jungle exceeded the 1974 spread by 60%.

Under the right conditions, none of the plants studied can match water hyacinth for sheer proliferation. Beginning with few scattered plant nuclei in early April of 1974, the plant mat filled 50% of the Jungle (approximately 600 ha) in a matter of 8 to 10 weeks.

The plant's establishment of equilibrium boundaries has already been discussed; it seems unable to extend into an area of deep water, strong current, or perhaps a combination of the two. Its spread is definitely inhibited by presence of mature submersed plants. During 1974 and 1975, in the west end of the Jungle, it did not spread into an established coontail area. In 1975, at the southeast Jungle tip, it seemed trapped in small, tight clumps by underlying hydrilla. This was also true in Brushy Creek during 1974 and 1975.

The standard approved herbicide for hyacinth control in Texas is 2,4-D. Application in maximum repeated dosages in small areas, as in 1975 in the upper White Rock Creek section of Lake Livingston, can totally control hyacinth expansion. When there is substantial areal hyacinth spread (as in the Jungle in 1974), the same level of effort produces a haphazard impact, temporarily restraining growth in the scattered areas of 2,4-D application.

Evidence that less than total herbicide application can be counterproductive exists. Herbicide application to a hyacinth mat is invariably followed by early regrowth within the cleared area. The sprayed portion of a senescing mat can thus revert to a vigorous, youthful growth stage. The low, active regrowth area is considerably more frost resistant than the tall, overmature mat adjacent to it. This showed clearly in the late 1975 and 1976 imagery of the Jungle.

*Potamogeton*

Some of the earliest plant growth seen on Lake Livingston is potamogeton. Although early growth is rapid, plant spread is slight. Potamogeton communities noted in 1974 were still essentially the same size in 1975 and 1976. In a head-to-head struggle with hydrilla for the same area of Brushy Creek habitat in 1975, potamogeton was slowly but surely the loser.

Its competitive disadvantage may be due to early loss of vigor. At a time of year when hydrilla is still rapidly increasing in extent and biomass, potamogeton as seen on color infrared film is turning the characteristic greenish brown of senescence. By late September or mid-October, the plant has begun its annual desiccation. By November, its image is usually gone, leaving the waters to more aggresive species.

*Duckweed*

Duckweed appears to decline relative to other aquatic plants in Lake Livingston.

This tiny, floating plant does not now seem a significant factor in the species makeup of Lake Livingston. In 1974 duckweed seemed to inhabit a quarter or more of the Jungle, cohabiting with coontail and water hyacinth. Its spring emergence in 1974 was as early as that of hyacinth, but its demise occurred in early November rather than late December, as was the case with hyacinth.

In 1975, when hyacinth spread in the Jungle was much reduced, duckweed emerged later, covered less area, and disappeared sooner. In 1976, its presence was limited to few areas.

*Watermilfoil*

Insufficient watermilfoil imagery prevented determination of its approximate life cycle. Indications at Lake Austin are that watermilfoil is ordinarily restricted to shallow depths. Watermilfoil is the dominant aquatic plant species in Lake Austin. Negligible amounts of other submersed species exist. In Lake Conroe, on the other hand, watermilfoil communities grow side by side with coontail. Relatively little algae was mixed with Lake Austin watermilfoil. At Lake Conroe, there was a good deal of algae overlying the plant.

*Coontail*

Coontail is the only noxious species studied that is actually a native of the U.S. It was found in varying quantities in most lakes visited during this study, with the exception of Lake Austin.

The plant is quite hardy and is found at all latitudes in the conterminous states. It does not reach a dense growth state in the observed lakes until early summer. It senesces quite early — ordinarily by mid to late October.

Unlike hydrilla, coontail spreads slowly. Observed in association with other submersed plants, it seems to hold its own in the short- to mid-term. By late summer, it invariably seems to have gained an overlay or intermixture with one or more algae species. It has a particular affinity for *Spirogyra*, a green algae. The coontail-spirogyra combination has an easily recognizable appearance on color infrared film. It resembles marshmallow sauce on chocolate ice cream.

*Hydrilla*

Hydrilla is one of the more interesting aquatic plant species. It was not available to this study until late 1974, when it was photographed in two marinas on Lake Livingston. Although the plant does not ordinarily reach the dense growth stage until early to mid-summer, it was found in lush stands during early June in the warm, high nutrient environment of Memorial Point Marina on Lake Livingston, a nearly enclosed basin which receives the effluent from a sewage treatment plant.

Hydrilla persists until well after the first frost. Total desiccation does not occur until water hyacinth turns to a dead mat, although the plant begins to lose density by late fall. It spreads rapidly and remains established year to year possibly because of its ability to generate tubers when stressed. The infestation at the Jungle's southeast tip is characterized by a steady spread, gravitating downstream in open water and along the shoreline. Once rooted in open water, new infestations spread in circular patterns, increasing in size throughout the late summer and fall growing seasons. Infestations along the shore seem to be more linear in growth direction and in community boundary establishment.

Hydrilla appears to have a significantly lower light requirement than other submersed plants. Unlike coontail and watermilfoil, hydrilla will grow in relatively clear water to depths beyond which color infrared film can detect its presence. In Lake

Livingston it has been dredged up live from depths as great as 12 m (40 ft). Shadowy plant clusters, probably hydrilla, have been seen from the air on rare days when Lake Livingston is only slightly turbid in areas where it had been dredged from the bottom.

Hydrilla is often associated with a variety of algae. The algae seem neither to hinder nor help hydrilla growth. To date, a consistent association of hydrilla with a particular algae has not been found.

In cohabitation with water hyacinth, hydrilla exhibits a strong retarding effect on the normal hyacinth spread. In Brushy Creek, from observations of 3 years' photographic data, hydrilla seems to be steadily taking over areas previously dominated by water hyacinth. Although there are small amounts of coontail and watermilfoil in Brushy Creek, neither plant seems able to make inroads into areas in which hydrilla is established, let alone stake out a separate domain.

Hydrilla seems relatively unaffected by most herbicides except in the short term. Both the Texas Parks and Wildlife Department and The Trinity River Authority of Texas have had application programs on Lake Livingston. Neither organization has achieved lasting success. The extensive multiherbicide test (1975) conducted in Brushy Creek by the Texas Parks and Wildlife Department demonstrated very little lasting impact on hydrilla. Although late 1975 photography showed substantially diminished biomass within the test plots, 1976 mid-summer photography of the same area, substantiated by ground truth work, shows that there is no detectable difference in growth density between the test plots and the intervening buffer zones. In the same vein, the Trinity River Authority has been applying herbicides on hydrilla in several marinas on Lake Livingston since late 1974. Regrowth has invariably occurred in all cases, usually during the same year as the application was made. These characteristics amply demonstrate that hydrilla will continuously cause problems in Texas waters in the foreseeable future.

## SUMMARY OF RESULTS

It was determined that tandem photography with color film, color infrared film, and the proper filters provides a viable method for low-cost detection, identification, and monitoring of aquatic plants in Texas lakes. The major emersed species were relatively easy to detect and differentiate from one another. Their spectral reflectance characteristics, photographic image textures, and context within a given scene make them fairly easy to recognize.

Submersed species are more difficult to detect. In outline, texture, and locational context within a given lake, they often resemble one another. Their spectral reflectance characteristics due to changing depth, changing density, and changes in algal cover tend to further confuse the photoanalyst. Nevertheless, once it is realized that a given submersed species can assume several different appearances on color infrared film, and that confusion is dispelled by a small amount of ground verification, the problem becomes much simpler.

Spot verification is not a continuing need over the course of a growing season. Once established, there is little likelihood of a given species being supplanted by another submersed species in the short- or mid-term. Thus, the primary need for extensive ground truth work occurs during early summer when plants first appear in the submerged state. This is especially true on lakes where there has been no previous remote sensing activity. In any event, there is limited need for late season ground truth for discrimination alone. Table 5.1 summarizes the appearance features of the various aquatic plants studied as seen on color infrared film.

After frequent work with problems of species recognition on color infrared film, lay persons gained confidence in their ability to detect and discriminate a variety of

TABLE 5.1

**Typical Appearance of Aquatic Plants on Color Infrared Film**

| Aquatic plant | Approximate color | Texture | Shape of plant community | Usual location |
|---|---|---|---|---|
| Water Hyacinth | | | | |
| Youthful | Lavender | Smooth | Cumuloform | Adjacent to shore and in drowned brush and trees |
| Mature | Magenta | Mottled | Cumuloform | |
| Senescent | Olive | Mottled | Cumuloform | |
| Stressed | Light tan to tan-olive | Rough | Sharply defined spray borders | |
| Potamogeton | | | | |
| Youthful | Pink | Fibrous | Small, circular clumps | Just offshore or in shallows |
| Early-mature | Deep reddish brown | Fine | Clusters, dense outside, sparse inside | |
| Late-mature | Green-brown | Fibrous | Sparse outside, open inside | |
| Duckweed | Pink-white to white | Smooth | To fit shoreline | In shallows and coves, where blown by wind |
| Eurasian Watermilfoil | | | | |
| Youthful | Black submerged | Smooth | Often elongated, regular borders | Nearshore and adjacent to shore in shallows |
| Mature | | Smooth | Same as above | |
| At surface | Red-brown | Smooth | Same as above | |
| Airborne | Light red-brown | | | |
| Coontail | | | | |
| Youthful | Black submerged | Smooth | Regular boundaries | Nearshore and adjacent to shore in shallows |
| Mature | | Smooth | Regular boundaries | |
| At surface + *Spirogyra* | Cocoa brown<br>Gray-white on top | Mottled | Regular boundaries | |
| Hydrilla | | | | |
| Youthful | Black submerged | Smooth | Circular and cumuloform, sometimes elongated | Nearshore, adjacent to shore, in shallow and deep waters |
| Mature | | Smooth | Same as above | |
| At surface | Dark cinnamon brown | Smooth | Same as above | |
| Airborne | Light cinnamon brown | | | |
| Senescent | Gray-brown | Sparse | Indefinite | |

plant species. A brief experience with the imagery and with the photoanalytic instrumentation was all that was needed. Trinity River Authority personnel were trained in these photoanalysis and ground verification techniques during 1976 and are presently using the system.

The key to the viability of the system is the large amount of available information from color infrared and tandem color photography. The photoanalyst can simultaneously consider color vs. season; water level effects; location of a vegetation patch with respect to shoreline, marinas, or channels; indications of algal cover; the shape and orientation of a plant community boundary; and whether or not imagery is properly exposed. The analyst can then balance these subjective impressions against his own recollection of similar patterns previously analyzed. The human brain quickly sorts through these relationships and arrives at either a single conclusion or a very short list of alternatives.

As opposed to discrimination, detection of submersed vegetation is a much simpler matter. Detection is aided by overexposure of color infrared film. Particularly valuable is the use of tandem color film configured for maximum water penetration. Perhaps an ideal procedure using two cameras would call for use of color infrared film at half-stop overexposure (for optimum all-around discrimination) coupled with some sort of water penetration film used a half- to full-stop overexposure for maximum submersed vegetation detection and delineation.

In brief, the equipment and procedures tested in 1974, refined in 1975, and used operationally in 1976 seem an appropriate model for a viable low-cost, aquatic plant monitoring system.

## REFERENCES

1. **Pritchard, D.**, Texas Parks and Wildlife Department, personal communication, 1975.
2. **Benton, A. R., Jr.**, Monitoring Aquatic Plants in Texas, Rep. RSC-76, Remote Sensing Center, Texas A&M University, 1976.
3. **Rogers, H. H. and Davis, D. E.**, Nutrient removal by water hyacinth, *Weed Sci.*, 20, 423, 1972.
4. **Timmer, C. E. and Weldon, L. W.**, Evapotranspiration and pollution by water hyacinth, *Hyacinth Control J.*, 6, 34, 1967.
5. **Penfound, W. T. and Earle, T. T.**, The biology of the water hyacinth, *Ecol. Monogr.*, 18, 447, 1948.
6. **Gustafson, T. D. and Adams, M. S.**, The Remote Sensing of Aquatic Macrophytes, Rep. No. 24, Madison Institute of Environmental Studies, University of Wisconsin, 1973.
7. **Seher, J. S. and Tueller, P. T.**, Color aerial photos for marshland, *Photogramm. Eng.*, 39, 489, 1973.
8. **Anderson, R. R. and Wobber, F. J.**, Wetlands mapping in New Jersey, *Photogramm. Eng.*, 39, 353, 1973.
9. **Klemas, V., Daiber, F. C., Bartlett, D., Crichton, O. W., and Fornes, A. O.**, Inventory of Delaware's wetlands, *Photogramm. Eng.*, 40, 433, 1974.
10. **Reimold, R. J., Gallagher, J. L., and Thompson, D. E.**, Coastal mapping with remote sensors, in *Proc. Coastal Mapping Symp.*, American Society of Photogrammetry, Washington, D.C., 1972, 99.
11. **Fritz, N. L.**, Optimum methods for using infrared-sensitive color films, *Photogramm. Eng.*, 34, 1128, 1968.
12. **Fritz, N. L.**, New color films for aerial photography, in *Proc. 3rd Biennial Workshop on Color Aerial Photography in the Plant Sciences*, Gainesville, Fla., 1971, 34.
13. **Lockwood, H. E., Perry, L., Sauer, G. E., and Lamar, N. T.**, Water depth penetration film test, *Photogramm. Eng.*, 50, 1303, 1974.
14. **Harris, W. D.**, Planning and operation of a color aerial photographic mission, in *Manual of Color Aerial Photography*, American Society of Photogrammetry, Arlington, Va., 1968.
15. **Welsh, J. J.**, Eastman Kodak Co., personal communication, 1974.

Chapter 6

# ENVIRONMENTAL POLICY IN THE CORPS OF ENGINEERS*

## BACKGROUND AND HISTORY

The Chief of Engineers manages and executes the Civil Works Program of the Corps of Engineers under direction and supervision of the Secretary of the Army. The Corps of Engineers has responsibility for: investigating, developing, and maintaining the nation's water and related land resources; constructing and operating projects for navigation (Figure 6.1), flood control (Figure 6.2), major drainage, shore and beach restoration and protection, hurricane flood protection, hydroelectric power production (Figure 6.3), water supply, water quality control, fish and wildlife conservation and enhancement, and outdoor recreation (Figure 6.4), supporting emergency relief activities directed by other Federal agencies; administering laws for the protection and preservation of navigable water and for emergency flood control and shore protection; and conducting research and development related to Civil Works planning, design, construction, operation and maintenance, and regulatory activities.

The first civil responsibilities were delegated to the Army Engineers immediately following the Revolutionary War.[16] The new nation needed roads, railroads, lighthouses, bridges and many other substantial engineering works to support economic advancement. The Corps, the only available organized engineering force, was called upon to build them.

Beginning in 1815 when Lieutenant Zebulon Pike (after whom Pike's Peak was named) explored the Missouri Valley, the Army Engineers were the principal government agents in the exploration of the West to locate wagon and railroad routes. Lieutenant John C. Fremont surveyed the Oregon Trail and opened up California. Lieutenant A. J. Donelson surveyed the line of the Northern Pacific Railroad. Other Corps survey parties located the transcontinental routes of the Sante Fe, Union Pacific, and Southern Pacific Railroads. In 1871, Lieutenant George M. Wheeler commenced a survey of all the territory of the U.S. west of the 100th meridian, and completed it 13 years later. As part of its job, the Corps also surveyed and mapped the Great Lakes and marked all U.S. boundaries and most state boundaries as well.

During this period of exploration, the Army Engineers systematically catalogued wildlife and other natural resources in expedition areas. They fought for and conducted the first measures to preserve national treasures (the areas which are now Yellowstone, Yosemite, and Sequoia Parks) from private exploitation. The Corps administered the Yellowstone areas until the founding of the national park system, and administered the Yosemite and Sequoia areas from 1864 until California accepted them as State parks.

Since 1824, the Corps has also been the principal developer of the nation's water resources. In that year, Congress assigned it the task of clearing some snags and sandbars from the Ohio and Mississippi Rivers to facilitate navigation—the small beginning of its civil works program. This program was limited to improvements for navigation until 1879, when the Mississippi River Commission was created with flood control in the alluvial valley of the Mississippi as a function incidental to improvement of the river from the Ohio's mouth to the Gulf of Mexico to provide safe passage for increasing numbers of commercial vessels.

* By C. Grant Ash, Biologist, Corps of Engineers, Washington, D.C.

FIGURE 6.1. Navigation in the Cape Cod Canal. New England Division of Engineers Project. Bicentennial observance of "Tall Ships", July 1976. (Courtesy of the Corps of Engineers, James McDevitt, photographer.)

FIGURE 6.2. Flood control and water resource reservoir, Stockton Lake, Mo., June 1976. Kansas City District, Corps of Engineers. (Courtesy of the Corps of Engineers, R. E. Nelson, photographer.)

FIGURE 6.3. Hydroelectric power generators, McNary Dam Power House. Walla Walla, Washington, June 1976. Walla Walla District, Corps of Engineers. (Courtesy of the Corps of Engineers.)

FIGURE 6.4. Recreation boating, Belton Lake, Texas, July 1976. Fort Worth District, Corps of Engineers. (Courtesy of the Corps of Engineers, Rod Steiger, photographer.)

A few years later, the Corps was given responsibility for regulating hydraulic mining in California's Sacramento and San Joaquin Rivers and the institution of other measures to prevent creation of hazards to navigation as a result of mining. In 1917, flood control on the Mississippi was authorized, and the Corps was also directed to undertake similar work on the Sacramento River. Congress added coastal protection to Corp functions in 1930, and in 1936 its responsibilities were greatly enlarged to include nationwide flood control activities. Subsequently the construction of projects to provide multiple benefits, including, in addition to flood control and navigation, the generation of hydroelectric power, water supply, recreation, and conservation of fish and wildlife, was authorized.

During the last 200 years, the Corps has completed more than 4000 civil works projects. The Corps has participated in construction of 25,000 mi of inland and intracoastal waterways now in commercial use, 107 commercial port facilities, 400 small boat harbors, and 261 locks. Domestic commerce on these waterways moves 16% of the nation's tonnage and ton miles of intercity cargo, 60% of which is energy producing commodities. The Corps of Engineers has completed over 1000 flood control projects, a figure which includes both large dams and small local protection projects. In Fiscal Year 1975, for example, these projects prevented an estimated $15.9 billion in flood damages. Sixty-three of the Corps dams include facilities for generation of pollution-free hydroelectric power. Corps facilities annually produce 82.2 billion kilowatt hours of electrical energy. (Hydroelectric power is the most environmentally clean form of electricity since there is no air and water pollution as with burning of fossil fuels and no thermal pollution as with contemporary nuclear plants.) Recreation attendance at the Corps' 350 reservoirs has exceeded 350 million visits per year. The 11 million acres of land and water at these lakes are managed for recreation, fish and wildlife, and related purposes. In time of flood or hurricane, the Corps provides assistance in damage control, rescue work, and rehabilitation both under its own authorities and as engineering agent of the Office of Emergency Preparedness. Since 1963 it has been involved in more than 100 disaster operations, including flood and hurricanes.

## ENVIRONMENTAL MISSION

The environmental mission[36] of the Corps of Engineers is to carry out the mandate of the National Environmental Policy Act of 1969 (hereafter referred to as NEPA)[20] to "...encourage productive and enjoyable harmony between man and his environment; to promote efforts which will prevent or eliminate damage to the environment and biosphere and stimulate the health and welfare of man; to enrich the understanding of the ecological systems and natural resources important to the Nation;" and as further defined in Section 101(a) of the act " . . . to use all practicable means and measures, including financial and technical assistance, in a manner calculated to foster and promote the general welfare, to create and maintain conditions under which man and nature can exist in productive harmony, and fulfill the social, economic, and other requirements of present and future generations of Americans."

Implicit in this mission are four general environmental objectives:

1. To preserve unique and important ecological, aesthetic, and cultural values of our national heritage
2. To conserve and use wisely the natural resources of our nation for the benefit of present and future generations
3. To restore, maintain, and enhance the natural and man-made environment in terms of its productivity, variety, spaciousness, beauty, and other measures of quality

4. To create new opportunities for the American people to use and enjoy their environment

## ENVIRONMENTAL DECISIONS

In the early days of this country, Americans saw the natural environment both as the means to enchance their national well-being and as an obstacle to be overcome. Rivers and natural harbors presented opportunities for commerce and provided routes of communication. They also had to be defended. From the start, Army Engineers provided defenses needed to help protect American harbors and waterways from potential foreign enemies.

A gradual shift in national awareness eventually led to appreciation of our natural environment as a less than infinite resource. Thus, today in the Corps of Engineers, they discuss environment in terms formerly reserved for national economic development and growth syndromes. Economic competitiveness is still vital and will always be a factor, but growth-at-all-costs is no longer taken seriously. Environmental protection is now an equally important consideration in any agency decision and ranks high in national value.

A nation's first obligation is to insure its survival. At the survival stage of man's existence, concern for preservation of the natural environment is largely irrelevant. At the point where economic well-being is marginally achieved, environmental considerations have limited meaning and environmental effects are likely to be brushed aside. Not until a nation achieves a healthy economic situation and a self-realized portion of the "good life" will its citizens also manifest serious concern for the quality of their environment. But what is the proper balance between environmental and economic values?

NEPA was a dramatic statement on this problem. The feature of NEPA with which the public is most familiar is the requirement for preparation of an assessment of the environmental effects of future projects. This is displayed in the form of an Environmental Impact Statement (EIS) under the requirements of that Act's Section 102(2)(C). The EIS requires the decision maker to consider and evaluate all environmental aspects of a proposed action. He must also make all the factors considered matters of public record and conduct give and take sessions in the court of public opinion. After competitive views are discussed and needed decisions modified as appropriate, the entire record is placed in the hands of the Council on Environmental Quality (CEQ) and before Congress.

Perhaps the most important lesson learned in the Corps, as they have implemented NEPA, is that the environmental considerations of a project must be an integral part of the planning and decision making process in exactly the same way that social, economic, and engineering considerations are a part of that same process — a part of it, but only a part of it.

Decision making is a continuing and repetitive procedure dealing with matters of small consequence, as well as those of major significance. Decisions are made almost daily during the planning phase of a project, and become fewer as they go into the design, construction and operation, and maintenance phases.

The requirement to prepare an EIS has had the salutary effect of continually reminding the decision maker that environmental factors must be given adequate attention in all phases of planning. By its own terms, however, the main focus of the EIS is on a detailed examination of environmental factors of alternative actions, without commensurate attention to social, economic, and engineering factors that may be of lesser, equal, or greater importance depending on specific circumstances. For this reason, the Corps position has been that the EIS can serve as only one of the documents on which a decision about a major Federal action should be based.

Critics of the Corps[8,9] frequently charge that their decisions are arbitrary and capricious and that citizens are pushed aside as the Corps promotes construction. Those who make these claims are frequently poorly informed about the elaborate and complicated procedures that the Corps must follow for a project to be congressionally authorized for construction. Little publicity is given to the number of proposed projects found unacceptable to public interest or to the number of projects recommended for deauthorization.

When the people on a river, lake, or ocean have problems connected with the water, they can petition Congress for a study of remedial measures. They may be concerned about navigation, flood control, hydroelectric power, water supply, water quality control, recreation, fish and wildlife protection, or shore protection.

If Congress authorizes and appropriates funds for a study, the District Engineer investigates the situation thoroughly and tries to make a good plan to meet the people's needs. He will examine all practical alternatives and select the plan providing most benefit for least cost and with the least harm to the environment. The Corps gives special attention to minimizing ecological disruption during construction and to enhancing fish, wildlife, and recreation resources of the completed project.

The District Engineer will recommend a Federal project only when benefits exceed cost, the environmental impact is acceptable, and he finds the overall project in the public interest. In the course of his planning, he will hold public meetings and consult local, state, and Federal officials. He transmits his analysis and recommendations to Washington in a project report and environmental impact statement. In over half of such reports, District Engineers recommend against any Federal project.

In Washington, the Board of Engineers for Rivers and Harbors, the Chief of Engineers, the Secretary of the Army, and the Office of Management and Budget all review the report and the environmental impact statement to decide whether or not to recommend a project. Based on their recommendations, the President and Congress make the final decision on project authorization and funding. If their decision is favorable, Congress will pass a law authorizing the project, and in the years following, funds for design and construction will be included in the Annual Public Works Appropriations Act.

Throughout the planning and coordination process, the Corps pays a great deal of attention to views of the people whom the project will affect, the economic justification for the project, and its effect on the environment. Every plan necessarily involves some compromises. The Corps' planning staffs include experts in project design, economics and environmental preservation and enhancement. Since passage of NEPA, the Corps has intensified its efforts to make plans that balance the need for water resource development with that of environmental protection in the overall public interest.

In spite of these rigorous check and balance procedures, situations change and new priorities are established sometimes indicating that project reevaluation is appropriate. In accordance with Section 12 of the Water Resources Development Act of 1974,[40] the Corps of Engineers has reevaluated a large number of authorized projects under their jurisdiction and has recommended to Congress over the past 2 years the deauthorization of 395 projects. These projects are those authorized for at least 8 years, but not funded during the past 8 years and falling within any of the following five categories:

1. Lack of economic justification
2. Not adequate to meet current or prospective needs (obtaining adequate improvements would require additional funding authorization from Congress)

3. Significant and unacceptable adverse environmental impact
4. General opposition or lack of cooperation from local interests
5. Part or all of the project has been accomplished by other means or is no longer required

The following types of projects were recommended for deauthorization: 145 channelization; 160 levee, dike, or flood wall; 41 navigation structures; 37 dam and reservoirs; and 12 with other types of construction.Of the 395 projects recommended for deauthorization by the Corps, Congress has now deauthorized approximately 275 projects.

## ENVIRONMENTAL IMPACT STATEMENTS

The effect of NEPA on Federal agencies has been dramatic. Extensive modifications in point of view and procedure are evident. In the Corps of Engineers alone, over 2600 Environmental Impact Statements (EIS), draft and final, have been prepared and filed; total Draft EIS filed with Council on Environmental Quality (CEQ) 1970 through 1976 was 1527, and the total Final EIS filed for the same period was 1077 and does not include those prepared under regulatory authorities.

The Corps of Engineers evaluated effects of this Act on its current projects and found they had performed reasonably well when it comes to identifying adverse effects and eliminating the effect or otherwise solving the problem. In an assessment of over 500 projects in the construction and design stages, one third were modified to accommodate environmental considerations. Similarly, of 200 studies investigated, about one third of the final alternatives proposed had been significantly changed during the course of the study to minimize impact on the environment. In 43% of the 103 completed projects investigated in the operation and maintenance category, the operators had adopted new procedures to help protect the environment.

Based on these data, about one of every three active projects in Civil Works has been modified because of NEPA. The implementation has included a host of actions ranging from complete elimination of a project to minor project modifications: dams have been resited, elevation of pools adjusted, spoil areas relocated, fish and wildlife losses mitigated, fish passage structures added, highways relocated, levees set back to avoid channelization of natural streams, and greenbelt floodways established—just to name a few modifications involved.

The requirement to produce these statements is good! The Corps has re-oriented its thinking and now considers more alternatives and nontraditional alternatives as solutions to the Corps' mission problems. They work more closely with the public, civic minded groups, and elected officials. Their concepts and procedures have been modified and changed. Environmental considerations are now an integral part of planning and decision making in exactly the same way that the social, economic, and engineering considerations are part of that same process.

The current rules and regulations have been translated into a procedure that makes the EIS cost too much in terms of money, time, and other resources. Too much attention has been given to the procedural aspects. Environmental statements prepared today can generally be identified by their great bulk and diversity of information. Several hundred pages is not uncommon for an EIS and a thousand or more pages is somewhat unusual, but occasionally occurs. These statements are not necessarily bad. In fact, some are very good. They meet every procedural aspect of the law and reveal a tremendous amount of information. They meet the demands of the courts, but do they meet the needs of the public? Who reads these bulky documents? Perhaps more important is the question, what function do these environmental statements play in the decision process?

In this connection, I quote from a memorandum for heads of agencies, dated February 10, 1976, from Russell W. Peterson, Chairman of the Council on Environmental Quality (CEQ):

> Nevertheless, situations continue to arise in which the impact statement process has been more an appendage to or justification for decisions already made than an aid to decision making. Frequently these failures have been caused or aggravated by the inordinate and unnecessary length of EISs. Such documents at best obscure the intent of NEPA and can be extremely harmful to the environmental impact statement process . . . An unnecessarily large portion of many EISs has been devoted to descriptions of the proposed action and the existing environment. Frequently EISs follow lengthy, detailed outlines in order to assure that at least some treatment, however brief, is given to every subject conceivably relevant to the proposal. In following this approach, agencies make little or no attempt to rank and then analyze in depth the most significant environmental impacts . . . These reasons, however, ignore the precept that the EIS is not an end in itelf but is primarily intended to aid decision making. The statement does not achieve this purpose when it has such prodigious bulk that, while it may serve some academic purpose, no one at the decision making level in any agency will ever read it. Since its purpose is to clarify, not obscure, issues and to forecast and analyze significant impacts of a proposal and its reasonable alternatives, efforts must be made early in the EIS process to weed out unnecessary information. Then, by focusing effort and attention on meaningful analyses, the legal adequacy of an EIS will also be supported and enhanced.
>
> It is the Council's position, therefore, that descriptions of the existing environment and the proposed action should be included in an EIS only to the extent that they are necessary for a decision maker to understand the propoal, its reasonable alternatives, and their significant impacts. The EIS should explain how the scope of the statement and its level of detail have been carefully delineated in accordance with the significant environmental issues and problems foreseen by the agency. Data and analyses in an EIS should consequently be commensurate with the importance of the impact as determined by the agency's environmental analysis. Less important material should be summarized, consolidated or simply referenced.

President Carter,[6] in his environmental message to Congress, May 23, 1977, emphasized the importance of NEPA to the development of sound Federal decisions and to public involvement in government. He stated that the impact statement process had been of tremendous benefit in informing the public and in making agency programs more responsive to environmental concerns. "Nevertheless," he states "an increasing preoccupation with paper work and procedure in the EIS process has tended to obscure the Act's underlying policies." This has resulted in guidelines[4] that do not directly apply the environmental policy statements in NEPA. Instead, the current guidelines tend to focus only on the "action forcing" procedures and not enough on the substantive NEPA policies when conducting the decision process.

As a result, business and labor complain of the bureaucratic paperwork requirements. Environmentalists fear that the NEPA EIS process, which they value highly, is given a bad name by the concentration of paperwork. Suggestions have recently been made by several studies, including a report of the Federal Paperwork Commission, to improve Federal impact statement practices and to make them more consistent through issuance of regulations. Consequently, the CEQ is in the process of converting their guidelines into a regulation format.

Although Federal agencies must consult with each other before taking an action with significant environmental effects, interagency differences over the environmental impacts are not always resolved in a timely way. No procedures presently exist to encourage Federal agencies to anticipate conflicts and seek their resolution before significant decisions have been made. Effective use of the impact statement process to help resolve differences is still limited. For example, under Section 309 of the Clean Air Act of 1972, Congress required the Environmental Protection Agency (EPA) to review the impact statements of other Federal agencies and to refer agency proposals to CEQ if they are found to be environmentally unsatisfactory. Once such differences are known, the Federal government has no clear procedures to resolve them or to see that agencies take reasonable steps to assure that Federal proposals do not conflict with NEPA objectives.

The Council on Environmental Quality is responsible for administration of the environmental impact statement process throughout the Federal Government. It is the intention of the Council to revise the guidelines so as to reduce paperwork and concentrate on improving decision making. They wish to streamline the process without sacrificing protection of the public's environment.

Under the direction of Charles Warren, current Chairman of CEQ, public hearings were held in Washington, D.C. (June 6 to 8, 1977) as a first step in implementing President Carter's environmental message to Congress. The following problem areas were considered:

**Real issues instead of paper production**—How to concentrate less on paperwork and more on substance in implementing NEPA. Included are discussions of how to focus on real alternatives and on mitigation measures that offer decision makers real choices and less on bulky accumulations of background data. Make impact statements more useful.

**Ways to streamline the process**—How can project planning and the impact statement process be integrated to save time? How can statements, by covering whole programs, reduce the scope of needed coverage in subsequent site specific statements?

**Ways to eliminate conflicts and duplication**—What can be done to reduce conflict between agencies regarding which is to assume leadership? How can duplication be eliminated when there are both Federal and State impact statement requirements?

Other problem areas considered include impartial preparation, application to legislative proposals, insuring that mitigation is carried out, the consideration of economic and social factors, and environmental impacts outside the U.S. caused by Federal agency action.

These actions are typical of the constantly changing, dynamic process NEPA initiated. There are certainly more changes and refinements to come.

## PLAN FORMULATION AND EIS PROCESS

The requirements of NEPA and the U.S. Water Resources Council's established principles and standards[39] for planning water and related land resources, among other laws, necessitate an interdisciplinary planning approach[27] to identify and define planning objectives, develop creative alternative plans, and analyze a broad range of complex issues, including the likely economic, social, and environmental consequences of plan implementation. In the Corps of Engineers, this process is carried out in three distinct stages[28] with four tasks: problem identification,[29] formulation of alternatives,[30] impact assessment,[31] and evaluations.[32] Problem identification is the determination of the range of water and related land resources problems a study will address. It provides for establishing planning objectives that give direction to subsequent planning tasks, and its importance deserves special attention.

The building block of plan formulation is the development of viable alternatives, those that will solve the problems identified. Experience has shown that it is not practical or productive to produce alternatives for consideration which are not implementable or will not solve the problem authorized for investigation. The exception is the "no action" alternative which must be considered within any set of alternatives in order to comply with NEPA.

Once the alternatives have been formulated, an impact assessment is possible. The impact assessment provides for analyzing the significant effects of each alternative. These are the economic, social, or environmental consequences of an alternative likely to have material bearing on the decision making process. Impact assessment requires forecasting where and when significant primary, cumulative, and higher order effects could result from implementing a given alternative. This determination requires analyzing and displaying monetary and nonmonetary changes.[34]

Once the need for an environmental assessment is established, the first task is to define the objective and scope based on the purpose for which the information will be utilized. Failure to make these definitions clearly results in production of information insufficient to properly examine the viable alternatives essential to the assessment process. This makes justification for a given course of action almost indefensible. Critical to these definitions is determination of the level of study detail required. This varies depending on the planning process stage where data is required and on the complexity of the project. For example, if one is preparing environmental information dealing with 15 alternatives in preparation for a formulation stage public meeting, the data need not be as detailed and precise as that data used in describing the most viable three alternatives, the selection of the best alternative in preparation for the late stage public meeting, and the draft EIS. The establishment of useful environmental information to affect the decision making process requires both time and money. This also relates specifically to the level of detail and ultimately to the appropriate definition of objective and scope. Failure to recognize the significance of these influencing factors and the need for a well-defined work objective and scope has caused preparation of some less than satisfactory reports and put an unnecessary strain on our Contractor-Corps relationship.

The final task is to evaluate each plan's impacts against the "no action" condition and against the other plans. Systematic preparation and evaluation of alternative ways of addressing problems, needs, concerns, and opportunities under the U.S. Water Resources Council's principles and standards[39] is to develop and identify alternatives in turns relating to the two national objectives of National Economic Development (NED) and Environmental Quality (EQ). The process begins by establishing contributions of each alternative in relation to these two objectives. They are then ranked and arranged from the one with greatest NED to the one with greatest EQ. Judgments are made concerning the beneficial and adverse nature of the contributions of an alternative to establish its overall desirability. After this has been done for each alternative, plans that do not result in improvement over the "no action" condition will be eliminated. The process is then repeated in an iterative fashion, and with each iteration, a new presentation of alternatives from NED to EQ is designated. As plan formulation moves from one stage to the next, more and more detail is examined and assessments and evaluations are refined until ultimately a basis is established for selecting one detailed alternative plan as the best solution to problems identified.

Management of the plan formulation process and ultimate production of an EIS and a recommendation is related to preparation of material for three separate public meetings, review by the Office of Chief of Engineers (OCE) and the Board of Engineers for Rivers and Harbors (BERH), and coordination with Federal and State agencies and public groups. A study plan is developed by completion of Stage 1 in the plan formulation process and further refined by participation of interested individuals at an announcement public meeting where alternatives are solicited in preparation for development of intermediate plans in Stage 2 of the planning process.

An attempt is made to identify all possible alternatives during the second stage of the plan formulation process. By conducting several process iterations, the number of alternatives is reduced to the most feasible and implementable in preparation for the second meeting designated as the Formulation Public Meeting. While no formal reports are published at this point, an outline of the plan formulation report with its appropriate appendices (design and costs, evaluation, hydrology, etc.) should be available. Also, an environmental assessment of alternatives should have been prepared.

At Stage 3, the last iteration is made, the implementable alternative plans are completely developed and the District Engineer selects a plan for implementation. A summary is prepared and plan formulation appendices completed. The draft EIS is prepared and circulated for coordination 15 days before the third or Late Stage Public Meeting. Sometime following this meeting, a revised draft EIS is prepared incorporating all comments from state and federal agencies and other public groups. The revised EIS and the plan formulation reports are forwarded to the Board of Engineers for Rivers and Harbors (BERH) and the Office of the Chief of Engineers (OCE) for review. When the review is complete and the final recommendation of the Chief of Engineers determined, the final EIS is prepared, coordinated, and filed with CEQ. The final reports and EIS are forwarded to the Secretary of the Army and eventually to Congress for action.

The preceding description is an abbreviation of a complicated procedure which sometimes takes 4 or 5 years to accomplish. Only steps in the preauthorization or study phase of a project have been discussed. There are three phases in the postauthorization period: the engineering and design phase, the construction phase, and the operation and maintenance phase of projects. Allowing for the funding process and necessary engineering and construction activities, it is not unreasonable to conclude that the elapsed period between study authorization and completion of project construction may be as much as 15 to 20 years (average) depending on the project scope and urgency of the existant need. There are, of course, opportunities to shorten the process in specific cases if initial funding at various stages is immediately forthcoming, but this is largely a function of prevailing budgetary constraints and cannot be reliably predicted.

The overall process initiated when a water resources study authority is first granted to the Corps of Engineers to investigate problems in a single locality or in a broad region of the country, and which ultimately results in authorization and construction of water resources projects to solve the identified problems, is indeed a lengthy one. It reflects continuously increasing complexities of conducting activities in our complicated society. Try as the Corps does to improve the quality and efficiency of report production, there is little evidence to show that process time can be shortened or that major water resources projects, once authorized, can be operative much sooner than they are now.

## PUBLIC PARTICIPATION

Concern for the environment is not new to the Corps. Then what has changed? Many things, of course. Not only does the nature of a problem change over time, but the perception, understanding, and concerns of the people change as well. This is very important to an organization such as the Corps of Engineers. The Corps, as a public agency, must respond to the public interest. Though determination of the public interest is never simple, every public agency must meet this issue or lose the people's confidence and support. "Loss of credibility" is the term sometimes used.

The Corps is sensitive to public reactions.[18,21] Few agencies of government work with people so directly as it does through elected representatives or private citizens and at many levels of concern and involvement. Obtaining the public view is generally difficult. Reaching consensus is a task even more difficult. Responding to overall public interest within constraints and authorities imposed on a public agency becomes more difficult with each passing year.

Public interest can sometimes change rather rapidly. It depends to a certain extent where the individual sits at the time and the pressures imposed by his immediate surroundings. This is best illustrated by realizing that besides protecting the environment, the Corps of Engineers is foremost in protecting people. Every time the Corps looks away, Mother Nature reminds them not to turn their backs on flood protection or other natural disaster problems.

The Corps has traditionally provided a public forum for opinionation on proposed projects. This is most often a public hearing held during the preauthorization study.

To a great extent, this practice satisfies the procedural requirement for citizen participation in preparation of environmental impact statements under NEPA. The CEQ guidelines go beyond traditional Corps practice, however, in requiring early notification that an EIS is being prepared, that a draft of the statement be made available prior to the public hearing, and that all citizen comments be made part of the final statement. Nevertheless, one would expect that the Corps would follow a consolidating strategy concerning NEPA requirements for citizen participation. That would be to essentially continue with its standing practice of holding hearings to elicit the views of the public, but little more. Given the vagueness of NEPA on this point, this would be a reasonable strategy. This has not occurred. In fact, some of the strongest criticism of the Corps traditional policy of public participation, as well as innovative counterproposals, have come from within the Corps itself. The Corps' traditional citizen participation policy is depicted in-house as legally acceptable, but grossly inadequate.

The Chief of Engineers published the following policy and guidelines[36] for all civil works divisions and districts in the Federal Register, October 29, 1976:

> The Civil Works Program will be conducted in an atmosphere of public participation, trust, and mutual cooperation. As a public agency, the Corps responds to the public interest. That interest synthesizes many needs, desires, and aspirations. It finds expression in the views of individuals and groups and their representatives at local, State and Federal levels of government. The Corps of Engineers has an obligation to seek these views and to accommodate them insofar as possible. There is an equal obligation to provide information to those who express their views, so that they can participate effectively in the planning process.
>
> Public involvement in Civil Works requires a flow of information between planners and the public. It assures that Corps actions respond to public needs and preferences. To secure effective public involvement in Civil Works projects the Corps of Engineers will:
>
> 1. Open and maintain channels of communication with the public
> 2. Encourage public understanding of Federal, regional, State and local responsibilities, laws, authorities, and procedures in conducting water resources planning studies and implementing water resources programs
> 3. Present information which will promote public understanding of water resources problems, needs, opportunities, and objectives
> 4. Seek out the public's views and comments on problems, needs, alternative solutions, and related impacts, and possible Federal action
> 5. Give full consideration to public needs and preferences in all Corps activities

The Corps is aware of the gap between its traditional practices and the requisites of a meaningful program of public participation. A first step in closing this gap came in September 1970, when Corps policy was changed to require holding three public meetings throughout the pre-authorization planning of a project. A more concerted effort to broaden the "public" side of planning began in February 1971, when the division and district planning chiefs and their public affairs officers participated in a short course on "Public Participation in Water Resource Planning." In June of that year, the Corps' Institute of Water Resources established a Technical Assistance Program to promote new modes of public participation and offered consulting services to the divisions and districts. Also in May 1971, district engineers were authorized to increase their survey investigations costs by up to 10% to allow expanded citizen participation in project planning. Even further spending above this level was allowed, though it required approval by the Office of the Chief Engineer (OCE).

Encouragement of greater public involvement in district level planning has met mixed results. Several districts have taken advantage of the Technical Assistance Program, and some have undergone fundamental reorientation in their planning process. Some Districts have adopted a far-reaching program of hearings, workshops, and information dissemination regarding their studies and projects.

Tension is evident between the Corps' top echelon, which is pushing for greater citizen participation, and many divisions and district engineers more comfortable with the status quo. For the leadership, greater citizen involvement at the early planning stages is a rational method for reconciling competing local interests, thereby preventing later controversies, confrontations, executive and Congressional reprimand, and possible court suits. Yet some division and district engineers appear less eager to follow this changed orientation completely. From their viewpoint, they are being asked to expose their project plans to stringent and often hostile scrutiny of many potential community opponents. Asking this of any planning group requires sharp departure from past practices.

Although the Corps' headquarters is pushing for change, its complete adoption throughout the agency has not occurred. That it will occur is just a matter of time.

## INTERDISCIPLINARY TEAMS AND MANAGEMENT

During Fiscal Year 1977, the Corps of Engineers employed 209 full-time and 323 part-time employees to prepare and coordinate environmental assessments and impact statements. Included are a zoologist, botanists, ecologists, freshwater and marine ichthyologists, fish and wildlife managers, a few sociologists, landscape architects, and environmental engineers. Costs for labor, materials, and outside contractual services to assist in EIS preparation for this period amount to approximately 28 million dollars. Contractors were used in data preparation in approximately half of the EIS written by the Corps of Engineers. There have been approximately 900 separate contracts to support EIS preparation since enactment of NEPA.

It is imperative that there be a close working relationship between Corps environmental resource personnel and the staff of any contractor. In the most successful endeavors, the contractor is part of the planning process. The degree to which the contractor is isolated from the planning effort can usually be directly equated with the extent of dissatisfaction encountered with the final report.

From a management viewpoint, the basic problem is lack of effective communication.[3] This breakdown in communication frequently begins during initial contract negotiations. The organization requesting environmental assessment is frequently unfamiliar with subtleties of the environment but understand economic and engineering characteristics of the proposed action. On the other hand, organizations capable of performing environmental assessment are frequently naive about the economics, engineering, and operational structure of the client. As a result, the people involved in environmental assessments have an inadequate understanding of the proposed action.

Another problem is selection and management of an interdisciplinary team. Selection of the project manager is critical to the success of the contractor agency team. Management of such a team is no easy task. The manager must insure cross communication and production of information at compatible levels of detail. He must have the disciplinary inputs available to meet deadlines. Problems of interdisciplinary team management vary depending on the contractor (for example, university versus consulting companies). The Corps' experience has been a mixture of success and failure. Interdisciplinary team efforts at universities have many original contributions to environmental assessment approaches and methodologies, but are difficult to coordinate because of the complex organizational structure of most universities. Organizations such as consulting companies doing environmental assessments do not generally have problems as complex organizationally as does the university, but frequently do not have staff capabilities of the universities.

For proper management, all will agree that coordination is necessary, but this is sometimes difficult. The problem is when, where, and how much coordination. There are no simple answers. The basic guidance in the Corps is that the contractor personnel must be considered part of the planning team. When not working as an integral part of the team, the contractor and district staff must have frequent information exchanges. Results have been most satisfactory when data was exchanged informally on a weekly basis and formal coordination meetings were held monthly to deal with schedules and problems as they arose.

Certain criteria emerge as the most successful in selection of team managers and team members whether in-house or for the contractor. There is no one expert category needed more than another. Experience has shown that a specific biological background is not preferred over the ability to relate scientific information over a wide base of disciplines. Although formal training is important, competence is essential. A broadly trained, experienced professional who can apply himself to a range of problems and who is not afraid to admit limitations is superior to an expert. Highly technical input can be obtained from exogenous sources when needed. When developing environmental teams, the Corps looks to satisfying multidiscipline requirements by selecting individuals with a broad base of expertise and experience rather than those with a narrow specialty.

## APPROPRIATE METHODOLOGIES

Numerous methods[25] have been developed for quantifying environmental impacts of engineering activities and for displaying tradeoffs among planning objectives so that decision makers can wisely select the best alternative. Examples include those techniques developed by the U.S. Geological Survey, Universities of Georgia and Wisconsin, Bureau of Reclamation, Northwestern University, Battelle-Columbus, Battelle-Northwest, Stanford University, U.S. Environmental Protection Agency, USAE Tulsa District, U.S. Army Engineers (USAE), USAE North Pacific Division, USAE Construction Engineering Research Laboratory, and USAE Waterways Experience Station. Corps of Engineers specialists at the Waterways Experiment Station, Vicksburg, Mississippi have examined 54 available techniques as they apply to Civil Works activities and report that each method has certain advantages and disadvantages. The basic problem facing those conducting an environmental assessment is to determine which methodology or combination of methodologies are best suited for the situation.

There are subjective methods and numerical methods. Most environmental impact statements of the past and present utilize some form of a subjective method. They are characterized by a great deal of descriptive material in the form of written text, maps, and lists of specific parameters such as environmental conditions, biological species, or other factors which provide mostly a framework for an area inventory. When using either method, the assessment is generally accomplished by comparison of the assigned effects of development alternatives to a baseline of ''no action''.

The primary advantage of subjective comparisons is simplicity of concept. The reason for a judgment is stated and subject to public scrutiny and criticism. The importance of each effect is stated, and emphasis can easily be placed on the most important effects, but sometimes secondary effects are overlooked or only partially identified. The conclusions are based on trained professional subjective judgments and in the absence of well-defined, statistically valid data, there is no better substitute to do the job.

Primary disadvantages of the subjective comparisons are lack of uniformity in subjective analyses and nonuniform considerations of various types of environmental effects. Because of training, background, and experience, the subjective ranking of several alternatives can differ markedly if made by different persons. The amount of difference is usually a function of the amount, accuracy, quality, and availability of data. The more data, the more similar judgments. Team effort helps to resolve opinion differences, but where efforts fail to resolve these differences, more than one opinion should be reported. Another disadvantage to subjective methodology is a lack of continuity in the report. Any and all kinds of data can appear in the report, but it is sometimes impossible to determine what data is relevant and influential to project judgments and which might be superfluous. This situation occurs because fixed procedures and assessment factors are not used for evaluation of alternatives.

Primary advantages of the numerical methods are uniformity of application and completeness. They are uniform in application because they have a standard evaluation procedure. A systematic method, generally using some form of a matrix, is used to determine numerical values for each effect and overall numerical ranking for each alternative. The methods evaluate impacts by comparison to a format that plots in a matrix the degree and relative significance of a proposed development action or alternative against a range of specified environmental, economic, and social factors. In some matrices, a symbol or descriptive terminology system is used to rate impacts. In others, a point value rating system is determined and numerical values are assigned to each matrix cell where an interaction occurs. In either of these, magnitude and/or degree of importance may be indicated. Each alternative can be expressed in a two-dimensional matrix, but a network of multialternatives can be analyzed and compared using a system of multiple matrices. Weighting factors or multiplies are sometimes applied to individual parameters in an attempt to compensate for inherent differences in their magnitude or importance in relation to the final net effect. In any event, the total numerical values for several alternatives, when compared, will indicate the supposed best alternative.

The disadvantages of numerical evaluations are: (1) too complex for routine use, (2) difficulty in adopting parameters and weighting factors to local conditions and noneconomic resources, (3) subjective inputs can control results, (4) high cost in acquiring data and reducing it to proper form for calculation, (5) low understandability, and (6) inability to project public and community policies and preferences.

Matrix analyses are used extensively throughout the Corps, particularly for large complex projects. But use is limited mostly to early planning stages. In a few instances, the matrix analyses have been carried through to a point of assigning comparative numerical values to different alternatives. These figures assist district engineers in making their decisions, but these figures have not been the only factor influencing the final recommendation. There is no substitute for trained professional subjective judgments in evaluation of construction effects on the environment, nor is there likely to be for some time. However, matrices analyses should not be precluded from immediate use on a selective basis.

## MITIGATION

Ever since the Fish and Wildlife Coordination Act of 1958[21] (but particularly since enactment of NEPA[26]) environmentalists have sporadically, but frequently urged the Corps to provide extensive mitigation measures as part of projects. In some cases, the Corps regarded the scope of the measures as excessive and suspected the project was held hostage for what was really enhancement measures labelled as mitigation to avoid cost sharing. In other cases, obvious damages have gone unmitigated because of physical, biological, or institutional problems. In the past, the Corps has relied primarily on state agencies and the U.S. Fish and Wildlife Service to consider nearly completed plans and to recommend mitigative measures subject to negotiation and some public review. With adoption and implementation of the Water Resources Council's principles and standards[39] the Corps has begun giving full consideration to mitigation problems early in the planning cycle, surfacing the issues with the public, and exploring feasibility and acceptability of all reasonable alternative solutions. The result of these developments has been a growing need to review existing policies on mitigation, consider consequences of various options, and decide what policies to pursue.

The Fish and Wildlife Service is required by law[13] to recommend measures for fish and wildlife conservation and development for each project considered by the Corps of Engineers. These recommendations are the basis on which the Corps develops fish and wildlife features included in the project plan. Fish and wildlife features may be for either enhancement, mitigation, or both according to existing conditions at the project site. The impact of a project on fish and wildlife resources is currently given major consideration during development of Corps reports. If the proposed development results in overall loss to the fish and wildlife resources and there are feasible means of offsetting or minimizing them, the project plan will include measures for these purposes. The timing of the fish and wildlife report is most critical so that the services recommendations will be available during early stages of project formulation. Late receipt of this information causes "add on" features (usually very expensive) and seldom results in the most desirable plan.

The fish and wildlife agencies tend to resist evaluation of monetary loss and recently have been making an effort to negotiate in terms of habitat replacement rather than in animals lost or reduction in hunter days. On the other hand, Corps interpretations may have been too conservative by requiring that losses be quantified with assignment of monetary values wherever practical.

In all Corps reports to Congress, the fish and wildlife aspects are covered. Recommendations of the U.S. Fish and Wildlife Service are either accepted or rejected with justification. The Fish and Wildlife Coordination Act specifically gives the Corps of Engineers the responsibility to recommend only those mitigations found justified. If the report of the U.S. Fish and Wildlife Service does not contain sufficient justification to support a recommendation or sufficient information to support a favorable judgment on the expenditure of funds, and the Corps is unable to develop that justification, the recommendation is rejected. However, a discussion is presented in the report stating why the recommendation is rejected. All reports going to Congress where project-related fish and wildlife losses are recognized include a mitigation plan.

Some contend that the Corps contributes to delays in land acquisition for wildlife mitigation by following complicated and time-consuming procedures. The system developed over many years with Congress extends the issue over a considerable time period—for project completion, 15 or more years. Work on land acquisition cannot be initiated until the project is authorized and construction funds appropriated. General design plans must be approved and the real estate appraisal studies completed before acquisition can be initiated. In many situations, objections by land owners and Congressional delegations cause further delays. Sometimes curtailment of project appropriations necessitated allocation of available funds to contract work to avoid major increases in Government contract costs resulting from extension of the contract period.

Another aspect involves authorization for land acquisition. Mitigation is limited to what can be accomplished within the project boundary unless specific authorization is sought for land acquisition elsewhere. If the land is described in the authorizing document, recommended by the Corps of Engineers, and approved by Congress, acquisition can easily proceed. However, should subsequent studies indicate a major increase in the number of acres required or change in location of acreages, such as from one county to another, the legal and contractual arrangements to acquire lands can become extremely complicated.

In most situations, the extent of land acquisition required is based on certain assumed management concepts so that wildlife production will offset specified project-induced losses. Operation and maintenance costs for these fish and wildlife mitigation features, including land management, are joint costs and should be borne by project beneficiaries. In some instances, the Corps of Engineers will manage these lands and fund their operation and maintenance.

It is the Corps policy to offer lands having specific fish and wildlife values, including mitigation lands, to state or federal fish and wildlife agencies for use without cost; however, the Corps will not fund for expenditures of the managing agency. If the state or federal fish and wildlife agency does not elect to take the lands and fund for operation and maintenance costs, the lands will usually be retained under Corps control and managed according to a fish and wildlife management plan developed in cooperation with the appropriate fish and game agency.

It is the position of the Corps that all fish hatcheries, special acquisition wildlife mitigation lands for mitigation purposes, and associated funding responsibility should be transferred to one of the federal fish and wildlife agencies. Under this procedure, the agency receiving the facility budgets for the operations and maintenance requirements, and at the end of each year provides the Corps with a cost statement so that appropriate allocations can be made to the responsible purposes of the project.

Another problem relates to a determination of what constitutes mitigation as opposed to enhancement. Because mitigation becomes a project cost and enhancement must be cost-shared, there is a tendency in fish and wildlife circles to avoid classifying any features for fish and wildlife enhancement.

Should a project plan include features to mitigate a potential? This is a controversial question. Since a stream may have several natural resource potentials (hydropower, water supply, fishery), a river basin or project plan should give equal consideration to development of all potentials. These potentials must be weighed one against the other and a mix selected and recommended to provide the best balanced development according to economic and environmental objectives. In determining whether a potential should be classified enhancement or mitigation, it is Corps policy to consider as mitigation a potential existent only at project formulation time and for which there is an on-going plan for its realization. Otherwise development would be classified as enhancement.

The Corps can and does provide both mitigation and enhancement by implementation of the Fish and Wildlife Coordination Act[13] and the Federal Water Project Recreation Act.[11] The latter act has been amended to require only 25% cost-sharing by nonfederal interests for improvements of fish and wildlife resources, provided nonfederal interests stand the operation and maintenance costs. There is also a provision in the Recreation Act (Section 6e)[11] which allows full federal costs provided the feature is connected with an authorized federal program for conservation and development of fish and wildlife.

The National Environmental Policy Act[20] does, in a sense, call for a form of mitigation. The act requires the federal government to "attain the widest range of beneficial uses of the environment without degradation, risk to health or safety, or other undesirable and unintended consequences; preserve important historic, cultural, and natural aspects of our natural heritage; and maintain, wherever possible, an environment which supports diversity and variety of individual choice.[26]" The act further requires preparation of an Environmental Impact Statement utilizing "a systematic, interdisciplinary approach which will insure integrated use of the natural and social sciences and the environmental design arts in planning and decision making" and "identify any irreversible and irretrievable commitments of resources which would be involved in the proposed action."[20]

The effects of NEPA on mitigation have been twofold. First, the environmental dimensions subject to mitigation have been expanded from an examination primarily of fish and wildlife to an evaluation of the entire habitat and ecological system. This tended to make the analysis for mitigation more complex and difficult to establish a methodology applicable to various situations. Secondly, the requirements of NEPA have increased the work load not only of the construction agencies, but has increased demands for information and assistance from agencies such as the U.S. Fish and Wildlife Service and State Fish and Game organizations. In this regard, perhaps the latter groups have suffered more from a lack of sufficient funds and manpower to accomplish the new work load than the construction agencies.

Under Section 7, entitled "Interagency Cooperation", the Secretary of Interior is asked in the promotion of the purposes of the Endangered and Threatened Species Act[10] to utilize other federal departments and agencies authorized to accomplish purposes of this act. This act could relate to mitigation in that some mitigation actions support only small tracts of land, sometimes too small to manage efficiently and when considered by themselves will not contribute materially to protection of an endangered species. The Endangered Species Act would permit purchase of additional lands outside the project area, but under the same authority. Administratively, the actions may be separate but the results would place additional lands beyond those purchased for mitigation in federal ownership. This would tend to keep the mitigation procedure simple and defendable, an agreed position more attainable, and coordination easier.

## HABITAT EVALUATION

Mitigation of fish and wildlife or habitat losses due to engineering projects is not an easy task. Fish and wildlife agencies and water resource planners are reluctant to accept set procedures for determining wildlife losses or for selection of lands to mitigate these losses. Perhaps it is just as well to approach mitigation on a project by project basis; however, the goal should be development of acceptable uniform procedures.

The most widely used approach has been to express loses in terms of recreation usage.[11] This approach bases the fish and wildlife value of an area on utilization by hunters and fishermen. Damages are accounted and mitigation planned in terms of hunter and fishermen days lost. This approach is harmonious with traditional economic units and calculations of the benefit/cost ratio.

The concept of quantifying fish and wildlife values by this method has not been entirely satisfactory. It cannot be applied to situations in which the resource is not used and/or production does not occur. A resting area on a waterfowl flyway is a typical example. It is a necessary part of the system; however, procedures are not sufficiently refined to determine how many hunting days would be lost throughout the flyway if a particular resting area was removed from the system.

Another criticism of the user concept is the schedule of values. Most fish and game agencies believe some values are too low. This is considered particularly pertinent in the case of big game hunting. It is hard to justify the acquisition of large land areas considered necessary to offset loss using values set forth by the Principles and Standards.

Due to concept inadequacies, the Corps has deviated from normal procedures on a case by case basis. Decisions were made on judgment and negotiated compromise. In some situations, the Corps differs with fish and game agencies. The Corps agrees that complete mitigation results in the same level of fish and wildlife production as existed without the project. Nevertheless, in some situations it is almost impossible to justify expenditure of federal funds to obtain this objective.

For another example, consider a project that destroys pheasant habitat. If only one hunting day per acre of pheasant habitat is projected without the project, and provisions for a managed habitat will support 50 hunter days per acre, this could mean that 50 A of habitat outside the project could be mitigated by providing 1 A of managed habitat. If the objective is to preserve the habitat, the mitigation has been slight.

The approach used currently by the U.S. Fish and Wildlife Service destined to be the procedure of the future is one involving habitat replacement.[38] Following this approach, destroyed habitat will be replaced if possible in kind on an acre-by-acre basis without recourse to economic evaluation or a benefit/cost ratio calculation. Damages to a natural resource are considered based on the same principle as damages to public utilities, roads, or towns displaced by a project.

The habitat replacement concept has problems too. How do you replace a destroyed habitat?[2] Formerly, if wetlands were destroyed at one site, federal purchase of wetland at another site constituted mitigation. But this is not entirely consistent. The other wetlands were already there; merely transferring title of a land tract to the government does not mitigate damages on another land tract. To be sure, the acquired habitat may be made more productive by a game management program, but this can, with some justification, be challenged.

It has been a common practice in some areas to apply a factor of two to the value of any land acquired as mitigation and placed in government ownership as opposed to value of the same land under private possession.

Habitat Evaluation Procedures[38] developed for use by the U.S. Fish and Wildlife Service, Division of Ecological Services state as the objective, "...to provide a uniform, nationwide method for determining impacts on fish and wildlife and their habi tat arising from water development projects."

These procedures require a team of experienced field biologists to (1) delineate the major habitat types in the evaluation area, (2) determine the acreage of each, (3) select approximately 10 of the most biologically important wildlife species or groups of species, (4) develop or modify predeveloped key criteria for habitat values for each wildlife species or group of species, and then (5) by group consensus, rate a representative number of sites in each habitat, on a 1 to 10 scale for the selected wildlife species or group of species. The method works well with a team of experienced field biologists who have the time to examine in detail an adequate number of sites. However, any team gains experience as they do the job, so it may become difficult to compare results from the first sites with those of the last. Also, the individuals on the team vary in their knowledge of certain species and in their ability to convince the rest of the team to see it their way. Further, it is uncertain if the same team (much less a different one) would arrive at exactly the same conclusions at different seasons. A major weakness of the U.S. Fish and Wildlife system is the difficulty of displaying the basis for values given each site so other biologists, professionals in other fields, and the public can understand them.

The U.S. Fish and Wildlife Service has recently been modifying and adding to their procedure some of the concepts developed by Whitaker, Roach, and McCuen.[41,42] Line charts to inventory important components of wildlife habitat and transformation charts to convert the inventoried characteristics to habitat values for specific species are being developed. In addition, line charts display the basis for all conclusions, provide a simplified base for determining future conditions, and are easily adapted to computer analysis. The value ratings from this system provide a firm foundation for assessing effects on wildlife habitat of water resource project alternatives and for planning any needed mitigating or compensation measures. With the addition of the Whitaker and McCuen concepts, even the more conservative Corps biologists and managers are willing to concede that the method has the potential of producing a useable product.

The Corps has indicated several areas of disagreement with these procedures. In addition, the Fish and Wildlife Service has identified potential conflict areas with the Corps' planning process that may reduce the quality and completeness of the Service's fish and wildlife report. Therefore, the Corps of Engineers and the Fish and Wildlife Service have agreed to mutually support a goal which will identify the depth and extent of these problems by practical application and establish a course of action for resolving such difficulties. The agreement is that at least one project per Corps Division will receive special emphasis to achieve the above goals. These projects have been selected and in most cases evaluation procedures initiated.

## ENVIRONMENTAL LITIGATION

As of July 1977, the Corps of Engineers had been in 75 major lawsuits filed against 49 Corps projects and 27 permit actions. All of this litigation involved environmental issues and occurred since passage of NEPH. There were 2622 water resources projects where an EIS had been filed with CEQ in at least the draft form. There were 134 additional EIS and hundreds of environmental assessments prepared for permit actions. Of the 49 lawsuits involving Corps projects, 26 were pending, the Corps had been enjoined on 12, and 11 had been dismissed in favor of the Corps of Engineers. Of the 27 lawsuits involving regulatory authorities[24,35,37] of the Corps, 15 cases were pending; the courts ruled in favor of the Corps' position in 10 cases and ruled against the Corps' stand in only one case and part of another.

These figures would indicate that only 2.7% of projects on which an EIS was filed were challenged in court. This percentage is about half that experienced over the whole nation.[5]

In completed cases against the Corps of Engineers' own projects, the courts ruled in favor the Corps in about 50% of the lawsuits. In cases involving the regulatory authorities, the Corps record for winning was about 90%.

In examining the cases in litigation involving water resources projects, it was not surprising to find that in all cases, the plaintiffs alleged violation of NEPA[20] or CEQ guidelines[4] and most often declared the EIS inadequate. In 25 to 50% of the cases, the following specific violations were alleged: (1) failure to consider appropriate alternatives, (2) violations of statutes other than NEPA (Fish and Wildlife Coordinations Act,[13] Federal Water Pollution Control Act[12] (FWPCA), Marine Protection Research and Sanctuaries Act[17] and Preservation of Historic and Archeological Data,[22] (3) federal actions were arbitrary and capricious, (4) illegal actions associated with the acquiring of lands, and (5) pollution of water or degrading water quality.

The following additional complaints were found in 15 to 25% of the cases: (1) lack of adequate coordination with other federal agencies or inadequate review of EIS,[33]

(2) specified loss of one or more specific natural habitats, (3) inadequate determinations and acquisition of appropriate mitigation measures to compensate for environmental loss, (4) improperly calculated benefit/cost ratio or challenge of the interest rate used in the determination, and (5) alledged Corps did not follow Congressional mandates or obtain Congressional consent.

Examples of some plaintiff allegations found only occasionally up to 15% are as follows: (1) inadequate assessment of secondary effects, (2) failure to quantify environmental amenities, (3) loss of historical or archeological sites,[33] (4) violation of Endangered Species Act,[10] (5) failure to adequately describe hazards from earthquakes, (6) potential for undesirable underground water movement, (7) failure to adequately assess the effects of economic development and other sociological impacts, (8) failure to adequately consider future and expected future needs (traffic, commodities, equipment), (9) inadequate assessment of impacts from dredging and dumping of dredged material, (10) creation of a salt barrier, (11) failure to consider energy needs of the country, (12) public rights and public hearings violated, (13) loss of esthetic beauty, and (14) encroachment of farming or project on wetlands.

The above lists illustrate the complexity and variation in issues associated with court cases involving water resources projects. Regulatory function cases were equally complex, but allegations tended to be less repetitive, more specific, and each case an issue unto itself. The majority of cases in the permit area alledged violation of the Federal Water Pollution Control Act[12] specifically Section 404, rather than NEPA per se.

To give an idea of differences found in regulatory cases as opposed to Corps project cases and provide an appreciation of the types of complaints, what follows is a list of allegations that appeared in at least two or more cases examined: (1) questioned the Corps' jurisdiction under navigable waters or waters of the United States, (2) EIS inadequate, and there was a contrary determination as a result, (3) failure to hold public meeting under Section 404 FWPCA of the Federal Water Pollution Control Act, (4) failure to conduct appropriate actions required by state laws,[26] (5) water quality affected by dredging and deposit of materials into waters of the U.S., (6) further violations under Sections 401, 403, or 406 of FWPCA, (7) prohibition of construction of trailer parks or housing units in wetlands, (8) questioned the Corps jurisdiction in relationship to tidal mean high or higher water mark under Section 10 of the Rivers and Harbors Act of 1899[24] and Section 404 of FWPCA, and (9) Section 404 regulations invalid because they attempt to regulate land use and activities unrelated to water quality.

The Corps of Engineers in responding to court cases has made every effort to comply promptly with judicial orders and may have over reacted by preparing excessively detailed EIS's. In controversial cases they may have resorted to filing voluminous statements in anticipation of legal challenges to comprehensiveness and adequacy in the courts. Given the circumstances, without a clearer legislative or judicial definition of what constitutes an adequate EIS, there appears to be no realistic solution to the time-consuming and lengthy EIS process for many proposed actions, especially actions expected to be controversial. The Corps experience in litigation indicates, however, that the Corps' EIS have met with substantial judicial approval, and it is not clear that less detailed EIS would have been equally successful.

## EVOLUTION OF EIS

The language in NEPA was purposefully written in broad and somewhat general terms. Even today certain sections are still interpreted and debated in both the courts and in Congress. To the individual who must prepare a statement, implementation of

NEPA was and remains a dynamic changing process fraught in many ways with frustration and uncertainties.

Surveying the last 7 years reveals three definite phases: (1) a period of interpretation, (2) a procedural or letter of the law phase, and (3) an integrated planning or intent of the law phase.[1]

The first phase, the period of interpretation, is characterized by much talk and not much action. Typical of this phase are long unproductive debates about what projects require an EIS in accordance with the act and preparation of rather meaningless reports based on hurried searches for libary data to be used in making rather cursory analysis of the environment. Typically, the data serve only as descriptive information in the EIS rather than the base for an assessment of the environmental impacts of the project.

It would be expected that at least some time would be spent debating issues in order to obtain clarification. This was done, but only for a short duration.

In the beginning, the Civil Works Directorate of the Corps faced a workload of over 2500 projects requiring an EIS, not counting the additional requirements under the permit program for which the Corps was then responsible. By the end of the first year, the Corps had prepared 105 impact statements. That first year was spent developing detailed guidance and procedures while either preparing statements for recommended and previously authorized projects involved in actions by Congress or for projects under litigation.

On June 4, 1970, when NEPA was just five months old, the 9th U.S. Circuit Court issued the first injunction against a Corps project, stopping channelization on the Gila River in Arizona. There had been only one EIS (dated March 1970) prepared up to that time to study dredging and water quality problems on the Great Lakes. By January 1971, 1 year after the passage of NEPA, the West Tennessee Tributaries Channelization project also had been stopped and President Nixon issued the order that stopped the Cross-Florida Barge Canal.

Court involvement in implementation and interpretation of NEPA rather subtly, but definitely, led the Corps into the procedural phase. In this phase, emphasis is placed on adhering to the letter of the law.

EIS prepared during this phase can generally be identified by quantity of information meeting every procedural aspect of the law. A general criticism, however, of EIS prepared by this concept is that data is hard to relate to impacts and alternatives. The documents are hard to read, and many times they are "after action" reports with no clear understanding of how all the facts relate to the decision making process.

To a certain extent, such an approach was undermining tried and proven planning procedures. An EIS prepared according to the procedural philosophy, for example, usually contained a social, economic, and environmental data and engineering. The EIS document duplicated much information contained in planning and engineering reports. Too often contents of the different documents were not correlated.

Neither the EIS nor the planning documents provided the decision maker with all essential and timely facts. The combined use of the documents was not completely satisfactory. Because of internal pressures, tight schedules, and the procedural syndrome, there was a tendency to assign preparation of the EIS to available staff (some with little training in the environmental area) or to a consulting contractor. This procedure was not bad except there was also a tendency to separate the group preparing the impact statement from another staff group preparing the project documents. Unsatisfactory results of this procedure became immediately evident to those with review responsibilities. Information presented in the documents differed considerably.

The Corps is now well along into the third and last phase in the scheme for preparing

EIS. The third phase, the integrated planning phase, emphasizes fulfillment of the real intent of NEPA. Under the integrated planning concept, adherence to NEPA for a particular project requires examination of at least three separate Corps documents: the Survey Report, the EIS, and a Statement of Findings.

The Statement of Findings is extracted from the Survey Report. It is a short, clear statement made by the responsible District Engineer. It summarizes the rationale for his recommendation based on professional and personal concerns for serving the public's best interest. It also notes his efforts to draw the soundest possible conclusions from data and information in the Survey Report including studies, observations, consultations and other factors bearing on the proposed action. This Statement of Findings accompanies all final EIS filed with CEQ and is distributed to all parties interested in Corps activities.

The importance of recognizing the significant engineering activities and relating them to environmental impacts at various stages of project completion was only slowly recognized and adopted. Engineering during earlier periods of EIS preparation was too often interpreted and reflected as an overall description of the project without identifying specific engineering methods and objectives with their impacts. The present EIS objective reflects a shift in emphasis and closer adherence to the intent of NEPA.

The role of the courts was probably the single most influential factor determining evolution of the EIS process and implementations of NEPA as it presently exists in the Corps of Engineers. Not that this is necessarily good. One would probably expect policy makers to have the greater influence perhaps reflected in the CEQ guidelines. But court decision consequences were felt directly at the field level, and therefore had the greater impact on field level behavior.

## SUMMARY

In the U.S. today, most citizens have developed at least a degree of concern for environmental quality. Public opinion polls regularly disclose that a majority of Americans want to breathe clean air, enjoy waters free of pollutants, have access to parks and open space, preserve wildlife, etc. Of course, most realize that these environmental protection goals must coexist with the functioning of our economic system, with measures to produce energy, house and feed our growing population, etc. However, most of us want to balance economic development with environmental quality. These beliefs follow from our realization that if we cannot have both a functioning economy and a livable environment, there can be little real value in an ever increasing gross national product; if the average American's quality of life must deteriorate as our economy and population grow, then growth can hardly be desirable.

Because some degree of this environmental consciousness is commonplace today, some citizens find it hard to understand why many institutions and government agencies have not always been closely identified with these currently accepted environmental policies. In fact, the more strident environmentalists[8,9,23] do not hide their distaste and contempt for large segments of American industry and for many governmental agencies which traditionally have emphasized economic development goals rather than environmental preservation.

Although the more extreme environmentalists' point of view is understood, institutions, economic systems, and agencies within democratic governments almost always reflect the predominant economic and social forces of their age. Very rarely indeed can a government agency give complete deference to values of a future generation in preference to those of the current generation. Environmentalism has become a truly powerful force in the U.S. only in recent times.

Today the environmental movement is becoming institutionalized at all levels of American government and in innumerable private organizations such as the Sierra Club, the National Wildlife Federation, the Natural Resources Defense Council, The Environmental Defense Fund, the Audubon Society, and Zero Population Growth. Nevertheless, the battle to save what's left of the American environment is far from won. Even today we have no truly effective government programs to preserve remaining farmland and wetlands from sprawling suburbs and recreational home developments, to reduce the rate of U.S. or world population growth, to enforce widespread conservation of energy, or to regulate strip-mining. Reviewing the history of the American environment, one can only conclude that the American environmental movement has been a long time coming and has not achieved even a minority of its objectives to date.

The Corps was assisting the conservation movement for many years before the term environmentalism had even been coined. Nevertheless, one must recognize that environmental preservation was never a dominant priority for the U.S. for the first half of the 20th Century, any more than it had been during the 19th. Thus, it was not made a paramount mission of the Corps. Instead, this nation concentrated far more of its resources and attention on economic growth, improvement of the material standard of living, and development of natural resources in addition to waging major wars and a long "cold war."

If one merely recalls the more noteworthy events and trends of each decade of the 20th century prior to 1970, one must agree that economic values and concerns of the 19th century still predominated. Most new developments were hardly beneficial to environmental preservation since most entailed rapid and large-scale acceptance of new technologies which consumed energy voraciously and polluted air, water, and the land itself. The mechanization of transportation via trucks and private autos used up most of our petroleum reserves and inspired the mushrooming of suburbs which covered the countryside. The technological revolution in agriculture brought tractors, extremely costly machinery, and an "agribusiness" founded on chemical pesticides and massive energy consumption. American manufacturing industries began to use and discard thousands of new chemicals that polluted water and air and used natural resources rapidly. New packaging and marketing techniques plus a growing population caused problems in disposing of solid wastes. These are a few examples of 20th century trends hostile to environmental preservation.

Unquestionably, the Corps of Engineers performed many vital missions from 1900 to 1970, but most were developmental rather than environmental in nature. From Corps efforts in World War II to Corps contributions to the U.S. space program; from continued work for flood control, water supply, and navigation to creation of hydroelectric energy, the Corps built the U.S. economy. However, because public interest priorities were focused primarily on development, the Corps was less frequently ordered to preserve the American environment during this time than to develop and utilize its resources.

The growing environmental movement achieved a signal victory in 1969 with passage of the National Environmental Policy Act - NEPA.[20] That famous statute created the Council on Environmental Quality (CEQ) and required preparation of a thorough environmental impact statement for every major federal action possibly having significant impact on the human environment.

An objective evaluation of the Corps of Engineers' record during the 1960's and 1970's demonstrates conclusively that the Corps not only accepted environmental policies adopted by the federal government, but provided environmental leadership.[14,15,18,19,21] Before and after enactment of these federal environmental laws, the Corps worked to redirect national attention to programs balancing objectives of development and conservation, rather than merely emphasizing development.

In this chapter, an attempt has been made to explain some ways in which the Corps of Engineers has accepted and advanced new national goals and policies of protecting the environment, while continuing to perform the Corps' traditional missions of developing and using wisely our country's natural resources.

## REFERENCES


1. **Ash, C. G.**, Three-years evaluation, *Water Spectrum,* 5(4), 28, 1973.
2. **Ash, C. G.**, Wetlands engineering projects, *Mil. Eng.,* 66(430) 113, 1974.
3. **Ash, C. G.**, Corps Problems in Preparing EIS, paper presented at 40th North American Wildlife and Natural Resources Conf., Pittsburgh, March 16 to 19, 1975.

3a. **Ash, C. G.**, Hydropower projects and their associated environmental impacts, in *Weed Control Methods for River Basin Management,* Gangstad, E. O., Ed., CRC Press, West Palm Beach, Fla., 1978, Ch. 4.

4. Council on Environmental Quality, Guidelines for statements on proposed federal actions affecting the environment, *Fed. Reg.,* 38(147), 1, 1973.
5. Council on Environmental Quality, Environmental Quality—1976, 7th Annu. Rep. Stock No. 041-010-00031-2, U.S. Government Printing Office, Washington, D.C., 1976, 123.
6. Council on Environmental Quality, The President's Environmental Program, Stock No. 040-000-00384-3, 1977, U.S. Government Printing Office, Washington, D.C.
7. Council on Environmental Quality, Memorandum for Heads of Agencies, Subject Environmental Impact Statements, Letter from Executive Office of the President, February 10, 1976.
8. **Douglas, W. O.**, The public be dammed, *Playboy,* 16(7), 143, 1969.
9. **Drew, E.**, Dam outrage: the story of the army engineer, *The Atlantic Monthly,* 225(4), 51, 1970.
10. Endangered Species Act of 1973, Public Law 93-205 (87 statute 884), December 28, 1973.
11. Federal Water Project Recreation Act of 1965, Public Law 89-72, as amended (16 U.S. Code 668aa-668ee), July 9, 1965.
12. Federal Water Pollution Control Act Amendments of 1972, Public Law 92-500 (66 Statute 816), October 18, 1972.
13. Fish and Wildlife Coordination Act of 1958, Public Law 85-624, as amended (16 U.S. Code 661 et sequel), August 12, 1958.
14. **Florman, S. C.**, Hired scapegoats, *Harpers,* 254 (1524), 26, 1977.
15. **Gangstad, E. O.**, *Weed Control Methods for River Basin Management,* CRC Press, West Palm Beach, Fla., 1978.
16. **Gordon, R.**, Engineering for people—200 years of army public works, *Mil. Eng.,* 68(443), 180, 1976.
17. Marine Protection, Research and Sanctuaries Act of 1972, Public Law 92-532 (86 statute 1052), October 23, 1972.
18. **Mazmanian, D. A. and Lee, M.**, Tradition be dammed. The Corps of Engineers is changing, *Public Admin. Rev.,* 35(2), 166, 1975.
19. **Morris, J. W.**, A time for reflection, *Water Spectrum,* 7(3), 1, 1975.
20. National Environmental Policy Act of 1969, Public Law 91-190 (83 Statute 852; 42 U.S. Code 4221 et sequel), January 1, 1970.
21. **Nienaber, J.**, Bureaucracy, Policy, and Change: the Impact of Environmentalism in the Corps of Engineers, Board of Engineers for Rivers and Harbors, Resident Scholar Program, Research Paper No. 4, U.S. Government Printing Office, Washington, D.C., April 1975.
22. Preservation of Historic and Archeological Data, (88 Statute 174) Amending Public Law 86-523, The Reservoir Salvage Act of 1960 (74 Statute 220), May 24, 1974.
23. **Reuss, H. S.**, Needed: an about-face for the Corps of Engineers, *Readers Digest,* p. 129, November 1971.
24. Rivers and Harbors Act of 1899 (30 Statute 1151, 33 U.S. Code 403), (30 Statute 1152, 33 U.S. Code 407), March 3, 1899.

25. **Soloman, R. C., Colbert, B. K., Hansen, W. J., Richardson, S. E., Canter, L. W., and Vlachos, E. C.,** Water Resources Assessment Methodology (WRAM) — Impact Assessment and Alternative Evaluation, U.S. Army Engineers Waterways Experiment Station, Technical Report Y-77-1, February 1977.
26. Title III, Management of the Coastal Zone, Public Law 92-582 (86 Statute 1280), October 27, 1972.
27. U.S. Army Corps of Engineers, Planning process: multiobjective planning framework, Engineering Regulation 1105-2-200, U.S. Government Printing Office, (33 code Federal Regulations 290), Washington, D.C., November 1975.
28. U.S. Army Corps of Engineers, Planning Process: Plan Development Stages, Engineering Regulation 1105-2-210, U.S. Government Printing Office (33 Code Federal Regulations 291) Washington, D.C., November 1975.
29. U.S. Army Corps of Engineers, Planning Process: Problem Identification (Task 1), Engineering Regulation 1105-2-220, U.S. Government Printing Office, (33 Code Federal Regulations 292), Washington, D.C., November 1975.
30. U.S. Army Corps of Engineers, Planning Process: Formulation of Alternatives (Task 2), Engineering Regulation, 1105-2-230- U.S. Government Printing Office (33 Code Federal Regulations 293) Washington, D.C., November 1975.
31. U.S. Army Corps of Engineers, Planning Process: Impact Assignment (Task 3), Engineering Regulation 1105-2-240, U.S. Government Printing Office (33 Code Federal Regulations 294), Washington, D.C., November 1975.
32. U.S. Army Corps of Engineers, Planning Process: Evaluation (Task 4), Engineering Regulation 1105-2-250, U.S. Government Printing Office (33 Code Federal Regulations 295) Washington, D.C., November 1975.
33. U.S. Army Corps of Engineers, Preparation and Coordination of Environmental Statements, Engineering Regulation 1105-2-507, U.S. Government Printing Office, (33 Code Federal Regulations 209.410) Washington, D.C., April 1974.
34. U.S. Army Corps of Engineers, Feasibility Reports: System of Accounts, Engineering Regulation 1105-2-921, U.S. Government Printing Office (33 Code Federal Regulations 393), Washington, D.C., November 1975.
35. U.S. Army Corps of Engineers, Permits for activities in navigable waters or ocean waters, *Fed. Reg.,* 40(144), 31320, 1975.
36. U.S. Army Corps of Engineers, Environmental policies, objectives, and guidelines; revision, *Fed. Reg.,* 41(210), 47676, 1976.
37. U.S. Environmental Protection Agency, Navigable waters procedures and guidelines for disposal of dredged or fill material, *Fed. Reg.,* 40(88), 19766, 1975.
38. U.S. Fish and Wildlife Service, Habitat Evaluation Procedures, Division of Ecological Services, U.S. Department of the Interior Special Publication, July 1, 1976.
39. U.S. Water Resources Council, Establishment of principles and standards for planning, *Fed. Reg.,* 38(174), 24778, 1973.
40. Water Resources Development Act of 1974, Public Law 93-251 (88 Statute 12), March 7, 1974.
41. **Whitaker, G. A. and McCuen, R. H.,** A proposed methodology for assessing the quality of wildlife habitat, in *Ecological Modeling, Vol. 2,* Elsevier, Amsterdam, 1976, 251.
42. **Whitaker, G. A., Roach, E. K., and McCuen, R. H.,** Inventorying habitats and hating their value for wildlife species, paper presented at 30th Annu. Conf. Southeastern Association of Game and Fish Commissioners, Jackson, Miss., October 24 to 27, 1976.

# *Part II*
# *Ecosystem Studies of Impounded Water*

Chapter 7

# COMPARATIVE CONCENTRATIONS OF NUTRIENTS IN AN AQUATIC ECOSYSTEM*

## INTRODUCTION

Adequate evaluation of the aquatic environment of any water body requires thorough knowledge of its many facets, not the least important of which is the relative concentration of nutrients (elements) within its various components. Previous research has been primarily concerned with determination of the roles of one or more of these elements within a single component of an aquatic environment. This chapter summarizes information on relative concentrations of calcium, magnesium, iron, manganese, copper, and zinc in filtered waters and suspended matter from three reservoirs on the Chattahoochee and Flint rivers and the comparative concentrations of these elements in filtered water, suspended matter, hydrosol, plants, and fish.[1-3]

## DESCRIPTION OF RIVER SYSTEM

The Chattahoochee River begins in the Blue Ridge foothills of North Georgia (Figure 7.1). The uppermost impoundment is Lake Sidney Lanier which has a surface area of 15,390 ha and a drainage area of 2693 $km^2$. The next major impoundment, Bartlett's Ferry Reservoir, is approximately 250 km downstream. It has a surface area of 2369 ha and a drainage area of 11,033 $km^2$.

Lake Eufaula is the 10th downstream impoundment on the Chattahoochee River, with a surface area of 18,250 ha, an average depth of 6.2 m, and a drainage area of 19,320 $km^2$. The averaged rate of flow since its impoundment in 1963 has been 10,900 $m^3/hr$.

Lake Seminole, formed by a dam at the mouths of the Chattahoochee and Flint Rivers, is the 12th impoundment on the Chattahoochee River. It has a surface area of 15,188 ha, an average depth of 3 m, and a total drainage area of 44,420 $km^2$.

The drainage area for Lake Sidney Lanier is located entirely within the Blue Ridge foothills physiographic area. The portion of drainage area between Lake Sidney Lanier and Columbus, Georgia, is Piedmont plateau. The drainage area immediately adjacent to Lake Eufaula is primarily Coastal Plains, but a small portion of the drainage area on the reservoir's western side is Black Belt soil. The drainage area adjacent to Lake Seminole is lower coastal plains.

Dispersed throughout this drainage area are numerous towns and cities discharging domestic and industrial wastes in varying degrees of stabilization into the Chattahoochee River. Thus, by the time the river reaches the upper end of Lake Eufaula, it has received effluent from over 3/4 million people and their associated industries.

## DATA COLLECTION

Samples of raw lake waters, for chemical and suspended matter analyses, were collected at the surface and at intervals of depths periodically from March through October during 1967 and 1968 from several stations within each of the lower three reser-

* Updated and abstracted from The Summary Report, Contract DA-01-076-CIVENG, Mobile District, U.S. Army Corps of Engineers, and the Agricultural Experiment Station, Auburn University, Auburn, Ala., 1973.

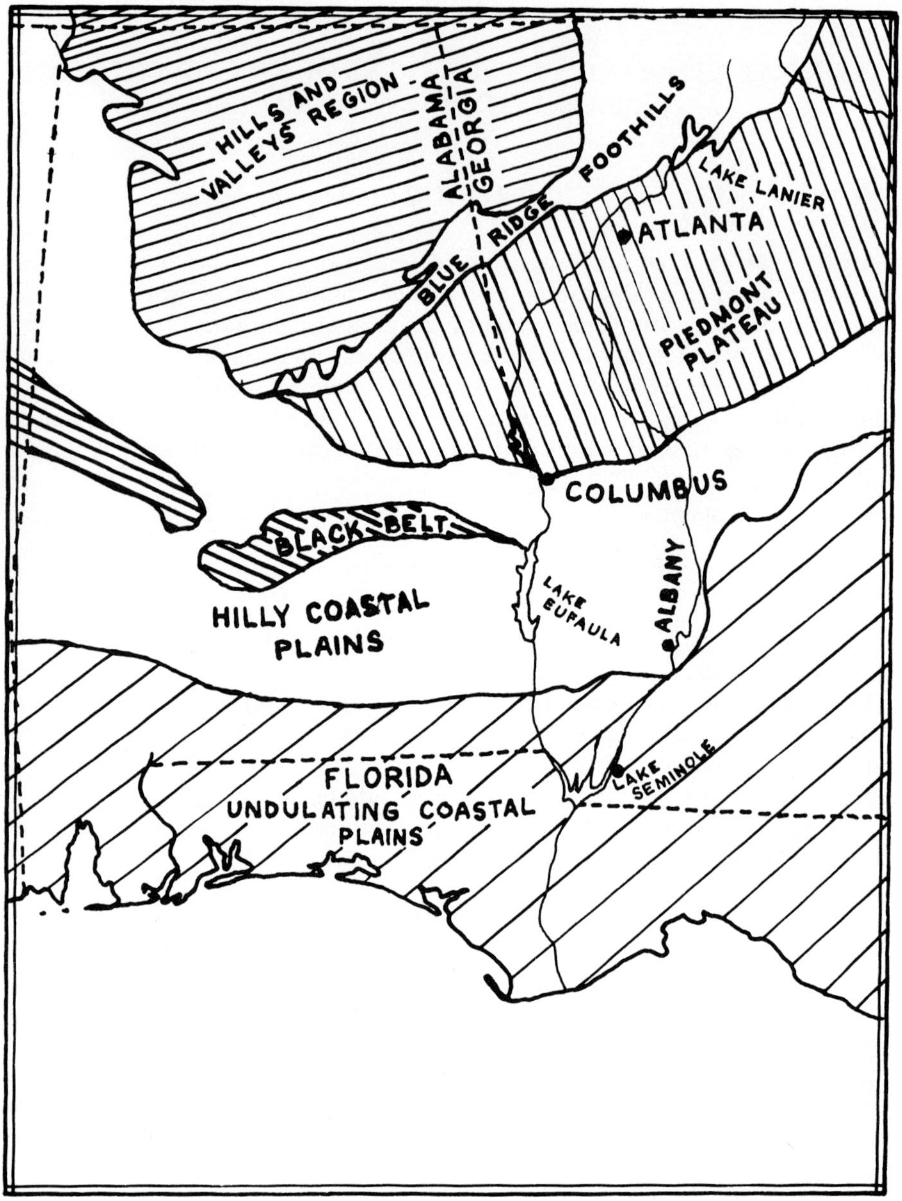


FIGURE 7.1. Portions of Georgia, Alabama, and Florida that comprise the drainage areas of Chattahoochee, Flint, and Apalachicola Rivers with indicated locations of physiographic areas and municipalities.[3]

voirs. Suspended matter was separated by passage of raw lake water through a 0.8 $\mu$m Millipore filter®. Residue on the filter was ashed and then dissolved in aqua regia, dried, and redissolved in dilute HCl. The concentrations of elements in this residue were reported as parts per million based on the volume of water from which it was removed.

Samples of the interface area adjacent to water and bottom mud were collected in 1967 and 1968 by a plastic hydrosol sampler. An area of 0.3 $m^2$ was included in each sample. The entire sample content was dried over infrared lamps, weighed, ground to pass 40-mesh screen, and stored for analyses. The neutral ammonium acetate was utilized for extracting these hydrosol samples. Elemental concentrations were reported as parts per million of dry weight of soil.

Samples of selected species of rooted aquatic plants were collected at various times and places throughout the study period. Plant samples were transported to the laboratory in plastic bags, washed in tap water, and freeze-dried. Dried samples were ground to pass 40-mesh screen and stored in a desiccator until analyzed. Weighed plant sample was ashed, dissolved in aqua regia, dried, and redissolved in dilute HCl. Elemental concentrations were reported as parts per million of plant dry weight.

Samples of selected fish species were collected by various techniques at different times and places in Lake Eufaula. Briefly, methods of collecting fish included seining, electric shocking, and by rotenione treatment. Immediately following collection of fish, they were stored on ice until they reached the laboratory where they were separated by species and their total lengths, total depths, and weights were determined individually before they were freeze-dried. The dried fish were ground to pass a 40-mesh screen. Weighed sample of dried fish was ashed, dissolved in aqua regia, dried, and redissolved in dilute HCl. Elemental concentrations were reported as parts per million of dry weight of fish.

Concentrations of calcium, magnesium, iron, manganese, copper, and zinc in prepared samples of filtered water, suspended matter, hydrosol, plants, and fish were determined by atomic absorption spectrophotometry.

## RESULTS AND DISCUSSIONS

Discussion of the comparative concentrations of the six elements in the filtered water and suspended matter from three reservoirs on Chattahoochee River is presented separately from discussion of comparative element concentrations in filtered water, suspended matter, hydrosol, plants, and fish from Lake Eufaula.

## FILTERED WATER AND SUSPENDED MATTER COMPOSITION

Practically all published analytical procedures for determination of chemical composition of water specify that the sample be filtered prior to analyses, but there is no mention of how this residue should be analyzed and there is no consideration of its composition in reporting the water analyses. In the Chattahoochee River system study it was felt that this suspended matter had to be separated and analyzed to give a better understanding of the phytoplankton composition and the composition of particles and colloids transported by these waters.

The averaged composition of filtered water and suspended matter in three reservoirs and some tributary creeks is summarized in Table 7.1. Several differences in concentrations of elements in both the filtered water and suspended matter are immediately obvious.

Calcium salts are among the most commonly encountered substances in water. They are leached from soil and other natural sources by runoff waters and their concentrations are increased by sewage and industrial wastes. These data indicate that the upper drainage areas of the Chattahoochee River are poor in calcium salts, but the concentration in waters increases in the lower coastal plains areas particularly where large limestone springs contribute to the stream runoff.

The calcium content of suspended matter apparently does not increase in proportion to the soluble content in the water. Magnesium comprises 2.1% of the earth's crust, and its carbonates are 10 to 20 times as soluble in water as calcium carbonate. Mag-

TABLE 7.1

**Comparative Averaged Chemical Composition of Filtered Water and Suspended Matter from Raw Water Samples Collected from Bartlett's Ferry Reservoir, Lake Eufaula Reservoir and Tributary Creeks, and from Lake Seminole Chattahoochee River Arm, Flint River Arm, and Spring Creek Arm[a]**

| Element | Component (ppm)[b] | Bartlett's Ferry Reservoir | Lake Eufaula | | Lake Seminole | | |
|---|---|---|---|---|---|---|---|
| | | | Reservoir | Creeks | Chattahoochee River | Flint River | Spring Creek |
| Ca | water | 2.525 | 3.681 | 5.767 | 5.743 | 14.754 | 22.107 |
| | SM | 0.109 | 0.111 | 0.101 | 0.178 | 0.159 | 0.125 |
| Mg | water | 1.175 | 1.343 | 1.206 | 1.271 | 1.074 | 0.656 |
| | SM | 0.050 | 0.043 | 0.067 | 0.045 | 0.022 | 0.009 |
| Fe | water | 0.226 | 0.121 | 0.193 | 0.394 | 0.261 | 0.086 |
| | SM | 0.671 | 0.522 | 0.524 | 0.487 | 0.349 | 0.142 |
| Mn | water | 0.119 | 0.078 | 0.172 | 0.023 | 0.031 | 0.070 |
| | SM | 0.064 | 0.073 | 0.064 | 0.085 | 0.049 | 0.037 |
| Cu | water | 0.016 | 0.012 | 0.010 | 0.017 | 0.013 | 0.014 |
| | SM | 0.025 | 0.022 | 0.016 | 0.014 | 0.017 | 0.017 |
| Zn | water | 0.168 | 0.158 | 0.190 | 0.245 | 0.264 | 0.131 |
| | SM | 0.051 | 0.044 | 0.056 | 0.063 | 0.068 | 0.040 |

[a] Study conducted during period March to October, 1967 and 1968.
[b] SM = suspended matter.

nesium analyses of Chattahoochee River waters indicate lack of soluble salts of this element in soils of this watershed. These data also indicate that the underlying limestone formations on the Flint River and Spring Creek arms of Lake Seminole are even poorer sources of this element. The abundance of magnesium in the suspended matter seemingly was more closely related to the concentration in the water than was the case with calcium.

Iron is an abundant element in the soils of the upper watershed on the Chattahoochee River. Its chief source in these river waters appears to be suspended soil particles. This is supported by the fact that the iron content of the suspended matter is several times that of filtered water. However, since the filtered water was passed through a 0.8-$\mu$m filter, it is very probable that a portion of the iron reported in water was in colloidal rather than soluble form. The changes in soil types within the downstream drainage areas are indicated by decreased iron content in waters, particularly from the Spring Creek arm of Lake Seminole.

Manganese is an element closely associated with iron in soils and is apparently rather abundant in the upper watershed of the Chattahoochee River. However, its abundance in filtered waters and suspended matter throughout this system does not follow a particular pattern. Thus, in the redland area of this watershed, the filtered waters indicate an inflow of this element either in colloidal or soluble form. Downstream in the sandy land areas, data indicate an outward movement of manganese in the form of suspended matter. Copper is found in trace amounts in all plant and animal life, and it is assumed that these are primary sources of this element in filtered waters throughout this reservoir system. It is assumed further that the source of copper in suspended matter is the amount assimilated in the microscopic plant and animal life present in these waters.

Soluble or colloidal zinc is fairly abundant in the soils of this and other watersheds in the southeastern U.S. Agronomists do not agree with the above statement since addition of zinc in fertilizers increases production of some crops throughout this area.[1] An apparent explanation for abundance of zinc in these river waters is its introduction by industrial and domestic sewage. This latter statement is partially true, but analyses of some fish ponds within the watershed having no external source of zinc indicate a high zinc concentration in runoff waters. From data presented in Table 7.1 it is evident that a major portion of zinc was found in filtered water. This indicates that zinc was present in either a soluble or colloidal form. However, in most surface water samples, the zinc content of suspended matter was higher than indicated by the overall averages given. This suggests that a portion of the zinc present is in a soluble form that can be absorbed by the phytoplankton.

As previously mentioned, most analytical techniques ignore those concentrations of elements in waters that are incorporated into suspended matter. In analyses for calcium and magnesium, this technique apparently would not have a marked effect upon total concentration data. However, in heavy metals such as iron, manganese, copper, and zinc, presence of the contents of the residues would markedly affect total concentration data. More importantly, analyses of this suspended matter probably reflect a more realistic picture of the movement of these heavy metals within a reservoir or any flowing water body than does any other currently available method.

Accumulation of data presented in this chapter was only possible as a result of continuous adaptations and developments of new sampling techniques occurring during the first 3 years of this study. This resulted in as complete a set of compositional data as are available to date on some of the components of a reservoir's aquatic environment. It must be remembered that even these data are inadequate for a full description of the situation, but they do demonstrate some facts previously ignored in an aquatic environment evaluation. Thus far, five facets of the environment (water, suspended matter, hydrosol, rooted aquatic plants, and resident fish) have been sampled and analyzed.

The hydrosol includes that fraction of the bottom often referred to as the interface area, but generally might be called the bottom ooze. It is assumed that this is the site of heaviest concentration of precipitated elements in both colloidal and particle form. This area has previously been relatively inaccessible to sampling since it is semifluid and would escape from most bottom soil samplers. The sampler finally devised was 2 years in development, but has operated more efficiently in recovering this muck than was anticipated. In fact, the heavier the muck deposit on the reservoir bottom, the more effective the sampler was in retrieving the sample. Normally a layer approximately 0.5 cm in thickness was recovered. The dry weight of a square foot of such bottom materials varied from 5 to 500 or more grams, but averaged 265 g.

Since practically no science or aquatic soil technology exists, the extraction technique used in this study was that employed by terrestial soil scientists and may or may not be the most applicable to this situation. Further research on soil extraction will be done to determine what system produces the greatest recovery.

Rooted aquatic vegetation has never developed in Lake Eufaula to a measurable extent. While the nutrients nitrogen, phosphorus, and potassium were present in sufficient amounts to support a measurable population of aquatic plants, the winter draw-down technique employed on Lake Eufaula has apparently been instrumental in eliminating over-wintering propagules of troublesome plants. Only three aquatic species, alligator weed (*Alternanthera philoxeroides*), smartweeds (*Polygonum* spp.), and a grass, *Paspalum fluitans*, have appeared in sufficient numbers to provide plant samples.

The average calcium content for each species was: alligator weed 7155 ppm (range 3630 to 18,500 ppm), smartweeds 8687 ppm (range 4400 to 16,500 ppm), and *P. fluitants* 1500 ppm (range 1100 to 1800 ppm). The average magnesium content was: alligator weed 5979 ppm (range 3300 to 10,000 ppm), smartweeds 4605 ppm (range 550 to 7500 ppm), and *P. fluitans* 3133 ppm (range 605 to 5300 ppm).

The average iron content for each species was: alligator weed 1575 ppm (range 265 to 4600 ppm), smartweeds 2074 ppm (range 400 to 4500 ppm), and *P. fluitans* 897 ppm (range 430 to 1400 ppm). The average manganese content was: alligator weed 528 ppm (140 to 950 ppm), smartweeds 599 ppm (range 300 to 890 ppm), and *P. fluitans* 1076 ppm (range 310 to 2560 ppm).

The average copper content for each species was: alligator weed 80 ppm (range 32 to 230 ppm), smartweeds 65 ppm (range 10 to 250 ppm), and *P. fluitans* 33 ppm (range 16 to 46 ppm). The average zinc content was: alligator weed 135 ppm (range 60 to 360 ppm), smartweeds 142 ppm (range 85 to 206 ppm), and *P. fluitans* 199 ppm (range 170 to 240 ppm).

The average calcium content of fish was 37,753 ppm (range 20,540 to 50,436 ppm), and for magnesium it was 1939 ppm (range 1401 to 2569 ppm). The average iron content of fish was 290 ppm (range 23 to 2500 ppm), and for manganese it was 57.5 ppm (range 6 to 138 ppm). The average copper content of fish was 59.7 ppm (range 11 to 330 ppm), and for zinc it was 211.5 ppm (range 40 to 940 ppm).

The Chattahoochee River drains one of the more impoverished calcium-magnesium watersheds in the southeastern U.S. Consequently, the calcium and magnesium content of the river water, which totals less than 7 ppm, categorizes it in the lowest 5% of all world surface waters. Such low concentrations of these two elements in water should be reflected in the composition of suspended matter and hydrosol, and this appeared true in the Lake Eufaula environment. The calcium and magnesium content of rooted aquatic plants and fish from this impoverished environment do not appear materially different from the contents of the same plant and fish species removed from an environment with an adequate calcium supply. Apparently, higher plants and fish are adapted to absorb and assimilate these two elements under what might be classified as a physiological stress condition.

The iron and manganese contents of waters and associated suspended matter of the Chattahoochee River also reflect nutrient contributions from the watershed. Since each element is practically insoluble under aerobic conditions, it is obvious that a majority of the concentrations of both elements in filtered waters was in colloidal form, and that the suspended matter contained both particulate and colloidal forms. On the other hand, each element becomes reduced and soluble under anaerobic conditions existant in deeper Lake Eufaula waters for varying periods during hot weather. Thus, a portion of the concentration of these elements might be considered dissolved from the hydrosol and recirculated within the lake environment.

Extractable iron and manganese from Lake Eufaula hydrosols gave a greater quantity of manganese than iron. This was the reverse of most other data obtained elsewhere in this environment. Needless to say, determination of the actual iron-manganese ratio in the hydrosols of lakes requires additional research.

Concentrations of iron and manganese in broad leaved aquatic plants were in approximately the same ratio (4:1) as that found in raw waters. However, the one species of grass analyzed had a ratio (0.8:1) of iron to manganese which was similar to the hydrosol. Concentrations of iron and manganese in fish were in approximately the same ratio as that found in raw reservoir waters.

Concentrations of copper and zinc in filtered water and suspended matter of Lake Eufaula are believed to reflect abundance of these elements in the watershed. All available data indicate that the zinc concentration is derived primarily from the watershed and only a minor portion is contributed by sewage and industrial pollution.

Extraction of hydrosols yielded quantities of copper and zinc in the approximate ratio found in raw waters. The broad leaved aquatic plants apparently accumulated twice as much copper as was found in the grass species. The grass species accumulated more zinc than either broad leaved species. Fish were apparently able to assimilate copper and zinc in approximately the same ratio as found in raw river waters.

## SUMMARY

The chemical composition of water and suspended matter in reservoirs on the Chattahoochee River apparently reflected the abundance of soluble elements in the watershed. The abundance of the six elements in raw waters occurred in the following order: Ca>Mg>Fe>Zn>Mn>Cu. The abundance of the six elements in broad leaved aquatic plants occurred in the following order: Ca>Mg>Fe>Mn>Zn>Cu, but in the grass species they occurred in this order: Mg>Ca>Mn>Fe>Zn>Cu. The abundance of six elements in fish occurred in the following order: Ca>Mg>Fe>Zn>Cu>Mn.

## REFERENCES

1. **Fried, M. and Broeshart, H.,** *The Soil Plant System in Relation to Inorganic Nutrition*, Academic Press, New York. 1967.
2. **Keup, L.,** Phosphorous in flowing waters, *Water Res.*, 2, 373, 1968.
3. **Lawrence, J. M.,** Dynamics of Chemical and Physical Characteristics of Water, Bottom Muds and Aquatic Life in a Large Impoundment on a River, Zoological-Entomological Department Series, Fisheries No. 6., Auburn University Agricultural Experiment Station, 1968.

Chapter 8

# INPUT, STORAGE, AND OUTPUT OF PLANT NUTRIENTS*

## INTRODUCTION

All water contributing to the surface runoff in a river basin is the solvent and vehicle for transporting varying amounts of the elements that are essential nutrients for aquatic plant and animal growth. The concentration of these elements in runoff waters and eventually in river waters depends on the types of soils and agricultural operations occurring in the drainage area, plus amounts of domestic sewage and industrial effluents that may also be discharged within this area.[1-9]

Within the impoundment, a multitude of events may occur determining the fate of input nutrients. A portion of the nutrient content of an impoundment is always present in soluble form. Some of these soluble nutrients originate from resolution, from bottom muds, and from waste and decomposition of plants and animals. A portion of the nutrients may be precipitated as colloidial matter directly into the bottom muds which may be a temporary or permanent type of storage. Another portion of input nutrients may be used in growth and reproduction of bacteria, fungi, algae, or rooted aquatic plants which in turn may be consumed by some animal form, or the plants may die and deposit their nutrients in the muds. Animals will eliminate most nutrients consumed as waste and retain only a small portion for growth. This retained nutrient portion may be removed from the environment if the animal flies, walks, crawls, or is bodily removed from the impoundment. If the animal remains in the impoundment, it eventually dies and nutrients may return to bottom muds or become food items for another animal whose waste then returns to the bottom muds. Finally part of the input nutrients pass out of the impoundment into the tailwaters where they are then classified as outlet nutrients.

These outlet nutrients may occur as soluble forms, bacteria, fungi, algae, rooted plants, various animal forms, organic materials, and soil colloids. All forms of these nutrients move downstream and combine with additional drainage area runoff and eventually become input nutrients for the next impoundment where retention-storage process is repeated until the river flows into the ocean.

This process might well be called an abbreviated nutrient cycle for an impoundment. To utilize this cycle to man's best advantage requires knowledge of the quantity of each nutrient involved in each niche described for the environment and of rates of partial and permanent retention or storage. Such information would make it possible to determine the element or elements responsible for overproduction of noxious plants and so to isolate and eventually correct the problem.

Since the nutrient cycle of impoundments is intimately related to eutrophication, and since a moderate degree of eutrophication is essential for fish production in river impoundments, a tolerable nutrient loading is beneficial. In those areas where there are excessive nutrients, the use of seasonal aquatic plants as possible retention sites for excess nutrients during hot weather and low stream flow and cold weather and frost provides a slow release mechanism when there is a higher rate of stream flow.[4-9]

By 1970, scientists, pollution control experts, health authorities, and informed laymen all recognized that one cause of deteriorating water quality was introduction of

* Updated and abstracted from The Summary Report, Contract DA-01-076-CIVEN G, Mobile District, U.S. Army Corps of Engineers, and the Agricultural Experiment Station, Auburn University, Auburn, Ala., 1973.

excessive quantities of nutrients (nitrogen and phosphorus) into streams and lakes. These nutrients stimulated algae production. Because of accumulations of these minute plants, the water became more expensive to filter, unpleasant to swim in, and (in some cases) so short of oxygen that production of some game fish was inhibited. Also, excessive quantities of nitrates in water were a potential health hazard.

The Water Pollution Control Act of 1972 (PL 92-500) focused additional attention on these issues. This act required development of information, including, "(1) guidelines for identifying and evaluating the nature and extent of nonpoint[1] sources of pollutants and (2) processes, procedures, and methods to control pollution resulting from . . . agricultural and silvicultural activities, including runoff from fields and crop and forest lands ...." The legislation forced recognition of the need for identification and evaluation procedures for "nonpoint" sources. It underlined the need for clear data defining the contribution agriculture was making to the nitrogen and phosphorus content of water.[10-12]

A historical review of scientific literature on lake eutrophication shows a rather dramatic focusing in recent years on phosphorus supply as the critical factor in the process. Initially, this was partly due to the realization that of nutrients essential for plant growth (carbon, nitrogen, and phosphorus) phosphorus was the only one that could feasibly be controlled in most cases. Increasing evidence that phosphorus is indeed the substance limiting productivity in most temperate latitude lakes has accumulated.

There are three arguments concerning the unique importance of phosphorus. First, relative to needs of aquatic plants for growth supporting substances, phosphorus is the element of greatest geochemical rarity. It is enhanced in the aquatic environment by insolubility of many phosphorus compounds and the tendency of phosphates to be adsorbed to soil particles thus settling to lake bottoms where they become available to algae. The second argument addresses questions of nitrogen and carbon sources available to algae. Some algae (the blue-greens) can use nitrogen directly and hence draw on the limitless reserves of this gas in the biosphere. Many algae can use the bicarbonate ion as a carbon source. This and the high solubility of carbon in water suggests that it is never likely to be more than a temporary limitation to growth of algal communities. Finally, as shown in the present study and its later extension to lakes in world temperate regions, algae are present during the summer at concentrations closely proportional to the annual inputs of dissolved phosphorus per water volume.[10-16]

Phosphorus in runoff occurs in three general forms: dissolved organic, dissolved inorganic, and particulate (most of which is attached to soil particles). With which, then, should researchers and water quality managers be concerned — with total phosphorus or with only one or two of the chemical forms? This study and information already in the scientific literature led to the important conclusion that dissolved phosphorus (organic and inorganic) has a far more important influence on algal growth than has the phosphorus attached to soil particles. Dissolved phosphorus controls the quantities of algae occurring in the most temperate latitude lakes during summer.

The study of 13 New York lakes led to quantitative expressions of the relations between phosphorus loading and concentration of algae. Furthermore, the fish yield from lakes is a function of the algae crop and so is indirectly dependent on dissolved phosphorus loading. This research project emphasized sources and transport of dissolved phosphorus because this phosphorus is of biological importance to receiving lakes.

Recently, biologists and others have pointed out that household laundry detergents containing phosphates are important contributors to sewage phosphorus content. In June 1973, New York State banned use of such detergents. A similar ban was imposed earlier in Erie County, New York. A study of three large sewage plants near Buffalo indicated a reduction of 50 to 60% in effluent phosphorus (most of which was dissolved) as a result of the ban. If a similar reduction occurred at other New York sewage plants, the ban would result in approximately 30% reduction in the total amount of dissolved phosphorus.[7]

## LOCATION AND DESCRIPTION OF THE STUDY AREA

The three mainstream impoundments studied are located on the Chattahoochee River between West Point, Georgia and Chattahoochee, Florida and on the Flint river between Bainbridge, Georgia and Chattahoochee, Florida (Figure 8.1).

The Chattachoochee river arises in the Blue Ridge foothills of northern Georgia. The uppermost impoundment is Lake Sidney Lanier having a surface area of 15,400 ha and a drainage area of 2690 km$^2$. The next major impoundment, Bartlett's Ferry Reservoir, is approximately 240 km downstream. It has a surface area of 2369 ha and a drainage area of 11,000 km$^2$. This is the upstream reservoir of the three impoundments included in this study.

Lake Eufaula (Walter F. George Reservoir) is the 10th downstream impoundment on the Chattahoochee River, with a surface area of 18,225 ha, an average depth of 6.7 m, and a drainage area of 19,320 km$^2$.

Dispersed throughout this upper Chattahoochee drainage area are numerous cities and towns that discharge both domestic and industrial wastes in varying degrees of stabilization into the water. By the time the river reaches the upper end of Bartlett's Ferry Reservoir, it has received effluent from over 1/2 million people. By the time it reaches the upper end of Lake Eufaula, it has received the effluent of 3/4 million people and their associated industrial wastes. Agricultural wastes in this portion of the drainage area are believed negligible since less than 10% of these lands are in cultivation and pastures.

Approximately 64 km downstream from Columbus, Georgia a paper mill with a capacity of 1500 t/day began discharging its treated effluent, not to exceed 1 t of BOD per day, into Lake Eufaula in 1967. This mill waste is released 12 m below the water surface at a point on the reservoir coinciding with the uppermost area where oxygen depletion in deeper water was detected in both 1965 and 1966, but did not occur this far upstream in 1967, 1968, or 1969.

Approximately 19 km downstream from this paper mill installation on Lake Eufaula, runoff wastes from a swine production operation (whose pastures included an area excessive of 400 ha) enter the impoundment via two small streams.

Lake Seminole (Jim Woodruff Reservoir), formed by a dam at the mouths of the Chattahoochee and Flint rivers, is the twelfth impoundment on the Chattahoochee and the third impoundment on the Flint. This reservoir has a surface area of 15,187 ha, an average depth of 2.8 m and a combined drainage area of 44,418 km$^2$.

Incoming waters into Lake Seminole were more varied in chemical composition than were those entering Lake Eufaula. The Chattahoochee waters are very soft (15 to 18 ppm $CaCO_3$ equivalent), the Flint waters are soft (45 to 55 ppm $CaCO_3$ equivalent), and the tributaries between the two rivers have moderately hard waters (80 to 100 ppm $CaCO_3$ equivalent). These latter wastes are discharged from large limestone springs originating in an area of lime sinks.

Another paper mill with a capacity in excess of 2700 t/day is located on the Chattahoochee river approximately 11 km below Columbia Dam. This mill releases 9 t of BOD per day of partially stabilized waste into the river. A depresssion in dissolved oxygen (DO) content of water has been observed for 40 km downstream from this release point.

On the Flint arm, Bainbridge, Georgia released untreated domestic sewage into the upper end of Lake Seminole. As a result of its large shallow water areas, excessive

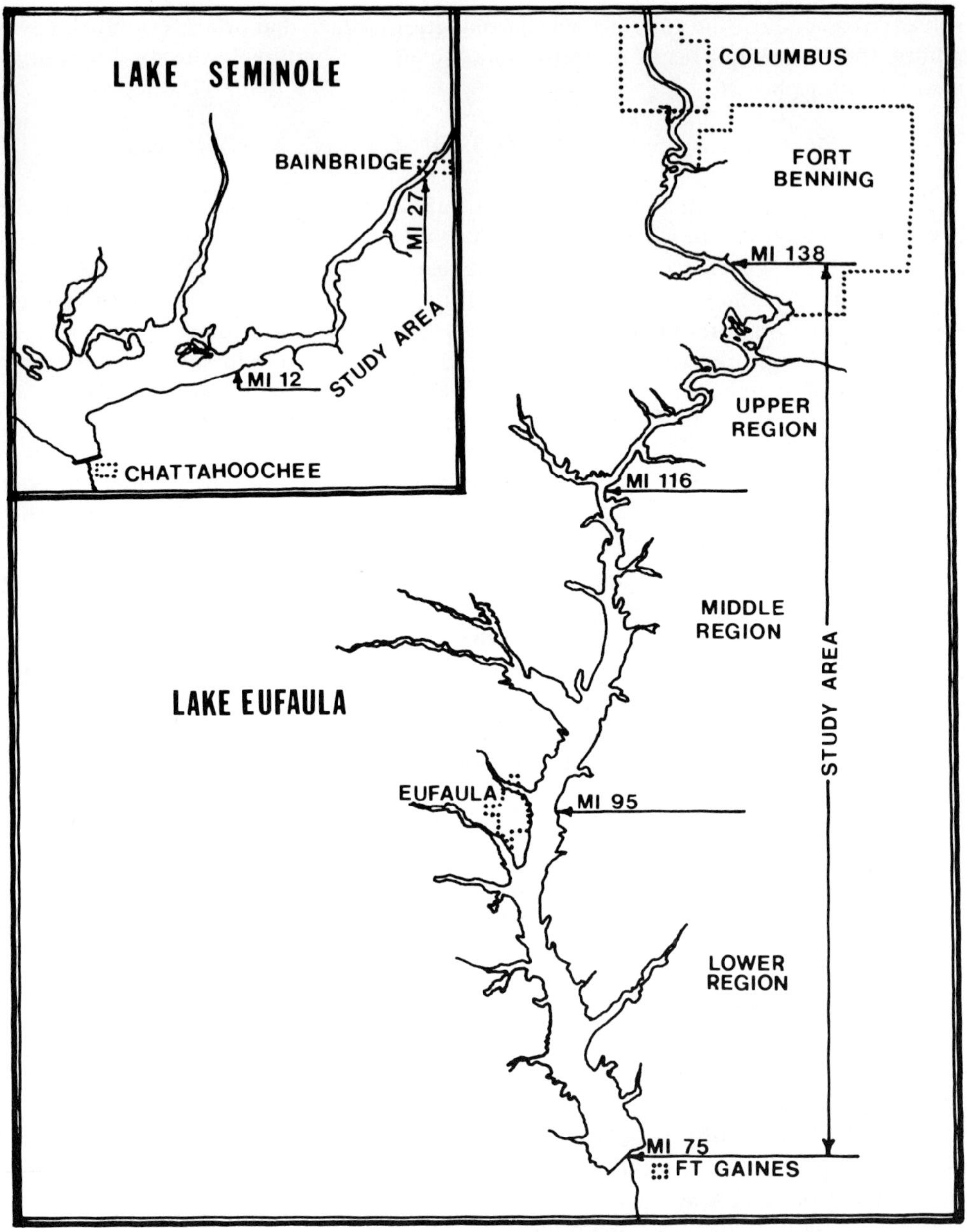


FIGURE 8.1. Locations of study areas on Lake Seminole and Lake Eufaula.

nutrient loading, and swamplike lake bottom, Lake Seminole has supported a population of aquatic weeds consisting of as great a number of different species as can be found in any other impoundment in the U.S. Many species were present in sufficient quantities to permit a reliable estimate of their nutrient retention-storage-release potential in an impoundment situation.

## DATA COLLECTION TECHNIQUES

Details of sampling water, suspended matter, hydrosol, plants, fish, and the analytical techniques for determining the concentrations of major and minor elements are described by Lawrence.[6] Samples of water and suspended matter (mainly phytoplankton) were collected at intervals throughout the study period. Chemical analyses of nutrients included were mde on both filtered water and suspended matter. Sampling of rooted aquatic plant populations was performed to obtain estimates of the wet and dry weights per acre of each major species occurring within the study area. Samples of each dried plant species were analyzed in the laboratory for their nutrient composition. Fish population estimates were obtained by rotenone sampling of given areas in the impoundments. Chemical analyses of dried samples of the various fish species were made in the laboratory to determine their elemental composition.

## RESULTS AND DISCUSSION

This chapter presents data on the input-assimilation and retention-output of the inorganic nutrients (nitrogen and phosphorus) in a multiimpoundment system on a large river, and ascribes the sources of these nutrients within each investigated area.

The soluble nutrient supply included only those fractions of the two elements detectable in the water after it passed through an 0.8 μm Millipore filter®. The suspended matter was material present in raw waters retained upon the 0.8 μm Millipore filter®. Nutrient input for each impoundment was estimated from average concentrations of each element in soluble form and in suspended matter of the inflowing waters. In Bartlett's Ferry Reservoir, total inflow included the quantity contributed by the Chattahoochee River and by the Osanippa and Halawakee Creeks. Output was determined by release through Bartlett's Ferry Dam. Estimates of inflow waters in Lake Eufaula were more difficult since the total flow increased 50% from the upper end of the lake to the dam. To compensate for this increase, the lake was divided into an upper region, middle region, and lower region as indicated in Figure 8.2. The output flow was that quantity of water passing through Walter F. George Lock and Dam. The input flow for Lake Seminole was also complicated and had to be calculated for the Chattahoochee arm, the Flint arm, and the Spring Creek arm. The output data was determined by that quantity of water passing the Jim Woodruff Lock and Dam during the study periods.

The assimilation data include those quantities of nutrients included in the fish and rooted plant population in each impoundment. Retention data may include differences between input quantities. Hydrosol data are also included as retention data. These hydrosol data represent the total nitrogen content and that portion of phosphorus content that could be extracted with aqua regia. Information on surface area, drainage area, average depth, water volume, and flow data for each impoundment are given in Table 8.1. Tabular summaries for nitrogen data on water and hydrosol in the three impoundments are given in Tables 8.2 and 8.3 and for phosphorus data in Table 8.3.

An examination of these data reveals on both nitrogen and phosphorus that extreme variations in total quantities of each element existed in these river systems in relation to time and to location. It might be assumed that the high quantity of nitrogen and phosphorus passing through Bartlett's Ferry Reservoir was primarily due to domestic sewage since the sewered population above this input was approximately 600,000. However, 500,000 people dispose wastes into the river at points 100 or more miles upstream from this reservoir, and it would be reasoned that most of this nitrogen would dissipate before reaching this point. If we assume that each person eliminated 3 kg of nitrogen, as raw sewage, then during this study period, the nitrogen quantity entering this reservoir in 1968 was 20% of this possible domestic sewage load. In 1969, it was over 80% of this sewage load.

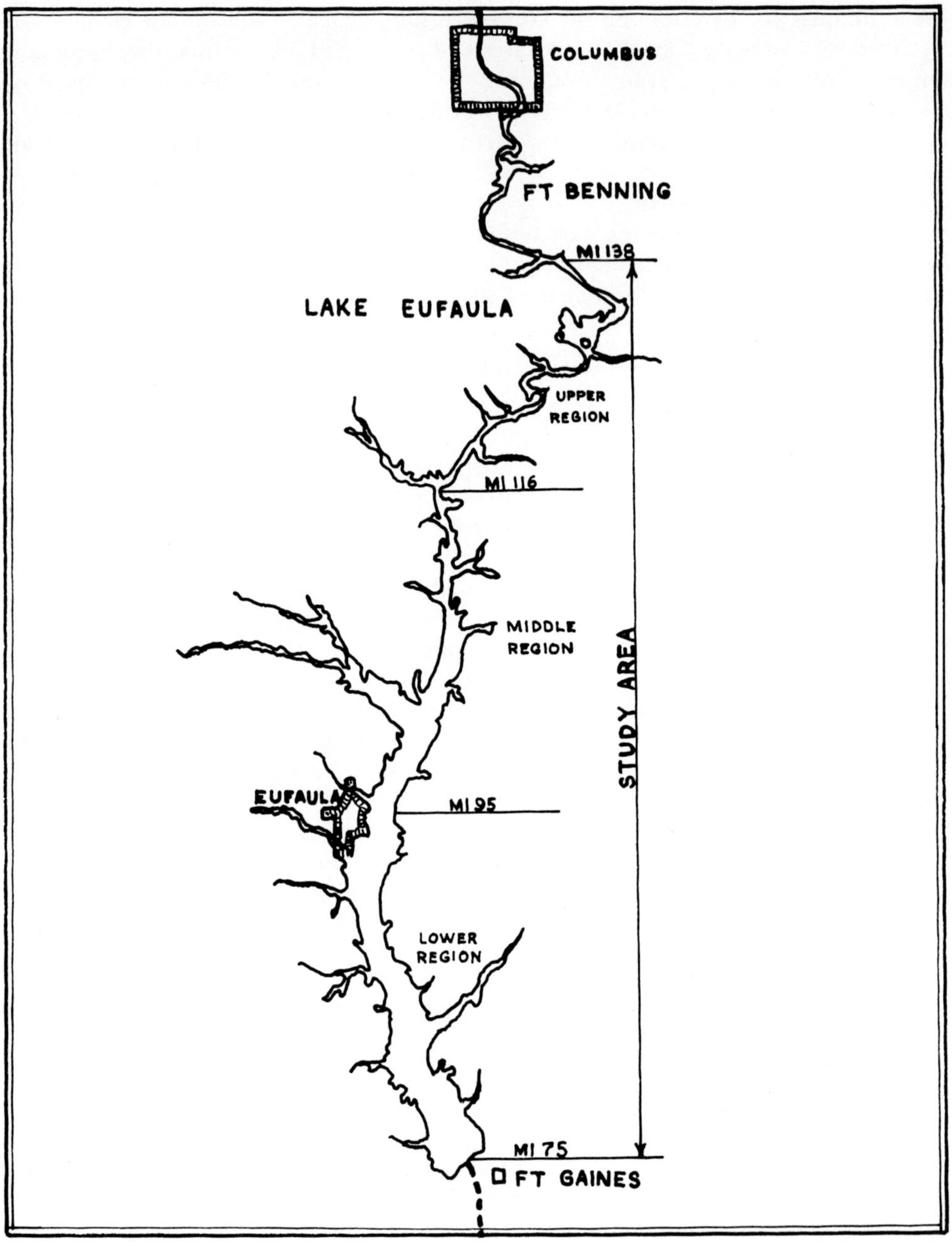


FIGURE 8.2. Study area of Lake Eufaula showing the upper, middle, and lower regions.

If lands included in this drainage area contained 1% organic matter, then the nitrogen content of the upper 6-in. layer would contain 454 kg nitrogen per hectare. This means that each hectare centimeter of soil contains 73 kg nitrogen, and erosion less than 0.0254 cm of soil could contribute an amount of nitrogen equal to the calculated quantity entering Bartlett's Ferry Reservoir in 1969.

The increase in output nitrogen content over input content can be attributed to lakeshore septic tanks plus contributions from sewage lagoons of Opelika, Alabama and from a large Pepperell Manufacturing Company plant near Columbus, Georgia.

TABLE 8.1

**Pertinent Information on Total Flow, Total Drainage Area, and Total Surface Area of Bartlett's Ferry Reservoir and Lakes Eufaula and Seminole[a], [12]**

| Location | April-October flow (ha-m) 1968 | 1969 | Drainage area (km²) | Surface area (ha) |
|---|---|---|---|---|
| Chattahoochee River | | | | |
| Bartlett's Ferry | | | | |
| Upper end | 157,269 | 161,152 | 10,360 | 2,369 |
| Dam | 174,289 | 179,032 | 10,948 | — |
| Lake Eufaula | | | | |
| Ft. Benning Bridge | 199,076 | 253,258 | 12,950 | 2,695 |
| Cottonton | 220,373 | 276,315 | 14,224 | 5,722 |
| Eufaul Bridge | 265,197 | 332,417 | 17,116 | 9,673 |
| Walter F. George Dam | 297,039 | 371,460 | 19,172 | — |
| Lake Seminole | | | | |
| Columbia Dam | 474,108 | 486,912 | 21,433 | 5,467 |
| River mile 3 | 488,888 | 502,078 | 22,102 | — |
| Flint river | | | | |
| Lake Seminole | | | | |
| Bainbridge Bridge | 289,866 | 297,646 | 19,275 | 5,873 |
| River mile 3 | 307,062 | 315,278 | 19,917 | 3,037 |
| Spring Creek mouth | 29,284 | 300,852 | 2,185 | — |
| Apalachicola river | | | | |
| Jim Woodruff Dam | 825,235 | 847,564 | 44,204 | 14,378 |

[a] For period April through October 1968 and 1969.

These same nitrogen sources existed throughout Lakes Eufaula and Seminole on the Chattahoochee Rivers, and to a lesser degree on the Flint arm to Lake Seminole. The Spring Creek drainage must be assumed to be entirely from agricultural origin since no town dumps sewage in this stream. Certainly the nitrogen data on Lakes Eufaula and Seminole appear more reasonable than those for Barltett's Ferry Reservoir. It must be remembered also that differences in quantity of nitrogen in 1968 and 1969 may have been largely due to inclusion of suspended matter, nitrogen plus soluble nitrites, and nitrates in the 1969 data.

The phosphorus data for the Chattahoochee River system present a real problem in interpretation since our knowledge of the phosphorus sources is limited. If we assumed that all sewered domestic waste entering the river above Barlett's Ferry Reservoir was moved through river mile 190, then 83% of the estimated phosphorus load can be accounted for. If we added a detergent phosphorus load, the combined phosphorus input during this period would be 300% the estimated phosphorus load passing river mile 190 in 1968, and 150% of the 1969 load.

It has been shown that 30% total phosphorus is retained by normal sewage treatment, but as much as 90% may be retained by additives such as $CaCO_3$ or $Al_2(SO_4)_3$. Our 1969 data show that a possible 70% of the phosphorus entering Bartlett's Ferry Reservoir was retained during the 7-month period, and data on Lake Eufaula (1968) show approximately 65% phosphorus retention for a like time period.

This suggests that the source of this phosphorus loading comes mainly from particulate phosphorus originating at all points above river mile 190. It might also be speculated that much of this particulate loading could have resulted from poultry production and processing, and that particles high in phosphorus content entered the river

TABLE 8.2

**Averaged Concentration of Nitrogen in Water[a], [12]**

| Location | Concentration of nitrogen in water (ppm) | | Standing crop of nitrogen suspended (kg/ha) | |
|---|---|---|---|---|
| | 1968[b] | 1969 | 1968[b] | 1969 |
| Chattahoochee River | | | | |
| Bartlett's Ferry Reservoir | | | | |
| Input | 0.266 | 0.780 | 21.5 | 74.8 |
| Output | 0.314 | 0.994 | — | — |
| Lake Eufaula | | | | |
| Ft. Benning Bridge | 0.205 | 0.687 | — | — |
| Cottonton Bridge | 0.173 | 0.511 | 14.0 | 51.7 |
| Eufaula Bridge | 0.181 | 0.472 | 10.6 | 36.3 |
| Walter F. George Dam | 0.194 | 0.388 | 11.1 | 25.3 |
| Lake Seminole | | | | |
| Columbia Dam | 0.180 | 0.239 | — | — |
| River mile 3 | 0.216 | 0.293 | 5.5 | 10.3 |
| Flint River | | | | |
| Lake Seminole | | | | |
| Bainbridge Bridge | 0.186 | 0.395 | — | — |
| River mile 3 | 0.212 | 0.327 | 5.5 | 10.2 |
| Spring Creek mouth | 0.154 | 0.184 | 4.3 | 5.2 |
| Jim Woodruff Dam | 0.216 | 0.174 | 5.9 | 4.0 |

[a] Parts per million averaged standing crop of nitrogen in impounded waters, April through October 1968 and 1969.

[b] 1968 data does not include values for nitrogen content of suspended matter.

through surface runoff and were transported with time through the entire river and into the Gulf of Mexico. An examination of the river profile certainly suggests this possibility. Also, the river flow is regulated so that almost daily flushing actions occur and the sediment is never permitted to stop moving.

The movement of nitrogen and phosphorus through the Flint River arm of Lake Seminole is presented separately for two reasons. First, this is a different river system originating in a different soil type and with a more gentle river profile. Second, the drainage area is predominantly an agricultural region with a total sewered population of only 140,000.

The average quantity of total nitrogen entering Lake Seminole at Bainbridge, Georgia in 1968 or 1969 is unknown. The difference, however, in 1968 and 1969 data on soluble nitrogen is believed to be due to inclusion of $NO_2$ and $NO_3$ analyses. Too few estimates of nitrogen content of suspended matter are available for inclusion in the 1969 data. From these data, however, it appeared that the suspended matter nitrogen could double or even triple the total nitrogen load.

As mentioned previously, domestic sewage would not be expected to be an abundant source of nitrogen in this arm of Lake Seminole. Yet, if we compare the urbanized population equivalents (PE) of the upper region of Lake Eufaula (730,000) to that at the upper end of the Flint arm of Lake Seminole (140,000), we would expect the nitrogen loading to be less than 50% of what was estimated. This indicates that other sources had to contribute. In this case, major sources would be agricultural (peanuts, soybeans, shade tobacco, and livestock). Whatever the sources of nitrogen, its origin

TABLE 8.3

**Average Concentration of Phosphorus in the Water and Standing Crop of Phosphorus in Impounded Waters[a, 12]**

| | Concentration of phosphorus in water | | Standing crop of phosphorus (suspended) | |
|---|---|---|---|---|
| Location | 1968 (ppm) | 1969 (ppm) | 1968 (kg/ha) | 1969 (kg/ha) |
| Chattahoochee River | | | | |
| Bartlett's Ferry Reservoir | | | | |
| Input | 0.283 | 0.336 | — | — |
| Output | 0.215 | 0.176 | 18.4 | 19.2 |
| Lake Eufaula | | | | |
| Ft. Benning Bridge | 0.984 | 0.333 | — | — |
| Cottonton Bridge | 0.304 | 0.425 | 47.9 | 28.6 |
| Eufaula Bridge | 0.063 | 0.203 | 10.9 | 19.1 |
| Walter F. George Dam | 0.079 | 0.086 | 4.2 | 8.7 |
| Lake Seminole | | | | |
| Columbia Dam | 0.104 | 0.069 | — | — |
| River mile 3 | 0.122 | 0.087 | 3.1 | 2.2 |
| Flint river | | | | |
| Lake Seminole | | | | |
| Bainbridge Bridge | 0.098 | 0.442 | — | — |
| River mile 3 | 0.15 | 0.076 | 3.5 | 7.3 |
| Spring Creek mouth | 0.094 | 0.132 | 2.5 | 3.7 |
| Jim Woodruff Dam | 0.150 | 0.136 | 3.4 | 3.8 |

[a] April through October in 1968 and 1969.

was above Bainbridge, Georgia as indicated by sampling at river mile 35 and by preponderance of $NO_3$ ion over $NH_3$ ion in the water.

The phosphorus situation in this Flint River arm was similar to that indicated for nitrogen with one exception. The soluble phosphorus content steadily declined throughout the length of Lake Seminole, and its concentration near the dam was 1/5 that at the input point. The nitrogen and phosphorus content of hydrosol in this Flint River sector was approximately the same. The low element content and a low rate of deposition over the entire inundated area indicates that the flushing action was operative on this segment of the reservoir also.

The total nitrogen content of hydrosols from Bartlett's Ferry Reservoir and from Lakes Eufaula and Seminole were generally of the same magnitude for all locations and for both years. There does not appear to be any particular relationship between the nitrogen quantity passing through an area and the quantity of nitrogen present in the bottom muds. Thus, if the source of soluble nitrogen was from particulate matter, this material was apparently being transported at a higher level in the river channel than was included in the hydrosol samples.

The identifiable assimilation of nitrogen and phosphorus by the biomass in Bartlett's Ferry Reservoir and Lake Eufaula was restricted to those quantities present in the fish populations. The averaged quantity of nitrogen present in the standing crop of fish population was 1.7 kg/ha. The total standing crops of nitrogen and phosphorus in the fish populations in Bartlett's Ferry Reservoir are 12,800 kg nitrogen and 40,102 kg phosphorus. In Lake Eufaula, crops are 98,473 kg nitrogen and 30,849 phosphorus.

## SUMMARY

A study was conducted from April through October, 1968 and 1969 in three reservoirs on Chattahoochee and Flint rivers in which the input-storage-output of nitrogen and phosphorus were determined from the drainage area.

Bartlett's Ferry Reservoir, located on the Chattahoochee River between miles 180 and 190, has a drainage area of 10,360 $km^2$, an impounded area of 2369 ha and an average depth of 7.6 m. The input nitrogen for the 1968 study period was 269 kg/$km^2$. The storage of nitrogen in hydrosol was 3.4 kg/ha/0.1 cm layer. In the 1969 study period, the nitrogen input was 971 kg/$km^2$. The storage of nitrogen in hydrosol was 1.19 kg/ha/0.1 cm.

The phosphorus input in 1968 was 285 kg/$km^2$, and the output was 228 kg/$km^2$. The hydrosol content was 1.4 kg/ha/0.1 cm layer. The 1969 input of phosphorus was 658/$km^2$ and the output was 193 kg/$km^2$. The 1969 hydrosol content was 2.2 kg/ha/0.1 cm.


## REFERENCES


1. **Fried, M. and Broeshart, H.,** *The Soil-Plant System in Relation to Inorganic Nutrition,* Academic Press, New York, 1967.
2. **Keup, L. E.,** Phosphorus in flowing waters, *Water Res.,* 2, 373, 1968.
3. **Lawrence, J. M.,** Control of a branched alga, *Pithophora,* in farm fish ponds, *Prog. Fish Culture,* 16(2), 83, 1954.
4. **Lawrence, J. M.,** Weed control in farm ponds, *Prog. Fish Culture,* 17(3), 141, 1955.
5. **Lawrence, J. M.,** Preliminary results on the use of potassium permanganate to counteract the effects of rotenone on fish, *Prog. Fish Culture,* 18(1), 15, 1956.
6. **Lawrence, J. M.,** Recent investigations on the use of sodium arsenite as an algacide and its effects on fish production in ponds, *Southeastern Conf. Game Fish Comm. Proc.,* 11, 281, 1957.
7. **Lawrence, J. M.,** Further investigations on the use of Delrad as an algacide in fish ponds, *Prog. Fish Culture,* 20(2), 89, 1958.
8. **Lawrence, J. M.,** Methods for Control of Aquatic Weeds in Fish Ponds with Emphasis on the Use of Chemicals, Auburn Univ. Agric. Exp. Stn. Prog. Rep. No. 69, 1958.
9. **Lawrence, J. M.,**Fertility and Biomass in Southeastern Impoundments, Auburn Agricultural Experiment Station, Auburn, Ala., 1969.
10. **Siedel, K. and Kichutlo, R.,** Biological treatment of phenol containing sewage with the help of rush plants, *J. Wass. Tech.,* 6, 209, 1967.
11. **Wagner, J. J. and Andreassen, D.,**Propagation of *Scirpus* Species in Hydropondic Cultures, Lifascruise Research Laboratory, Allied Chemical Corporation, Morristown, N.J., 1971.
12. **Wagner, J. J.,**A Biological Water Purification System Based on Aquatic Weeds, Division of Environmental Science and Planning, Huntington - Laison Architects, Springlake Heights, N.J., 1979.

Chapter 9

# AQUATIC PLANT SUCCESSION IN LAKE SEMINOLE*

## INTRODUCTION

Plant growth and their successive stages in an aquatic habitat are natural phenomena that have occurred for millions of years evidenced by present geological formations. Thus, one anticipates that small and large impoundments will eventually support a growth of aquatic plants. The kind of plant succession within a given impoundment depends on water depth, turbidity, ferility, composition of plant species, and presence or absence of plant pests and pathogens. Recently, man has by various means attempted to inhibit or manage obnoxious aquatic plant populations.

One example of widespread management of aquatic plant populations is use of a combination of deepened pond edges and inorganic fertilization to produce and maintain a growth of phytoplankton in farm fish ponds. This system coupled with diligent partrol of the pond edge has resulted in absence of rooted aquatic plants. However, after a few years, the fertility of such managed ponds has promoted development of dense blue-green algae blooms which indicates excessive nutrient enrichment.

Unfortunately, large impoundments such as Lake Seminole were not designed for deepened edges or controlled water flows which would utilize nutrient fertility and produce phytoplankton blooms of sufficient density to adequately shade the bottom and prevent establishment of rooted aquatic plants. Neither have edges of such large impoundments been cleared so that marginal water weeds could be controlled by mechanical methods, such as periodic mowing.

About 150 species of aquatic and semiaquatic marginal plants were weed problems in one or more aquatic situations in the U.S. and Canada. The most recent available U.S. Census data reveal that there are more than two million ponds and small reservoirs. Surveys have shown that all or most of these aquatic areas are infested by or susceptible to infestation by aquatic or bank weeds which interfere with various uses of the water resources. Weeds are particularly serious in shallow areas less than 3 to 4 m deep in ponds and lakes and in sluggish stretches of rivers and streams.

Extensive use of herbicides has been necessary for many years to replace or supplement mechanical methods of controlling aquatic and bank weeds and preserve agricultural, navigational, and recreational uses of our water resources. In 1957, a careful survey by the Agricultural Research Service and Bureau of Reclamation revealed that 63% of the total area of irrigation canals in 17 western states was infested with aquatic weeds. More than 75% of the ditchbanks were infested with bank weeds. In that year, 54% of the weed-infested canals and 80% of the weed-infested ditchbanks were treated with herbicides at a total cost of about $8 million. Those weed control operations were estimated by irrigation companies and districts to have prevented losses of irrigation water and resulted in savings in labor, canal and structural damages, and crop losses valued at about $16 million. The net productive value of the water saved was estimated to be $39 million, nearly five times the cost of weed control. An extensive weed control program has been continued since 1957 on western irrigation systems. In 1969, more than six million pounds of seven herbicides was used on irrigation systems in Idaho, Oregon, and Washington, only 3 of the 17 western states. Aquatic weed problems are

* Updated and abstracted from The Summary Report, Contract DA-01-076-CIVENG, Mobile District, U.S. Army Corps of Engineers, and the Agricultural Experiment Station, Auburn University, Auburn, Ala., 1973.

especially critical in lakes, streams, and navigation channels in Gulf and Atlantic Coast states. A survey conducted in 1963 by the U.S. Army Corps of Engineers showed total infestations of 65,000 ha of water hyacinth, 40,000 ha of alligator weed, and 85,000 ha of submersed weeds.[6]

## TOLERABLE SPECIES CONCEPT

Any suitable aquatic habitat is subject to invasion and occupancy by numerous types of rooted plants. In most cases such plants are considered noxious to use of or destructive to the habitat. In mainstream impoundments, varying areas of shoreline are suitable for such growths, and the extent of plant development is dependent largely upon available nutrient supply. It does not appear feasible at present to control or eliminate such noxious aquatic plant growths by herbicides; neither are sufficient biological controls known or suspected to exist to adequately combat this menace. Thus, another approach, that of tolerable plant species, needs to be evaluated to more adequately define proper control measures. This concept is based upon the premise that some types of aquatic plants are tolerable to varied water uses, i.e., these plants will occupy the area and their presence will produce the least physical hazards, they can be economically eliminated, and they will reduce the probability of habitat elimination more than any others.

The first step in developing this concept into a management technique is to determine what species possess those characteristics necessary to classify them as tolerable to varied water uses. Such a classification can be approached by determining the good and bad characteristics of various plant species occupying varied aquatic habitats. A very important second step is to determine the species' susceptibility to thinning or elimination by approved economical means. The third step would involve techniques that can be applied to a habitat to assure the establishment of the selected tolerable species.

Lake Seminole, inhabited by practically all known noxious aquatic weeds present in the southeastern U.S., provides a model habitat for preliminary investigations into the feasibility of accomplishing the first two steps of the proposed concept.

During the last two growing seasons (1968 and 1969), two plant species, alligator weed and water hyacinths (considered by many as the most troublesome water weeds in subtropical and tropical countries), have received extensive study.[6] Actually, these two plants have been under observation for almost 20 years on the Flint River arm of Lake Seminole. To understand their influence on this aquatic habitat, one must review the history of their existence in this area.

Lake Seminole, formed by a dam at the mouths of the Chattahoochee and Flint rivers, was impounded in 1957. Then, no growth of either plant species was present in the Flint River arm of the impoundment. However, there was another upstream impoundment, Lake Blackshear, that did have a population of water hyacinths. Strays began to drift downstream into Lake Seminole during 1958 and 1959. Since this area was newly impounded and also contaminated with raw sewage, the invader hyacinth growth became rampant, and by 1960 covered approximately 2500 ha of this river arm. At this time, it was impossible to maintain navigation on this arm of the reservoir. Rafts of hyacinths would drift against the powerhouse trash racks and interfere with power generation. Thus, in a period of approximately three summers, this plant speices became an economic hazard to the operation of the reservoir. Chemical control was instigated in 1960 and continued through 1961 and 1962. By 1963, practically every hyacinth had been killed; however, during 1961 and 1962, an invader, alligator weed, had moved into the niches relieved of hyacinth growths. By 1963, this new pest occupied approximately 120 ha of the marginal area of this arm of the reservoir. Alligator weed plants were robust as a result of bountiful nutrients supplied by raw sewage. Interestingly, the stand of alligator weed stabilized at slightly above 125 ha for the next 5 years. This density of alligator weed did not interfere with navigation and did not reduce fish production or eliminate spawning areas. Nor did it provide a suitable habitat for extensive mosquito breeding. In fact, this growth of alligator weed stabilized the marginal aquatic plant population of this area for 5 years.

In 1968, a population of flea beetle (*Agasicles* sp.) which had been introduced in 1967, became abundant and devoured extensive quantities of the terrestial alligator weed growth. The damage was so severe in 1968 that more than 50 ha of alligator weed were eliminated. However, nature reversed itself and water hyacinths began to refill the niches vacated by alligator weed destruction. In 1969, the early spring occupancy by alligator weed was some 60 ha. After the beetles became active in June, they devoured sufficient "gatorweed" to reduce the stand to approximately 20 ha by November.

During the early (1968) destructive stages of alligator weed, the water hyacinth population began to expand. It is estimated that expansion within this study area in 1968 was from 1 to 2 ha in May to some 2030 ha in November. One factor that acted as natural control for this hyacinth population was a 1 m drawdown of reservoir water level in the fall and the maintenance of this low level for a sufficient period to allow frost and freezing weather to kill over 80% of the plants.

## DESIRABLE SPECIES CONCEPT

The program of aquatic plant control is further complicated by the content and concentration of elements in the input waters, particularly amounts of industrial waste water and domestic sewage containing a complex of toxic organic compounds, heavy metals, and plant nutrients that interfere with water use. Recent research on the biology, biochemistry, and metabolism of different plant species demonstrates that different plant species are able to degrade various toxic substances present in the input waters. Plant species able to do this efficiently are most desirable for reservoir population.

German scientists started research work in the early sixties at the Max Planck Institute in Ploen, Schleswig-Hollstein and demonstrated that certain weeds (bullrushes) known to prefer clear water actually created clear water by a metallic process and adapted to utilization of various inorganic and organic pollutants.

Aquatic plants which are effective in slime dispersion and antifoaming action are: common reed (*Phragmites communis*), giant weed (*Arundo donax*), and marsh reed (*A. polyonium*). Plants having high capacity for removal of inorganic and organic compounds are: European bulrush (*Scirpus lacustrio*), Hard-stem bulrush (*S. acutus*), American bulrush (*S. validus*), and common cattail (*Typha angustifolia*). Plants specifically effective in removing organic and synthetic compounds including halogenated phenols such as pentachlorophenol are: European bulrush, soft rush (*Juncus effusus*), and needle rush (*J. roemericans*). Root exudates of a plant such as European bulrush are also known to dissolve the cell membrane and kill pathogenic microbes — salmonella, enterococcus, and parasitic worm eggs.

Reed lagoons have been operating as primary, secondary, and tertiary water cleaning systems for several years in Germany, Holland, Hungary, Poland, Yugoslavia, Finland, Israel, and Italy. There is reason to plan and design similar treatment facilities in areas of the U.S. to improve water quality. Marginal aquatic plants such as the bulrush have their own oxygenation system and contribute to the oxygen level of the

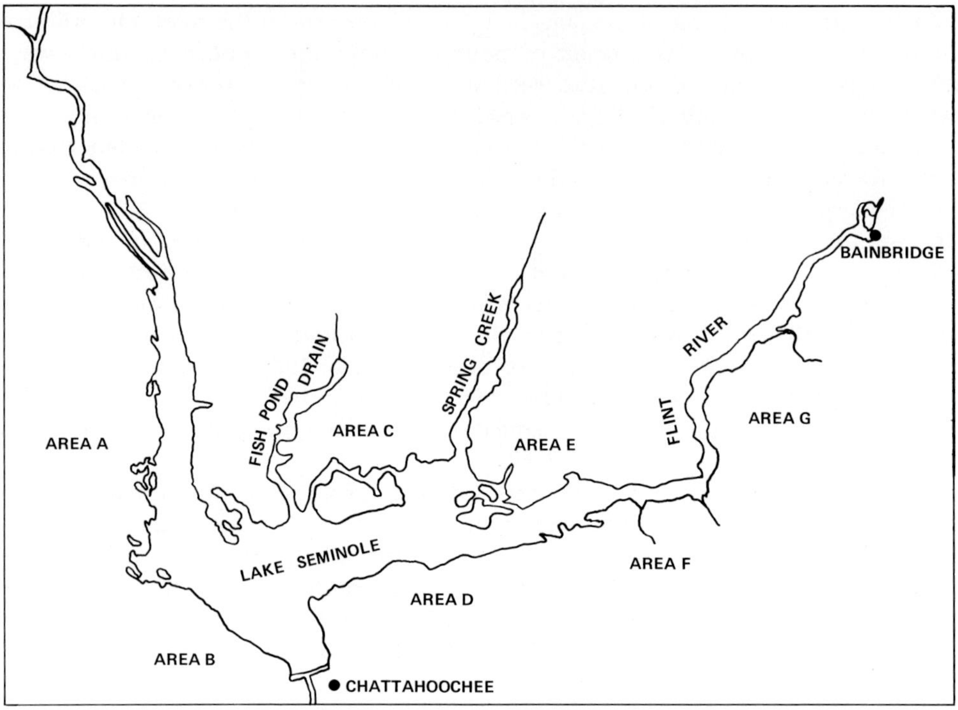


FIGURE 9.1. Distribution of major aquatic weeds in Lake Seminole, 1965.

water resource by stimulating aerobic microbial degradation in water and soils where they grow counteracting build-up of slime and sludge from anerobic decomposition. They are able to accumulate high amounts of phosphorus, nitrogen, and other nutrients of the input water. They propagate from root stocks and rhizomes and do not invade the total water area.

Bulrushes and similar plants of eastern North America are described in detail by Hutchkins,[16] Research Biologist of the Patuxent Wildlife Research Center in Patuxent, Maryland. Bulrushes are important wildlife plants. Seeds provide duck food, and root stocks of some types provide important goose and muskrat foods. The plants also provide valuable shelter for ducks and muskrats. Species prevalent in the North die back in the winter but those adapted to the South stay green year round.

## AQUATIC PLANT SURVEY 1957 TO 1965

Lake Seminole supports as great a diversity of aquatic weed species as can be found any place in the southeastern U.S. It is also probable that development of various plant associations in Lake Seminole has been more closely observed over the years of the existence of the impoundment than any other large impoundment. A brief resumé of the major patterns of observed plant population developments follows.

Soon after the inundation of Lake Seminole in 1957, the major portion of Area G (Figure 9.1) was infested with water hyacinths. This situation was corrected by use of 2,4-D sprays, and by 1963, water hyacinth was virtually eliminated from this reservoir. The population has been under maintenance control since 1965.

As water hyacinths were being eliminated in Area G (Figure 9.1), a growth of alligator weed established itself in this Flint River area. This plant reached maximum growth about 1964 when it occupied more than 200 ha of marginal water areas.

During 1963, Area E (Figure 9.1) had some growths of *Potamogeton* spp. ranging from occasional plants to plots almost impossible to penetrate with an outboard motor boat.

In 1966, the entire Area E (Figure 9.1) was filled with Eurasian watermilfoil. The exact invasion time was not detected, but latter 1965 was suspected. Since that time, the Eurasian watermilfoil has been confined to Area E, and quite recently was found in Area C, where water hardness was greater than 50 ppm $CaCO_3$ hardness.

Flooded lime-sinks (Areas D and F, Figure 9.1) have changed very little during the past 10 years (1965 to 1975) except that marginal growths of cattails and giant cutgrass have eliminated some very shallow water areas for any type of fishing or wildlife use. Though areas occupied by other species varied from year to year, the composition of species remained fairly constant. Two new species were found in Area B (Figure 9.1) in 1968. These plants were identified as *Limnophla* sp. and *Hydrilla* sp.

Area A, actually a river-run habitat, has been free of aquatic weeds throughout the history of this impoundment. Likewise, Area H, the open water portion of Lake Seminole, has never been infested with anything more than limited straggler, transient aquatic plant growths.

Area C, a tributary impoundment with several lime-sinks, has been filled almost completely with a mixture of submersed and emersed weeds since early days of inundation.

The cattails and giant cutgrass have continued to spread lengthwise along the shore into new lake areas. Cutgrass expansion has also occurred lakeward in Areas B, D, F, and G. This expansion of anchored, floating, giant cutgrass is believed to be potentially the most menacing plant of the future.

These plants have been present in the impoundment as scattered marginal growths ever since its inundation in the late 1950s.

## AQUATIC PLANT SURVEY 1965 TO 1975

For a period of approximately 10 years, there was limited lateral spread of both species into the shallow moist edge. Then cutgrass growth suddenly spread into deeper waters where many stands exist as floating mats. Invasion of cutgrass into noninfested areas continues at an alarming rate. Giant cutgrass completely dominates the cattail at all locations on Lake Seminole. The area of infestation has increased from 250 ha in 1965 to 1000 ha in 1975. It continues to increase (see Figures 9.1 and 9.2).

In 1965, alligator weed growths were confined primarily to the Flint River arm of Lake Seminole and were dominant and robust. In 1971, these growths were present in both the Flint and Chattahoochee River arms and were confined mainly to undergrowth in cutgrass, cattail, and water hyacinth stands. The infestation area has decreased from 200 ha in 1965 to 50 ha in 1975 (see Figure 9.2).

In 1965, only a few isolated plant clumps were present throughout the reservoir. In the spring of 1971, water hyacinth growth was greatly retarded from what was anticipated in the autumn of 1970 when there was an estimated 260 ha of plants present. However, during the late winter of 1971, high waters apparently washed a majority of these plants downstream into the Gulf of Mexico. Consequently, insufficient brood stock was left to produce nothing more than a widespread scattering of water hyacinths in 1975. Water hyacinths have been controlled by spraying with 2,4-D, and current acreage is limited to isolated colonies (see Figure 9.2).[3]

As mentioned previously, no Eurasian watermilfoil was known to be present in 1965, but by mid 1966, it was present in an estimated 600 ha of the Spring Creek area. In 1975, the growth was still confined to the same area and occupied approximately 2000 ha (see Figure 9.2).[4]

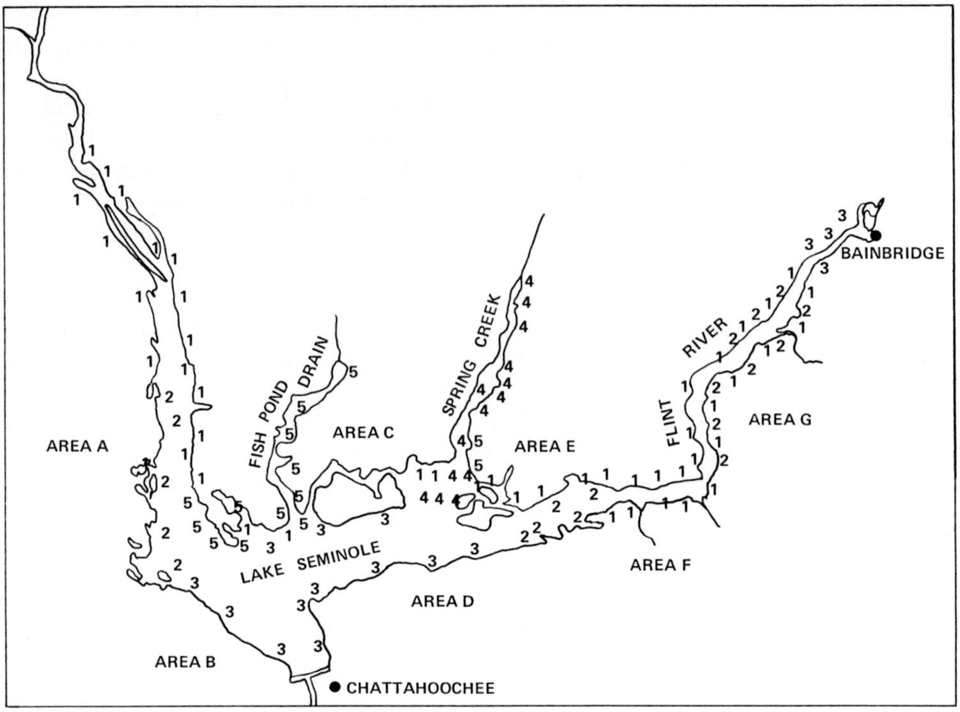


FIGURE 9.2. (1) Cattails and cutgrass; (2) alligator weed; (3) water hyacinth; (4) Eurasian watermilfoil; (5) water lily and water shield.

Water lillies (*Nymphaea*) and watershield (*Brasenia*) occupied approximately the same area (250 ha) in 1975 as they did in 1965 (see Figure 9.2).[5]

## AQUATIC PLANT CONTROL OPERATIONS

The progression of aquatic plant growths in large water bodies has been of academic interest to plant scientists for over 100 years. Only within the last few decades have scientists become interested in how these aquatic plant growths in man-made impoundments could be controlled. During the past 25 years, practically every chemical and mechanical method man could devise has been tried to reduce the nuisance problems caused by aquatic weeds in streams, lakes, and reservoirs. Due to recent public concern about addition of herbicides to surface waters, the chemical control method has been largely abandoned and replaced by biological control techniques. However, use of some promising organisms has been criticized.

The current approach to aquatic weed management on Lake Seminole is to utilize the habitat barrier (i.e., water hardness) to confine Eurasian watermilfoil to the Spring Creek area, to utilize the flea beetle to reduce the stand of alligator weed, to rely on winter kill and flooding to check water hyacinths, and to utilize underwater weed cutters to keep the shallow boat channels open. To date, no technique is known to retard or eliminate growth of giant cutgrass other than removal by dragline.

The 1972 status of this program indicated that Eurasian watermilfoil infestation was still confined to the Spring Creek area where water hardness was 50 ppm $CaCO_3$ or higher. Other areas with lesser hardness possibly had springs of Eurasian watermilfoil floating in the water, but no rooted specimens could be found.

The flea beetle apparently reduced growth of alligator weed to a point at which it would no longer be considered a dominant noxious plant in Lake Eufaula in 1972. This is not meant to imply that alligator weed is no longer present, it is as widespread along the margins today as it was 8 years ago. However, instead of being the dominant plant in an area, it is often the shoreline fringe in a water hyacinth stand or the understory in a giant cutgrass stand. Where alligator weed still existed in a fairly pure stand, the flea beetle attacked it early in the season with the result that regrowth was stunted for the remainder of the year. In fact, by August, the entire alligator weed population in Lake Seminole was so dwarfed in growth and in stem size that the reproductive rate of the flea beetle had practically dwindled to zero.

In the fall of 1971, only a moderate population of water hyacinths existed on Lake Seminole. There was some fall drawdown in water level permitting some soil rooting of hyacinths in the shallow edge. These rooted hyacinths, plus a mild winter and low water flows, allowed a sizable stock of viable plants to overwinter. Thus, there was a considerable water hyacinth population by late summer of 1972. There was a considerable 1972 fall drawdown in Lake Seminole water level permitting many hyacinths to root in the soil. With a large floating population and a sizable rooted hyacinth population, the only hope for next year is a cold winter with heavy frosts before the water level returns to normal followed by one or more spring floods. If winter weather conditions are as mild and dry as in 1971 to 1972, there will be a tremendous overwintering population of hyacinths. Consequently, troubles could occur in summer of 1973.

Submersed and emersed weed growths in shallow boat channels continued to cause problems even though dredging in some areas occurred in 1972. In most instances, these channels are too narrow to permit sufficient deepening to eliminate an aquatic weed habitat and thus keep the boat lanes open without use of underwater weed cutters.

In 1972, the eelgrass infestation was noted to have spread considerably, especially into previously uninfested areas. The cause of this plant's widespread infestation was the spawning habits of common carp. Prior to 1972, the location of small, widely scattered eelgrass beds was most easily accomplished during carp spawning season. Carp preferred this plant as an attachment site for their eggs. However, when many of these fish congregated in such an area, their vigorous activity uprooted the entire eelgrass plant. By looking for floating eelgrass in early spring and noting where considerable fish activity occurred, locating plant beds could be accomplished with minimum effort. Thus, carp uprooting of entire eelgrass plants was probably responsible for spread of the plant in Lake Seminole.

Giant cutgrass continued to spread at a rate of 50 ha per year and is present in some degree of density on 75% of the lake margin. To date, no economical control method is available. This is potentially the most serious noxious plant problem encountered on Lake Seminole. In many areas where cattails and giant cutgrass were present, cutgrass has virtually eliminated the cattail stand. Only one plant, *Peltandra* sp., is apparently able to dominate giant cutgrass. This situation probably occurred due to the mild winter (1971) which did not kill the plant's top growth and allowed it to spread more rapidly in the summer of 1972.

To summarize the aquatic plant situation in Lake Seminole, it might be said that the lake area inhabited by weeds continues to increase by a slow, but steady rate. As biological control reduces one species of plants, another species is present to replace it.

Water level fluctuations during the growing season may benefit mosquito control, but it is doubtful that it helps reduce aquatic plant problems. Rather, it permits movement of submersed plants into deeper water and this results in increased total acreage of these plants. It is probable that the drawdown also permits giant cutgrass to gain better bottom rooting conditions and thus increase its spread.

## REFERENCES

1. **Lawrence, J. M.,** A Tolerable Species Concept, Auburn University Agricultural Experiment Station, Auburn, Ala., 1970.
2. **Siedel, K. and Kichutlo, R.,** Biological treatment of phenol containing sewage with the help of rush parts, *J. Wass. Tech.*, 6, 209, 1967.
3. **Wagner, J. J. and Andreassen, D.,** Propagation of *Scirpus* Species in Hydropondic Cultures, Lifascruise Research Laboratory, Allied Chemical Corporation, Morristown, N.J., 1971.
4. **Wagner, J. J.,** A Biological Water Purification System Based on Aquatic Weeds, Division of Environmental Science and Planning, Huntington - Laison Architects, Springlake Heights, N. J., 1979.
5. **Hatchkins, N.,** Bulrushes and Bulrushlike Plant of Eastern North America, Circ. No. 221, Fish and Wildlife Service, Bureau of Sport Fisheries and Wildlife, U.S. Department of the Interior, Washington, D.C., 1965.
6. Expanded Project for Aquatic Plant Control, House Document No. 251, U.S. Government Printing Office, Washington, D.C., 1965.

Chapter 10

# ECOLOGY OF *HYDRILLA VERTICILLATA**

## INTRODUCTION

A reservoir will support varied types of vegetation. The type and density of vegetation depends upon the morphology of the reservoir, the nature of the bottom sediments, plant nutrients and light conditions. New reservoirs undergo succession with subsequent changes in species composition and abundance as they age. The species of aquatic plants endemic to an area are capable of dispersal into a newly formed reservoir. The major factors that determine the establishment of a given species in one area of a reservoir are temperature, light, current, wave action, water depth, soil, and water chemistry. The competitive interactions between various species is also important in determining the development of a dominant species.

We can state that the aquatic plant flora of a reservoir is determined by environmental and biological interactions with species that invade or occur in the system. Problems arise when the aquatic vegetation interferes with the human activities on the lake. In many cases, the problem is simply the result of the reservoir having an extensive littoral zone for aquatic vegetation.[4-7]

## RESEARCH OBJECTIVES

The ecological study conducted in Lake Ocklawaha had the following objectives:

1. Determine the distribution, composition, and density of the aquatic plant community.
2. Determine water quality and its effect on plant growth.
3. Determine the levels of selected nutrients in aquatic plants, and the relationship to water quality.
4. Determine the change of light intensity as effected by other environmental factors in the lake.
5. Determine the accumulation of organic debris on the lake bottom as related to growth and distribution of aquatic plants and their control.
6. Determine the effect of water drawdown on growth and distribution of hydrilla (*Hydrilla verticillata*).

## HISTORY OF AREA

Lake Ocklawaha, which was filled in 1968, covers approximately 5275 ha at a water elevation of 6.1 m msl. The lake is approximately 40 km long by 5 km wide. The reservoir was partially cleared in the area where the 4 m deep Cross Florida Barge Canal traverses the lake. The trees were entirely cleared away along a margin and in other areas the trees were selectively cleared. A total of 2230 ha was totally cleared with the remainder being partially cleared.

In 1969, a study of aquatic vegetation in Lake Ocklawaha showed that 47 species of plants were growing in the lake. Water hyacinth (*Eichhornia crassipes* (Mart.)

* From Blackburn, R. D., Factors affecting the growth and distribution of submersed aquatic weeds in Lake Ocklawaha. Unpublished cooperative research reports of the Agricultural Research Center, Fort Lauderdale, Florida and the U.S. Army Corps of Engineers, Washington, D.C., 1974.

Solms) was the dominant floating aquatic, while southern naiad (*Najas guadalupensis* (Spreng.) Magnus) and egeria (*Egeria densa*) were the dominant submersed weeds. The other aquatic plants were found only in small areas of the lake.

In May 1970, a survey was made by a scientist of the U.S. Department of Agriculture Aquatic Weed Laboratory at Fort Lauderdale, Florida, to determine the extent of the aquatic weed problem in the lake. At that time, aquatic weeds were a minor problem. Spray operations by the U.S. Army Corps of Engineers had resulted in excellent control of the water hyacinth.

During late summer and early fall of 1970, the submersed weeds southern naiad and egeria became a problem in several areas of the lake; but by late fall, the submersed weed growth was still a minor problem. Local state agencies predict that because of the shallow depth of the lake and its high fertility, aquatic weeds would become a continuous and major problem.

In December 1970, the water level in Lake Ocklawaha was dropped 6 m msl exposing 15% of the lake bottom. The water level was scheduled to be returned to its normal level 6.1 m msl in the spring of 1971. This water level fluctuation was to be evaluated as a method of controlling aquatic weeds in Florida impoundments. Research in Louisiana had indicated that water level fluctuation would control submersed aquatic weeds in smaller lakes.

In August 1972, the water level in Lake Ocklawaha was dropped an additional 1.7 m msl exposing approximately 60% of the lake bottom. The water level was returned to its August level of 6 m msl in February 1973. The water level fluctuation during this period was for the purpose of forestry management, but it had a dramatic effect on the submersed aquatic vegetation.[1, 2]

## STUDY PROCEDURE

Permanent plots 60 m square were established in four different locations in the lake. These plots are permanently marked with metal stakes. The geographical location of each area has been marked on aerial maps of the lake and defined in research reports. The four plots were established in areas of the lake which have different water flow rates, water depths, soil types, and aquatic vegetation.

Biological conditions were studied by the standing crop or biomass method. The standing crop was measured by removing a 0.3 m square area of vegetation from the lake bottom. Three samples were removed from each permanently established plot. Samples were removed by scuba divers clipping the plants at the soil line. The divers placed the plants in plastic bags as they were collected in the sample area. Plants were removed from the plastic bags, separated by species, counted, measured, and weighed after draining for 5 min. The weighed samples were returned to the plastic bags, stored in ice coolers, and transported to the Fort Lauderdale Laboratory. At Fort Lauderdale, the plant samples were dried, weighed, and analyzed for nitrogen, phosphorous, and potassium.

A liter sample of water was taken from three locations in each of the four plots. A pH measurement, dissolved oxygen, alkalinity, temperature (water and air), and weather conditions were recorded at the time each water sample was collected. A direct-reading colorimeter was used for measuring water color and turbidity. The liter water samples were used for determining the quality of the water in each plot. Most of the water color and turbidity determinations were made by colorimeter, but periodic comparisons were made using other procedures.

The necessity of light for growth of submersed plants probably is one of the most important factors in their distribution. A limnophotometer was used to measure the surface and subsurface illumination from voltaic cells connected to a central control box. One cell was enclosed in a metal case with a window in the top and was used for surface measurement. The other cell, the subsurface unit, was mounted in a watertight case with a window to admit light. Because of the cosine error of the photovoltaic cells, a flashed-opal disk was fitted over the light-sensing area of each cell. Light readings were made by anchoring a boat in the center of the canal so that no shadows would interfere with the readings. Readings were taken with the submersible cell at the surface, 0.1, 0.3, 0.5, and 1 m depths, and at 1 m intervals thereafter to the lake bottom. Each time readings were taken with the submersible cell, they were taken also with the deck cell. Subsurface readings were recorded only when the surface readings were constant throughout a series of vertical measurements.

In August, 1972, an additional sampling station was established to study the competition between hydrilla and valisneria.[8, 9]

## DRAWDOWN EFFECTS

The aquatic environment in Lake Ocklawaha has changed tremendously since the lake was filled. Changing of the water level for long periods has had the most pronounced effect. Environmental factors and macrophyte population changed rapidly after each drawdown. Water drawdown has been used as a lake management tool in other southeastern states but the schedule of drawdown does not appear to be practiced for most Florida lakes.

### Water Quality

Water quality information collected at the various sampling stations is shown in Table 10.1. The fluctuations that occurred appear to be related to seasonal changes and macrophyte growth.

The levels of phosphate in the water have remained nearly constant throughout this sampling period. The reason for the sudden increase in the May sampling cannot be explained at this time. Since there had been little rainfall in the Ocklawaha basin during March and April 1974, most of the water flowing into the lake was supplied by the various springs in or around the lake. The water flowing into the lake from these springs contains high concentrations of phosphate.

Nitrate concentrations at the various sampling stations were directly related to aquatic plant density. Sampling Station 5 had the largest plant infestation in May and the smallest nitrate concentration; while the reverse was true at Station 4. Concentrations of nitrate also appeared to be related to seasonal changes which directly affects plant growth. The natural supply of essential nutrients in Lake Ocklawaha is considered to be adequate for vigorous growth of aquatic plants.

Dense growth of submersed aquatic plants has affected the water quality. Alkalinity and calcium carbonate ($CaCO_3$) have changed in relation to the photosynthetic rate of the submersed species. Hydrilla has been shown to affect both measurements by the rapid precipitation of marl ($CaCO_3$) during periods of rapid growth.

### Light

It is known that light intensity and spectral composition influences the distribution of submersed aquatic plants. They all show that rooted submersed aquatic plants may colonize suitable substrate down to a depth where the light intensity is only 0.5 to 3% of the average intensity at the surface. Lake Ocklawaha has light penetrating to the bottom in all areas of the lake. In May, Secchi disk readings were recorded to a depth of 4 m. The turbidity and color that was created when the lake was reflooded in February 1973 had disappeared by June 1973.

TABLE 10.1

**Water Quality of Five Areas in Lake Ocklawaha[a]**

| Sample number | pH | Total alkalinity | Total hardness | $CaCO_3$ | Nitrates | Total phosphorus |
|---|---|---|---|---|---|---|
| 1-1 | 7.3 | 97 | 166 | 100 | 0.45 | 0.04 |
| 1-2 | 7.9 | 100 | 128 | 64 | 0.50 | 0.01 |
| 1-3 | 7.3 | 110 | 96 | 46 | 0.02 | 0.04 |
| 1-4 | 7.3 | 90 | 170 | 112 | 0.45 | 0.08 |
| 1-5 | 7.3 | 95 | 78 | 36 | 0.19 | 0.06 |
| 1-6 | 7.2 | 110 | 126 | 80 | 0.10 | 0.05 |
| 1-7 | 7.2 | 110 | 176 | 112 | 0.16 | 0.06 |
| 2-1 | 7.6 | 100 | 128 | 86 | 0.28 | 0.02 |
| 2-2 | 8.2 | 125 | 126 | 70 | 0.33 | 0.00 |
| 2-3 | 8.1 | 102 | 108 | 48 | 0.07 | 0.12 |
| 2-4 | 7.4 | 97 | 190 | 116 | 0.33 | 0.06 |
| 2-5 | 7.7 | 110 | 46 | 20 | 0.12 | 0.08 |
| 2-6 | 8.1 | 97 | 120 | 90 | 0.08 | 0.05 |
| 2-7 | 6.9 | 110 | 180 | 110 | 0.10 | 0.18 |
| 3-1 | 6.9 | 75 | 120 | 72 | 0.42 | 0.04 |
| 3-2 | 7.0 | 92 | 116 | 68 | 0.03 | 0.00 |
| 3-3 | 9.3 | 133 | 74 | 30 | 0.13 | 0.15 |
| 3-4 | 7.3 | 100 | 160 | 110 | 0.50 | 0.20 |
| 3-5 | 7.0 | 82 | 70 | 30 | 0.18 | 0.10 |
| 3-6 | 7.0 | 100 | 140 | 110 | 0.18 | 0.08 |
| 3-7 | 7.0 | 110 | 174 | 112 | 0.13 | 0.13 |
| 4-1 | 6.9 | 130 | 152 | 100 | 1.26 | 0.05 |
| 4-2 | 8.8 | 82 | 78 | 40 | 0.06 | 0.04 |
| 4-3 | | | | | | |
| 4-4 | | | | | | |
| 4-5 | 6.6 | 105 | 60 | 30 | 0.17 | 0.09 |
| 4-6 | 7.6 | 144 | 164 | 122 | 0.16 | 0.10 |
| 4-7 | 7.1 | 110 | 198 | 150 | 0.13 | 0.15 |
| 5-1 | | | | | | |
| 5-2 | | | | | | |
| 5-3 | 7.3 | 125 | 200 | 145 | 0.38 | 0.19 |
| 5-6 | 7.6 | 97 | 164 | 112 | 0.47 | 0.22 |
| 5-7 | 7.0 | 115 | 42 | 22 | 0.21 | 0.10 |
| 5-8 | 7.3 | 150 | 240 | 150 | 0.18 | 0.09 |
| 5-9 | 7.1 | 135 | 204 | 150 | 0.18 | 0.09 |

[a] Each measurement is the average of three samples.

## Plant Composition and Density

The plant composition and density changed drastically after the August 1972 water drawdown. Water level manipulation is often considered as one of the best control techniques for aquatic plants. This tool should not have been recommended for Lake Ocklawaha. Plant species and time of year should have been considered before the drawdown.

Prior to the August drawdown, those submersed species having the greatest frequencies of occurrence were egeria, southern naiad, and hydrilla.[3] These same authors reported that the drawdown reduced the percent frequency of hydrilla by 72%. Their data did not agree with the data collected in this study. The frequency of hydrilla

TABLE 10.2

**Composition of Submersed Aquatic Plants at Five Locations in Lake Ocklawaha[a]**

| Sample number | Naid | Coontail | Egeria | Hydrilla | Vallisneria | Ludwigia | Total green weight (lb/A) |
|---|---|---|---|---|---|---|---|
| 1-1 | 20 | 50 | — | 28 | — | 2 | 11,500 |
| 1-2 | 27 | 45 | — | 27 | — | 1 | 9,500 |
| 1-3 | 25 | 50 | — | 24 | — | 1 | 5,250 |
| 1-4 | 20 | 35 | — | 45 | — | — | 7,800 |
| 1-5 | 20 | 20 | — | 60 | — | — | 10,550 |
| 2-1 | 3 | 94 | — | 3 | — | — | 15,600 |
| 2-2 | 5 | 90 | — | 5 | — | — | 12,500 |
| 2-3 | 7 | 90 | — | 3 | — | — | 6,350 |
| 2-4 | 5 | 91 | — | 4 | — | — | 8,000 |
| 2-5 | 6 | 80 | — | 14 | — | — | 11,250 |
| 3-1 | 5 | 40 | 5 | 45 | 2 | 3 | 14,825 |
| 3-2 | 4 | 43 | 3 | 48 | 1 | 1 | 11,750 |
| 3-3 | 7 | 50 | 7 | 35 | 1 | — | 9,250 |
| 3-4 | 4 | 40 | 5 | 47 | 3 | 1 | 9,100 |
| 3-5 | 2 | 30 | 2 | 64 | 1 | 1 | 13,750 |
| 4-1 | 4 | 1 | 40 | — | — | 55 | 5,250 |
| 4-2 | 6 | 2 | 50 | — | — | 42 | 3,100 |
| 4-3 | 9 | 3 | 50 | — | — | 38 | 2,500 |
| 4-4 | 10 | 5 | 50 | — | — | 35 | 3,625 |
| 4-5 | 10 | 5 | 45 | — | — | 40 | 3,250 |
| 5-1 | — | 15 | 25 | 53 | 5 | 2 | 16,700 |
| 5-2 | 3 | 10 | 25 | 55 | 6 | 1 | 14,275 |
| 5-3 | 5 | 20 | 20 | 50 | 4 | 1 | 10,650 |
| 5-4 | 3 | 9 | 15 | 65 | 6 | 2 | 12,780 |
| 5-5 | 1 | 4 | 10 | 80 | 4 | 1 | 13,950 |

[a] Each value is based on the average of three randomized samples removed from the bottom of the lake in a 600-ft circle.

occurrence in the lake increased with the drawdown and by the spring of 1974 hydrilla was the dominant submersed aquatic plant (Table 10.2). The reason for the variation in the two studies cannot be explained. A complete survey of the reservoir in November 1972 showed hydrilla establishing from Orange Springs Landing north to Rodman Dam and into the main canal.

A second survey was made on May 28, 1974 to record changes in the growth and spread of hydrilla in the lake. During the 3-week period, hydrilla density had increased and the spread seemed to have been phenomenal in several areas of the lake. The spread may not have been as rapid as recorded because it is assumed that part of the increase was missed during the survey on May 7. Regardless of the reason it is very evident at this time that hydrilla is the dominant submersed aquatic and will further increase in 1974 (Figures 10.1 and 10.2).

The growth characteristics and life cycle of the plant would indicate that drawdown may present problems instead of reducing them. A water drawdown schedule may be worked out for the lake but it would not be the one used in 1972. A fluctuating water level schedule would be more likely to be effective than a fixed drawdown period.

The information collected in this study makes it evident that hydrilla in Lake Ocklawaha was spread by the August 1972 drawdown. During the drawdown, plant frag-

(A)

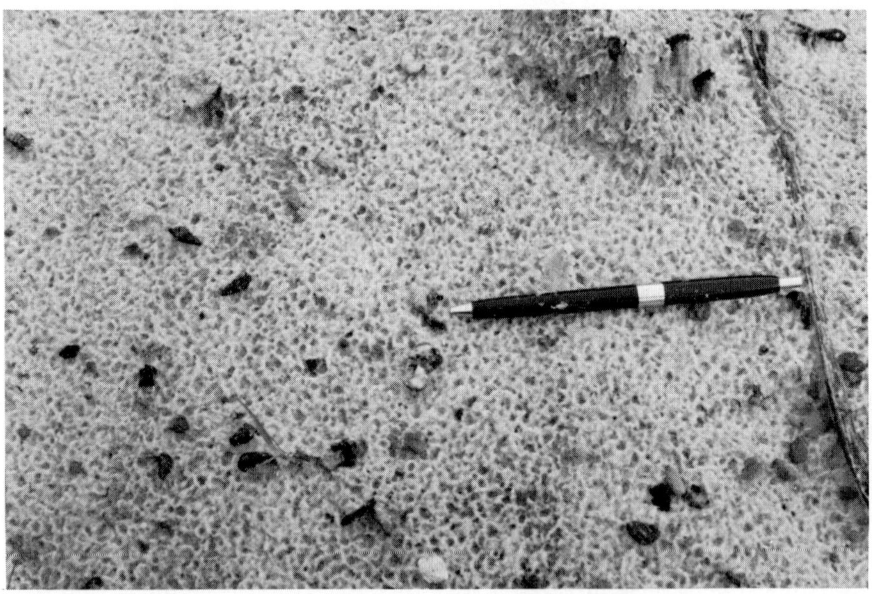

(B)

FIGURE 10.1. (A) Tubers of hydrilla in the soil along the shoreline near Buchman Lock, May 1972. (B) Sprouting of hydrilla tubers along the shoreline near Buchman Lock.

(A)

(B)

FIGURE 10.2. (A) Hydrilla fragments establishing near Orange Springs. Photograph taken in September 1972. (B) Same area one month later.

ments were found floating in the open water areas of the lake. The small infestation of hydrilla located near Orange Springs Landing, (see Figure 10.2) Recreational Site 2B, and Buckman Lock served as a source of infestation for the entire lake. Plant fragments broken off in these areas by wave action, water currents, boat propellers, and natural fragmentation floated throughout the main pool. During the past year hydrilla has spread upstream from Orange Springs.

The competitive relationship or coexistence of hydrilla with other submersed species is new in Lake Ocklawaha and will be important in future years. Once hydrilla is well established it will become the dominant submersed species. The water drawdown encouraged the establishment of *Ludwigia* in the dewatered areas. The drawdown also encouraged the germination of water hyacinth seeds in the dewatered areas.

### Plant Analyses

The productivity of a lake is often reflected in the macrophytes. Ranking growths of rooted macrophytes are usually associated with reservoirs having large areas of shallow water that receive good light penetration. Water analyses are commonly used in assessing the fertility of a lake. In large deep oligotrophic lakes such analyses may indicate the nutrient supply. However, in shallow euthrophic lakes, the store of soil minerals is probably of greater significance. Analyses, therefore, are of limited value in estimating anything other than the supply of solutes at the instant of sampling.

Nitrogen, phosphorus, and potassium levels found in various species of submersed aquatics are shown in Table 10.3. The nutrient levels varied with species, location, and time of year. The phosphorus levels have increased tremendously during this reporting period. We can only assume that the plants have assimilated phosphorus in excess of requirements. This could be considered as luxury consumption. Even with this excess consumption, phosphorus was still present in sufficient quantities to produce a luxurious growth of aquatic plants.

## SUMMARY OF EFFECTS

The submersed aquatic vegetation in Lake Ocklawaha has changed drastically since 1968. Submersed plants can now be found throughout the lake. The dominant submersed vegetation in 1970 was southern naiad (*Najas guadalupensis*), coontail (*Ceratophyllum demersum*) in 1973, and hydrilla (*Hydrilla verticillata*) in 1974. The water level drawdown during August, 1972 to February, 1973 played a major role in the ecological shift to hydrilla.

Hydrilla can now be found throughout the lake from Orange Springs Landing to Rodman Dam, and the navigational channel to Buckman Lock. During the drawdown, a small infestation served as the source of plant fragments and tubers to infest the entire lake. Hydrilla tubers established in the soil along the shoreline were washed into the lake after the drawdown by heavy rainfall. Hydrilla is now found in dense colonies in many areas of the lake. Water drawdown of this type should not be used as a control procedure for hydrilla.

Dense hydrilla growths in the lake has assisted in clearing the lake. Turbidity and other suspended organic materials are filtered out rapidly by this plant. This is substantiated by the greater light penetration into the water.

Lake Ocklawaha will continue to be a highly productive lake in the future. The abundant inflow of nutrients and the water clarity will assist in producing tremendous growths of hydrilla. A carefully planned management program could be developed for this shallow water lake using water drawdown and presently available control techniques.

TABLE 10.3

**Chemical Analyses of Submersed Aquatic Weeds Growing at Five Locations in Lake Ocklawaha[a]**

Sampling area 1

| Date | Plant species | N (%) | P (mg) | K (ppm) |
|---|---|---|---|---|
| 9/11/73 | Hydrilla | 2.51 | 3.35 | 49,280 |
| | Coontail | 3.27 | 4.38 | 43,050 |
| | Naiad | 2.14 | 2.50 | 34,650 |
| | Ludwigia | 2.27 | 2.89 | 9,240 |
| 11/8/73 | Hydrilla | 1.76 | 1.70 | 40,180 |
| | Coontail | 2.73 | 2.85 | 32,340 |
| | Naiad | 1.56 | 1.70 | 32,280 |
| | Ludwigia | 2.28 | 2.80 | 15,540 |
| 1/29/74 | Hydrilla | 1.94 | 1.65 | 37,550 |
| | Coontail | 2.79 | 3.30 | 34,780 |
| | Naiad | 1.54 | 1.85 | 31,650 |
| | Ludwigia | 2.05 | 2.75 | 12,780 |
| 3/20/74 | Hydrilla | 1.95 | 1.59 | 34,220 |
| | Coontail | 2.58 | 3.05 | 33,640 |
| | Naiad | 1.49 | 1.95 | 30,720 |
| | Ludwigia | 1.78 | 2.50 | 11,160 |
| 5/7/74 | Hydrilla | 1.52 | 1.42 | 30,690 |
| | Coontail | 2.08 | 2.75 | 46,740 |
| | Naiad | 1.41 | 2.05 | 37,720 |
| | Ludwigia | 1.44 | 2.15 | 6,000 |

Sampling area 2

| Date | Plant species | N (%) | P (mg) | K (ppm) |
|---|---|---|---|---|
| 9/11/73 | Coontail | 2.97 | 3.10 | 44,175 |
| | Coontail | 3.03 | 3.10 | 37,510 |
| | Naiad | 2.82 | 1.75 | 42,500 |
| | Naiad | 2.78 | 2.40 | 36,250 |
| 11/8/73 | Hydrilla | 3.07 | 3.40 | 36,890 |
| | Coontail | 2.89 | 2.00 | 34,100 |
| | Coontail | 2.91 | 2.09 | 35,250 |
| | Naiad | 3.37 | 4.35 | 32,000 |
| 1/29/74 | Hydrilla | 2.75 | 3.05 | 35,500 |
| | Coontail | 2.95 | 1.98 | 24,300 |
| | Coontail | 2.97 | 1.87 | 23,900 |
| | Naiad | 2.75 | 3.25 | 31,750 |
| 3/20/74 | Hydrilla | 1.68 | 2.95 | 36,980 |
| | Hydrilla | 1.98 | 2.87 | 35,650 |
| | Coontail | 1.93 | 1.97 | 22,780 |
| | Coontail | 1.84 | 1.95 | 27,230 |
| 5/7/74 | Hydrilla | 1.12 | 2.31 | 37,850 |
| | Hydrilla | 1.92 | 2.09 | 38,130 |
| | Coontail | 1.72 | 1.88 | 25,200 |
| | Coontail | 1.70 | 1.98 | 23,940 |
| | Naiad | 1.74 | 2.10 | 31,620 |

TABLE 10.3 (continued)

**Chemical Analyses of Submersed Aquatic Weeds Growing at Five Locations in Lake Ocklawaha[a]**

Sampling area 3

| Date | Plant species | N (%) | P (mg) | K (ppm) |
|---|---|---|---|---|
| 9/11/73 | Hydrilla | 2.26 | 3.10 | 47,560 |
| | Coontail | 3.34 | 5.40 | 55,335 |
| | Southern watergrass | 2.83 | 2.50 | 22.475 |
| | Ludwigia | 2.17 | 3.10 | 8,030 |
| | Naiad | 2.17 | 2.33 | 40,000 |
| | Milfoil | 2.98 | 3.00 | 17,980 |
| | Vallisneria | 2.36 | 2.07 | 80,000 |
| 11/8/73 | Hydrilla | 1.57 | 2.05 | 39,975 |
| | Coontail | 2.19 | 3.50 | 39,525 |
| | Ludwigia | 1.89 | 2.50 | 11,110 |
| | Naiad | 2.38 | 3.50 | 43,750 |
| 1/29/74 | Hydrilla | 2.01 | 1.95 | 41,360 |
| | Coontail | 2.78 | 1.83 | 37,550 |
| | Ludwigia | 1.79 | 2.25 | 10,650 |
| | Naiad | 1.85 | 1.75 | 44,680 |
| | Vallisneria | 2.05 | 1.96 | 50,200 |
| 3/20/74 | Hydrilla | 1.70 | 1.99 | 36,580 |
| | Coontail | 2.20 | 1.85 | 25,220 |
| | Ludwigia | 1.75 | 2.15 | 9,590 |
| | Naiad | 1.43 | 1.70 | 38,600 |
| | Milfoil | 1.40 | 1.92 | 14,380 |
| 5/7/74 | Hydrilla | 1.50 | 2.14 | 21,210 |
| | Coontail | 1.75 | 1.80 | 20,790 |
| | Southern watergrass | 3.01 | 3.37 | 21,420 |
| | Ludwigia | 1.63 | 2.41 | 7,740 |
| | Naiad | 1.12 | 1.65 | 24,570 |
| | Milfoil | 1.50 | 1.88 | 12,760 |
| | Vallisneria | 2.15 | 1.75 | 47,970 |
| | Potamogeton | 2.41 | 2.87 | 23,100 |

Sampling area 4

| Date | Plant species | N (%) | P (mg) | K (ppm) |
|---|---|---|---|---|
| 9/11/73 | Egeria | 2.97 | 3.60 | 27,590 |
| | Egeria | 2.55 | 2.17 | 18,700 |
| | Ludwigia | 1.99 | 3.00 | 14,300 |
| | Naiad | 2.90 | 4.30 | 50,840 |
| | Naiad | 3.05 | 4.53 | 63,550 |
| 11/8/73 | Egeria | 2.88 | 2.90 | 64,000 |
| | Ludwigia | 1.85 | 3.30 | 13,200 |
| | Naiad | 3.64 | 4.30 | 45,510 |
| 1/29/74 | Egeria | 2.73 | 2.50 | 26,540 |
| | Ludwigia | 1.99 | 2.95 | 12,500 |
| | Naiad | 3.04 | 3.25 | 38,650 |

TABLE 10.3 (continued)

**Chemical Analyses of Submersed Aquatic Weeds Growing at Five Locations in Lake Ocklawaha[a]**

Sampling area 4 (continued)

| Date | Plant species | N (%) | P (mg) | K (ppm) |
|---|---|---|---|---|
| 3/20/74 | Egeria | 2.61 | 2.45 | 25,200 |
| | Ludwigia | 1.78 | 2.70 | 11,580 |
| | Naiad | 2.95 | 2.70 | 25,680 |
| 5/7/74 | Egeria | 2.76 | 2.34 | 25,830 |
| | Ludwigia | 2.08 | 2.41 | 10,450 |
| | Naiad | 2.53 | 2.25 | 24,590 |

Source: Blackleurn, 1974.

Sampling area 5

| Date | Plant species | N (%) | P (mg) | K (ppm) |
|---|---|---|---|---|
| 9/1/73 | Hydrilla | 2.60 | 3.60 | 53,940 |
| | Coontail | 3.32 | 4.80 | 46,830 |
| | Ludwigia | 1.96 | 3.40 | 9,075 |
| | Egeria | 2.51 | 2.00 | 22,575 |
| | Vallisneria | 2.63 | 3.50 | 44,075 |
| 11/8/73 | Hydrilla | 2.26 | 3.40 | 36,580 |
| | Coontail | 3.00 | 3.93 | 32,185 |
| | Naiad | 2.95 | 4.07 | 35,875 |
| | Ludwigia | 1.81 | 2.80 | 9,185 |
| | Vallisneria | 2.19 | 3.83 | 38,010 |
| 1/29/74 | Hydrilla | 2.17 | 3.11 | 38,460 |
| | Coontail | 2.58 | 3.85 | 27,850 |
| | Naiad | 2.65 | 3.79 | 33,720 |
| | Ludwigia | 1.97 | 2.51 | 10,260 |
| | Vallisneria | 1.95 | 3.35 | 42,330 |
| 3/20/74 | Hydrilla | 1.98 | 2.75 | 38,790 |
| | Coontail | 2.03 | 3.55 | 43,520 |
| | Naiad | 2.41 | 3.55 | 30,550 |
| | Ludwigia | 1.95 | 2.34 | 11,220 |
| | Vallisneria | 1.91 | 2.41 | 39,780 |
| 5/7/74 | Hydrilla | 1.40 | 2.24 | 42,160 |
| | Coontail | 1.60 | 2.31 | 24,360 |
| | Naiad | 2.34 | 3.37 | 31,620 |
| | Ludwigia | 2.36 | 2.24 | 14,850 |
| | Vallisneria | 1.89 | 2.15 | 41,410 |
| | Egeria | 2.32 | 2.40 | 20,790 |

[a] Data collected at various intervals during the year.

Aquatic vegetation will probably continue to be a problem to all nonnavigation uses of reservoir waters. In the channel and canal area, water depths and vessel traffic will keep the channels open, but chemical or physical control measures are required in order to keep reservoirs suitable for multiple uses.

Water hyacinth and hydrilla are species that will present serious problems. If left to grow unchecked, they will displace desirable species and reduce the potential uses of the reservoir area. Herbicide, 2,4-D has been registered for use on water hyacinth in flowing water. The use of diquat to control hydrilla is limited to Florida and cannot currently be used in other states.

Various other used control methods are under investigation by the Corps of Engineers, including herbivorous fish, phytophagus insects, plant pathogens, environmental manipulations, mechanical control and integrated control.

## REFERENCES

1. **Achuff, P. L. and Zeiger, C. F.**, A survey of aquatic plants in the Cross Florida Barge Canal, *Hyacinth Control J.*, 8, 14, 1970.
2. **Boyd, C. E.**, The limnological role of aquatic macrophytes and their relationship to reservoir management, *Reservoir Fisheries and Limnology,* American Fisheries Society, Washington, D.C., 1971, 153.
3. **Hestand, R. S. and Carter, C. C.**, The effects of a water drawdown on aquatic vegetation in a shallow water reservoir, *Hyacinth Control J.*, 12, 9, 1974.
4. **Mitchell, D. S.**, The ecology of vascular hydrophytes on Lake Kariba, *Hydrobiologia,* 34, 448, 1969.
5. **Penfound, W. T.**, Plant communities of Oklahoma lakes, *Ecology,* 34, 561, 1953.
6. **Rantz, K. E., Davis, J., Hughes, J., and Schafer, H.**, Water level of fluctuation — its effect on vegetation control and fish management, *Proc. S. E. Assoc. Game Fish Comm.*, 18, 843, 1964.
7. **Shirley, H. L.**, Light as an ecological factor and its measurement, *Bot. Rev.*, 11, 467, 1945.
8. **Sculthorpe, C. D.**, *The Biology of Aquatic Vascular Plants,* Edward Arnold Ltd., London, 1967, 610.
9. **Ulrick, A. and Johnson, C. M.**, Analytical methods for use in plant anlysis, *Calif. Agric. Exp. Stn. Bull.*, 776, 25, 1959.

Chapter 11

# THE ECOLOGY OF *CABOMBA CAROLINIANA**

## INTRODUCTION

*Cabomba caroliniana* (Gray), commonly called fanwort, has become a serious submergent aquatic pest in lakes of northern Louisiana. It has reached harmful population levels in Black Lake, Saline Lake, and Lake Bisteneau, and it has been reported from all other major lakes in the area, including Toledo Bend Reservoir.

No lake in northern Louisiana typifies the extent of the spread of *Cabomba* more than Black Lake in northern Natchitoches Parish. Approximately 2000 of its 6000 ha are now infested with *Cabomba*, whereas it was hardly found in the lake 10 years ago. This species has the ability to grow in shallow or deep water. The economic impact of the infestation by *Cabomba* has been evident. Several commercial fishing camps have closed, others have faced marked decrease in gross income, private camp owners have sold their lake property, and many hunters, boaters, and fishermen have taken their recreation dollars to other areas.

Black Lake, the site of this study, is part of a lake complex formed by the construction of Allen Dam on Saline Bayou in 1930. Prior to this time there was no permanent lake in the area, although a temporary lake was formed in part of the area during the wet season. The damming of Saline Bayou created Black Lake, Saline Lake, and the prairie. The lake covers an area of 5265 ha, with an average depth of about 3 m. The deepest areas of the lake are in the channel, which was a bayou prior to the completion of Allen Dam. The lake is shallow enough that the entire lake is potentially susceptible to infestation by deep-ranging aquatic weeds such as *Cabomba*.[18]

The first major change to occur in the ecology of Black Lake is that the stream flow pattern of Red River changed as a result of a series of flood control projects upstream. Prior to 1960, flood waters of the Red River entered the Black Lake system by backing-up over the gates at Allen Dam and also over the spillway at Chivary Dam. This periodic flooding helped control submergent aquatic plants in Black Lake in three ways. First, murky water entered the lake and sometimes persisted for as long as 1 to 2 months. This usually occurred in the early spring at the time submergent plants would be initiating spring growth. The murky water reduced light penetration drastically and thereby prevented establishment in deeper areas. Secondly, the river water carried a heavy nutrient load. This probably resulted in an increased phytoplankton population, which in turn would result in a decrease in vascular plant populations due to the decrease in light penetration. Thirdly, the backwaters introduced new populations of buffalo, carp, and catfish into the lake. These species contributed to holding down aquatic plant populations by feeding directly on the submergent plants and by churning up the bottom, thereby increasing turbidity of the water.[18]

The second major change in the ecology of the area, which has led to conditions in Black Lake favorable for the growth of *Cabomba*, is a change in farming practices within the Black Lake Bayou watershed (the major stream feeding Black Lake). As late as the mid-1950s, much hill land in this watershed was actively farmed, being planted primarily in cotton and corn. Today, these hill farms are no longer being intensively farmed. Instead, they have been converted to pasture land or pine planta-

* By Dana R. Sanders, Sr., formerly Botanist, Northwestern State University of Louisiana, Natchitoches, Louisiana. Cooperative study with the New Orleans District, U.S. Army Corps of Engineers, and the Office of Chief of Engineers, Washington, D.C.

tions. Runoff water from this area no longer carries the heavy sediment load that it once did.

## DESCRIPTION OF THE PLANT

*Cabomba* (*Cabomba caroliniana* Gray) has finely divided opposite leaves that are dichotomously and trichotomously branched several times. Floating leaves are produced on flowering branches and are of firm texture, being borne alternately on the stem, with petioles peltately attached. Submersed leaves are slightly spatulate, 0.3 to 0.8 mm wide near the tip and narrower at the base. Flowers, which are borne on long axillary pedicels, are 7 to 11 mm long, with the three sepals and three petals slightly fused at the base. Perianth color ranges from white to yellow and purple. Pistil number is two to four, and there are normally six stamens. Each pistil contains one to three pendulous seeds, which are elongated and covered with elongate processes. A gelatinous slime coats the stems.[3-5, 9-15]

Herbaria of more than 200 colleges and universities in a 39-state area were asked to supply information on collecting stations for their specimens of *Cabomba*, especially those since 1950. Fassett[3] reported *Cabomba* from ponds and lakes from south central Texas to southern Florida, northern Kentucky, and, rarely, to southern Michigan. It was also reported from New York, southern Connecticut, and southeastern Massachusetts.[4,5,9] Results of the present study (Figure 11.1) indicate that a greater range now exists. New specimen locations were from Oklahoma, Kansas, Rhode Island, Pennsylvania, Georgia, and South Carolina.[18]

*Cabomba* seeds are produced in great numbers from April to December, with peak production occurring in May and October. Seeds, which are contained in long pistils, vary from one to three per pistil. The number appears to be environmentally related. Plants in one area may produce predominantly one seed per pistil in May and two or three in September. The time required for mature seed production is currently being investigated in a controlled environment. Pistils containing mature seed separate from the pedicel and fall to the bottom after a period of time. The pistil decomposes, leaving the seed on the hydrosoil surface. The pistil, while floating, provides a mechanism of seed dispersal. The pedicel soon drops off the determinate inflorescence, and finally the entire inflorescence separates from the branch.[1,4,15-17]

## PROGRAM OF RESEARCH

A series of five experiments was conducted to determine conditions conducive to germination of *Cabomba* seeds. Seeds from several sites and collecting dates were collected by hand. Only swollen pistils beginning to darken were selected, and these were placed in jars containing lake water. Seeds were considered mature when they had fallen to the bottom of the jar and were a dark color. It was necessary to collect seeds prior to field ripening, because field-ripened seeds fall from the pedicel soon after maturity. This could have had an effect on the low germination rates observed in this study. Seeds were subjected to approximately 15 treatments, with each being replicated twice. About 25% of seeds involved in Experiment I germinated within 2 months. This proved that *Cabomba* is capable of sexual reproduction in Louisiana and that a period of after-ripening is not necessary for germination to occur.[1,2] Riemer and Illnicki[8] found no sexual reproduction in *Cabomba* from New Jersey, although seeds were produced there. Germination rates (Table 11.1) were very low (10%), except for seeds stored in a greenhouse (85%). Field studies indicate that *Cabomba* seeds remain viable for more than 2 years after production. A study in which plots were established in September 1973 in Black Lake revealed seedling production in May 1974.

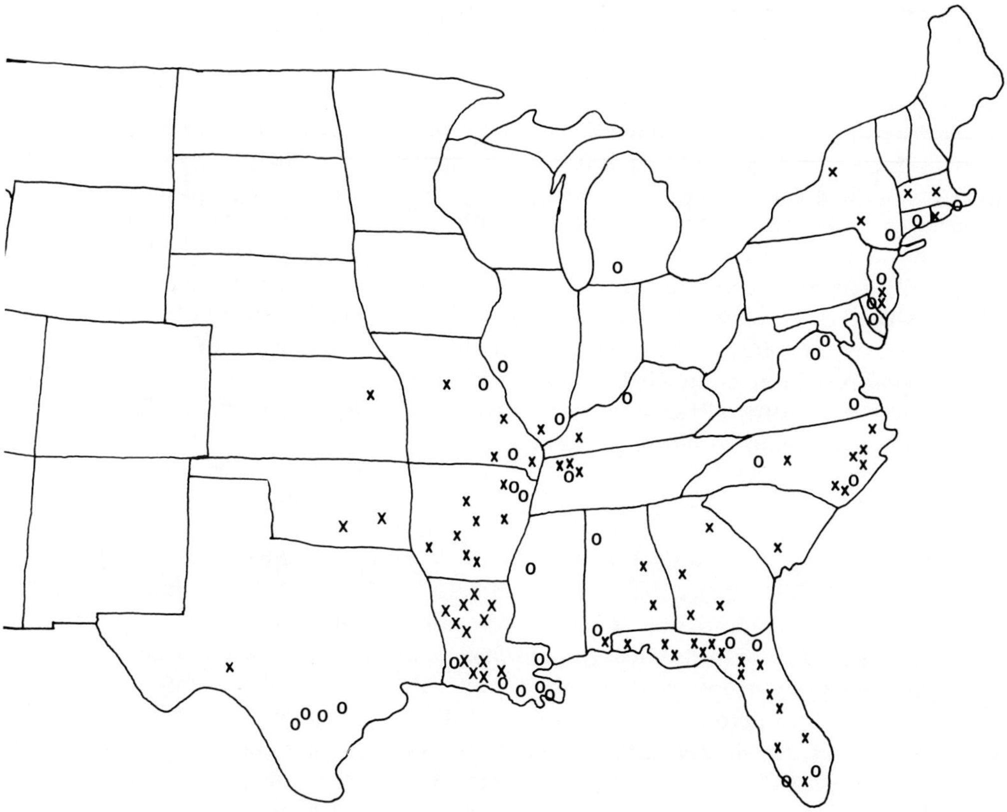

FIGURE 11.1. Distribution of *Cabomba caroliniana*. x = locations not reported by Fassett.[3]

TABLE 11.1

**Percent Germination From Three Collections of *Cabomba caroliniana* Seeds**

| Treatment | Fall 1972 Black Lake (% germination) | Spring 1973 Iatt Lake (% germination) A | Spring 1973 Iatt Lake (% germination) B | Spring 1973 Joyner's Pond (% germination) A | Spring 1973 Joyner's Pond (% germination) B |
|---|---|---|---|---|---|
| Placed in greenhouse | 86.0 | 100.0[a] | —[b] | 97.0[a] | — |
| None, placed in light | 6.0 | 2.0 | 0.0[c] | 6.0 | 8.0[c] |
| Air dried 14 days, placed in light | 6.0 | 2.0 | 0.0 | 2.0 | 0.0 |
| Dilute HCl, 70 sec | 6.0 | — | — | 0.0 | 0.0 |
| Heat, 14 days | 4.0 | 0.0 | 0.0 | 6.0 | 4.0 |
| Cold treatment, 30 days | 4.0 | 2.0 | 0.0 | 4.0 | 4.0 |
| Cold treatment, 14 days | 2.0 | 0.0 | 0.0 | 0.0 | 0.0 |
| 14-hr photoperiod, 6 weeks | 2.0 | — | — | — | — |
| Red light (650 nm), 5 min | — | 0.0 | 0.0 | 8.0 | 0.0 |

[a] Before red light treatment.
[b] Sample not available for study.
[c] After red light treatment.

**Water Level Manipulation**

Five seedlings were planted in 2.5 cm of hydrosoil in each of four aquaria filled with water from Black Lake and were allowed 2 weeks to become established. Seedling length (cm) was used to determine growth. The following treatments were made: (1) normal water level, (2) water barely covering seedlings, (3) all water removed, but hydrosoil saturated, and (4) all water removed, hydrosoil unsaturated. Treatments were maintained for 4 weeks, after which water levels were adjusted to normal for all aquaria. All seedlings survived in aquaria with standing water (Table 11.2). A mean survival rate of 53.3% was found in treatments where all water was removed but the hydrosoil remained saturated. Less than 10% of the seedlings survived when the hydrosoil was allowed to dry out. Under natural conditions, drying would likely be accompanied by temperature extremes that would make the treatment even more effective. *Cabomba* can be controlled by use of drawdown. However, seedlings located in areas where the hydrosoil remained saturated would have a good chance for survival. Therefore, water removal should be complete.

**Hydrogen Ion Concentration**

Seedlings were grown in Black Lake water in five aquaria at pH levels 4.0, 5.0, 6.0, 7.0, and 8.0. Two seedlings were planted in each of three small glass jars containing hydrosoil. Dilute HCl and 1.08 *N* NaOH were added dropwise to adjust lake water to the desired pH. The pH was checked colorimetrically and adjusted at 2-day intervals. Stem growth and branching were recorded. Mean elongation values from each pH level were converted into percentage growth that seedlings at each level contributed to total observed growth. Arcsin Percentage Transformations were used to produce values suitable for testing. An LSD test (Least Significant Difference) was performed and treatment means were found to differ significantly at the 0.05 probability level. Mean accumulated seedling elongation is shown in Table 11.3. Statistical analysis indicated that means of pH levels 4.0, 5.0, and 6.0 were not significantly different from each other, but were different from means of pH levels 7.0 and 8.0. Growth inhibition was observed at pH 7.0 and 8.0. Differences noted could be attributed to altered nutrient availability with a change in pH.

The optimum pH level for growth of *Cabomba* seedlings is between pH 4.0 and 6.0. The ability to grow only under acidic conditions could help explain the geographical distribution of *Cabomba*. It is usually found in acidic waters or in waters where the substrate would produce acidic conditions. Liming of small bodies of water infested with *Cabomba* would probably produce a shift in succession to less aggressive species. The net effect would be to eliminate *Cabomba* as a problem species in these areas.

**Turbidity**

Into each of four 19 ℓ aquaria, 4 cm of hydrosoil were placed, and enough water was added to bring water levels to within 5 cm of the tops of aquaria. Two seedlings were planted in each of 12 glass jars filled with hydrosoil (six seedlings per aquarium). After 1 week, the bottom hydrosoil was manually agitated with a glass stirring rod twice daily for 43 days. Turbidity levels of (1) Control-3-10 Jackson Turbidity Units (JTU), (2) Low-30-45 JTU, (3) Moderate-70-110 JTU, and (4) High-300-2350 JTU were used in this study. Stem length and branching were measured after 9, 22, and 43 days. Results shown in Table 11.4 indicate that *Cabomba* can grow well in turbid water. Growth at moderate turbidity levels was greatest, followed by growth at the high turbidity level.

TABLE 11.2

**Effects of Water Level Manipulation on Seedling Growth of *Cabomba caroliniana***

| Treatment | Replication I | II | III | Mean |
|---|---|---|---|---|
| None | 100.0[a] | 100.0 | 100.0 | 100.0 |
| Water only covering seedlings | 100.0 | 100.0 | 100.0 | 100.0 |
| No standing water; hydrosoil saturated | 60.0 | 60.0 | 40.0 | 53.3 |
| No standing water; hydrosoil unsaturated | 20.0 | 0.0 | 0.0 | 6.7 |

[a] Percent surviving seedlings after 8 weeks, based on five seedlings.

TABLE 11.3

**Mean Total Elongation of *Cabomba caroliniana* Seedlings at Five pH Levels**

| pH | Replication I | II | III | IV | Mean |
|---|---|---|---|---|---|
| 4 | 3.1[a] | 8.0 | 11.4 | 6.9 | 7.4 |
| 5 | 1.3 | 10.5 | 17.4 | 8.2 | 9.2 |
| 6 | 1.6 | 7.3 | 16.1 | 7.4 | 8.1 |
| 7 | −0.9 | 9.1 | 5.2 | 2.7 | 4.0 |
| 8 | −0.5 | 6.5 | 6.0 | 0.4 | 3.1 |

[a] Mean total elongation (cm) of six seedlings.

TABLE 11.4

**Average Elongation of *Cabomba caroliniana* Seedlings from Four Turbidity Levels**

| Treatment | Time in days 0 | 9 | 22 | 43 | Average total elongation |
|---|---|---|---|---|---|
| Control (3—10 JTU[a]) | 2.2[b] | 3.3 | 4.4 | 6.3 | 4.1[c] |
| Low turbidity (30—45 JTU) | 2.3 | 4.6 | 6.7 | 9.8 | 7.5 |
| Moderate turbidity (70—110 JTU) | 2.1 | 3.8 | 6.4 | 11.8 | 9.7 |
| High turbidity (300—350 JTU) | 2.1 | 3.9 | 6.3 | 10.9 | 8.8 |

[a] JTU = Jackson Turbidity Units.

[b] Average height (cm) of 18 seedlings.

[c] Values were obtained by subtracting initial height from height at 43 days.

## GROWTH AND REPRODUCTION CAPABILITY

Three sections of the following amounts of *Cabomba* were randomly placed in six 19 ℓ aquaria filled with Black Lake water: five, four, three, two, and one pair(s) of expanded leaves plus apical meristem; one pair of expanded leaves, no apical meristem; leaves only; and apical meristem only. Three 1-week replications were used. Any detached shoot with at least one pair of expanded leaves is capable of growing into a mature plant (Table 11.5). Many larger sections rooted at the nodes during the experiment. Consequently, when a motorboat passes through a *Cabomba* stand, hundreds of disseminules are created by the motor cutting action. This is probably the major method of infesting new areas within a water body.

### Productivity

On June 6, 1974, two transects were established in a pond containing a heavy infestation of *Cabomba*. Each transect consisted of two 9 m ropes placed on the pond bottom 0.4 m apart. Each rope was marked at 1.0 m intervals, thereby dividing each transect into nine 1.0 × 0.4 m plots. All plots were clipped 2 cm above the hydrosoil and rooted weights were determined (grams per square meter).

### Nutrient Uptake

A study was conducted to determine relative importance of roots and shoots in nutrient uptake in *Cabomba*. Root systems of *Cabomba* were placed in a solution containing labeled P, while stem sections were tested (after proper treatment) for presence of labeled P. Shoot sections were found to have taken up much more labeled P than did the root sections.

### Pollination

Observations of bees visiting flowers in June 1973 led to the hypothesis that *Cabomba* is entomophilous. Flowers that are to open on a given day emerge and begin to open at approximately 10:00 a.m. Flowers open for the first time characteristically have short stamens that are positioned at the base of the pistils. Pistils are divergent on the first day. Two-day flowers have long stamens (as a result of filament elongation) and upright pistils. At the end of day two, anthers are actually in contact with stigmatic surfaces of the pistil. From evidence gathered, it appears that *Cabomba* could only be facultatively entomophilous.

### Sexual Reproduction

The range of *Cabomba caroliniana* reaches from south central Texas north into Kansas, east to Michigan and upper New York, northern Massachusetts, and south to Florida. These studies indicate that *Cabomba* is capable of sexual reproduction in Louisiana. Nearly 100% germination was observed under greenhouse conditions. Factors believed to be of importance in *Cabomba* germination are red light, temperature, and high $CO_2$ levels. The seeds are found to have prolonged dormancy. Two-year-old or older seeds germinated under field conditions. Water level fluctuation did not render *Cabomba* seeds nonviable, but water level fluctuation effectively controlled *Cabomba* seedlings. *Cabomba* seedlings grow best at pH levels of 4.0 to 6.0 and are actually inhibited at pH levels 7.0 and 8.0.

Any stem section of *Cabomba* with a single pair of expanded leaves can produce a mature individual. The ability of *Cabomba* cuttings to grow under various pH levels appears to be similar to response to seedlings to different pH levels. It is found that nutrient uptake in *Cabomba* is primarily through leaves and stems. Experimental evi-

TABLE 11.5

**Effects of Size and Type of Plant Fragment on Branching and Rooting in *Colombia caroliniana***

| Plant section | % Branching | | % Rooting | | Week of earliest branching | Week of earliest rooting |
|---|---|---|---|---|---|---|
| | After 5 weeks | After 12 weeks | After 5 weeks | After 12 weeks | | |
| 5+[a] | 77.8 | 77.8 | 77.8 | 77.8 | 2 | 2 |
| 4+ | 33.4 | 33.3 | 66.7 | 66.7 | 4 | 3 |
| 3+ | 22.2 | 33.3 | 33.3 | 33.3 | 4 | 4 |
| 2+ | 33.3 | 33.3 | 33.3 | 33.3 | 2 | 4 |
| 1+ | 22.2 | 33.3 | 11.1 | 44.4 | 3 | 4 |
| Leaf | 0.0 | 0.0 | 0.0 | 0.0 | — | — |
| Apical meristem | 0.0 | 0.0 | 0.0 | 0.0 | — | — |
| 1−[b] | 66.7 | 77.8 | 66.7 | 88.9 | 2 | 2 |

[a] Number of pairs of expanded leaves plus apical meristem.
[b] One pair of expanded leaves, no apical meristem.

dence suggests that *Cabomba* is self-pollinated. This does not preclude occasional pollination by bees that frequent flowers from time to time.

## WATER FLUCTUATION STUDIES

Use of water level fluctuation or drawdown in management of Louisiana lakes has been a very controversial issue.[6, 7] Opponents have cited temporary loss of recreational use of lakes for fishing, boating, and hunting purposes, in addition to economic loss as valid reasons for not employing drawdown as a management tool. Proponents, primarily aquatic plant biologists, have suggested that water level fluctuation is the only economically feasible method available for controlling submersed aquatic plants in the many public lakes of Louisiana. They have also pointed out that increased desirable fish populations result when water fluctuation is utilized.

### Drawdown

The water level of Black Lake was lowered 1.8 m beginning July 20, 1972, at a rate of approximately 10 cm/day. It was interrupted due to low dissolved oxygen concentrations that occurred in September 1972. An additional 0.6 m of water was removed beginning October 1, 1972, making a total drawdown of 2.4 m below normal pool stage, which was scheduled to be maintained until January 15, 1973. Abnormally heavy rainfall, beginning in November 1972 and extending through the spring months of 1973, produced a rapid rise in water level to a maximum 2.1 m above normal pool stage in May 1973. This produced a total water fluctuation of 4.5 m within a 10-month period.

### Vegetative Analysis

Transects were established at eight locations, ranging from the upper end to the lower end of the lake. Another sampling procedure involved vegetational analysis in 24 permanent 0.4 ha plots located at various depth intervals.

Vegetative samples were collected at 30.5 cm depth intervals along each transect, with the first sample being taken at 15 cm. Samples were collected using a 61 cm × 61 cm sheet metal quadrat equipped with cutting edges along the bottom. The unit was hoisted above water level by a boom attached to a 4.3 m, V-hull aluminum boat. A quick-release device allowed the sampling unit to free-fall, severing all vegetation in its path. A nylon net sack was placed over the quadrat top to retain vegetation. A cutter inserted in a slot near the quadrat bottom was used to sever stems at the hydrosol surface. The unit was then raised above water and lowered into the boat, where all vegetation was removed and placed into a plastic bag. Plant samples collected each day were refrigerated until laboratory analysis. Plant samples were washed and sorted according to species, centrifuged at 1200 rpm for 2 min to remove excess water, and weighed.

The permanent plot study provided a means of testing reliability of the transect sampling method. Six permanent 0.4 ha plots were established in each of four depth zones: 0.1 to 1.8 m, 1.8 to 2.4 m, 2.4 to 3.1 m, and 3.1 to 3.7 m. Plots were established 2 years prior to drawdown initiation and thus were not specifically designed to demonstrate any particular effect of drawdown on aquatic vegetation. Frequency of species found in plots was determined for the summers of 1971, 1972 (prior to drawdown), and 1973. Frequency was calculated by dividing the number of samples containing a species by the total number of samples.[60] The bottom was checked for presence of vegetation using self-contained underwater breathing apparatus (SCUBA) prior to 1973 sampling, and no actual samples were taken in plots where no vegetation was observed.

### Chemical Analysis

Water samples were collected weekly during the 2-year period at 0.6 m below the water surface from 24 permanent 0.4 ha plots. Sampling time each week began 20 min before sunrise and was completed by 10:00 a.m. Six sample plots were located at each of four depth intervals: 0 to 1.8 m, 1.8 m to 2.4 m, 2.4 m to 3.1 m, and 3.1 m to 3.7 m.

Dissolved oxygen and temperature were recorded at each sample location using a YSI Oxygen-Temperature probe. Samples were immediately transported to the laboratory at Northwestern State University, where analyses of other physicochemical characteristics were made. Specific conductance was determined using a YSI Conductivity Bridge. Tests for ammonia, nitrite, nitrate, total phosphates, and orthophosphates were conducted using Hach Chemical Company procedures (which conform to APHA Standard Methods), and samples were analyzed using a Bausch and Lomb Spectronic 20 instrument. Color, turbidity, and percent transmittance were determined using a Hach testing kit, while pH was determined colorimetrically. All chemical concentrations were measured in milligrams per liter, while temperatures were recorded in degrees Celsius, specific conductance in mhos, turbidity in Jackson Turbidity Units (JTU), and color as units compared to an APHA Platinum-Cobalt Standard.

## DRAWDOWN EFFECTS

### Vegetative Analysis

Average freshweights of aquatic plant species per plot (grams per square meter) in Black Lake are shown in Table 11.6, arranged by depth intervals. Egeria, which had a high freshweight average at all depths in 1972, decreased 99.9% overall by August 1973. The only remaining egeria in August was found at 0 to 1.2 m depths and probably represented fragments that washed into shallow areas and became rooted. A similar response was noted in small Georgia impoundments for several species.[7]

*Cabomba*, the second most important species in Black Lake prior to drawdown, decreased 99.4% by August 1973 and showed essentially the same response as egeria

TABLE 11.6

**Average Fresh Weight (g/m²) and Percent Control of Aquatic Weeds in Eight Transects, Arranged According to Depth Before and After Water Fluctuation**

| Species | 0—1.2 m | | | 1.2—2.4 m | | | 2.4—3.7 m | | |
|---|---|---|---|---|---|---|---|---|---|
| | 7/72 | 6/73 | 8/73 | 7/72 | 6/73 | 8/73 | 7/72 | 6/73 | 8/73 |
| Egeria | 618.4 | 7.3 | 1.1 | 815.3 | 0.3 | — | 1073.0 | 10.2 | 1.1 |
| | | (98.8)[a] | (99.8) | | (99.7) | (100.0) | | (99.7) | (100.0) |
| Floating bladderwort | 60.8 | 50.6 | 82.6 | 111.1 | T[b] | 17.5 | 7.3 | 3.2 | T |
| | | (16.8) | (−35.8) | | (100.0) | (84.3) | | (55.6) | (100.0) |
| Cabomba | 11.0 | 1.6 | 2.4 | 102.8 | T | T | 275.2 | 1.6 | — |
| | | (85.4) | (78.0) | | (100.0) | (100.0) | | (99.4) | (100.0) |
| Snailseed pondweed | 7.0 | — | 3.2 | 10.2 | — | — | — | — | — |
| | | (100.0) | (54.0) | | (100.0) | (100.0) | | — | — |
| Coontail | 1.6 | 5.6 | 21.5 | 14.8 | T | T | 182.4 | 4.6 | — |
| | | (−350.0) | (−1333.0) | | (100.0) | (100.0) | | (97.5) | (100.0) |
| Stonewort | T | — | 105.4 | T | — | 4.3 | — | — | T |
| Southern naiad | T | — | T | — | — | — | — | — | — |
| Slender spikerush | — | 0.8 | 0.5 | T | T | 3.5 | — | — | — |
| Eastern bladderwort | — | T | T | 3.8 | — | — | 1.6 | — | — |
| | | | | | (100.0) | (100.0) | | (100.0) | (100.0) |
| Arrowhead | — | 0.8 | 3.2 | — | T | — | — | — | — |
| Water primrose | — | 8.3 | — | — | — | — | — | — | — |
| Southern water grass | — | 3.5 | T | — | — | — | — | — | — |
| Roundleaf bacopa | — | 0.3 | — | — | — | — | — | — | — |
| Variable watermilfoil | — | T | — | — | — | — | — | — | — |
| Smartweed | — | T | — | — | — | — | — | — | — |
| Frogbit | — | T | — | — | — | — | — | — | — |
| Total | 699.4 | 78.8 | 220.0 | 1058.0 | 0.3 | 25.3 | 1539.5 | 12.1 | 0.0 |
| | | (88.7) | (68.5) | | (100.0) | (97.6) | | (99.2) | (100.0) |

[a] Each number in parentheses represents degree of control achieved relative to 1972 values.
[b] T means that at least one plot contained the species, but less than 1 g in the plot.

with reference to depth distribution. The slight increase in abundance from June to August 1973 was due primarily to growth of *Cabomba* seedlings that germinated after the lake refilled.

Although coontail increased by 1300% at 0 to 1.2 m depths, freshweights before and after drawdown were low. Coontail was not abundant at 0 to 1.2 m depths prior to drawdown; therefore, the increase could be explained by the fact that coontail was not anchored in hydrosoil and drifted into shallow areas when the lake level returned to normal. Coontail was successfully eliminated as a problem species in deeper lake areas.

The response of floating bladderwort to water fluctuation was similar to that of coontail. The increase in bladderwort was related to surviving individuals in downlake areas being blown into shallow uplake areas where they became anchored. Indications are that bladderwort represents an earlier stage of aquatic plant succession than egeria or *Cabomba.*

Stonewort (*Chara vulgaris* L.) and slender spikerush (*Eleocharis acicularis* (L.) R. & S.) increased in abundance in shallow lake areas following water fluctuation. This also is indicative of a successional shift in lake vegetation. It is generally accepted that these species cannot compete favorably with more aggressive species such as egeria and *Cabomba* under stable water conditions.

Several primarily marginal species that were not present in samples taken prior to drawdown were found at 0 to 1.2 m depths in June 1973. These included arrowhead (*Sagittaria subulata* (L.) Buchenau), water primrose (*Ludwigia repens* Forst.), southern water grass (*Hydrochloa caroliniana* Beauv.) roundleaf bacopa (*Bacopa rotundifolia* (Michx.) Wettst.), variable leaf milfoil (*Myriophyllum heterophyllum* Michx.), smartweed (*Polygonum amphibium* L.), and frogbit (*Limnobium spongia* (Bosc.) Steud.). However, arrowhead and southern water grass were the only species of this group found 2 months later.

Southern naiad and eastern bladderwort (*Utricularia gibba* L.) were found in trace amounts in August 1973. Water fluctuation did not eliminate these species, but they were restricted to shallow areas following fluctuation.

Frequency data obtained by sampling permanent plots (Table 11.7) show the same trends among plant species as was determined by transect analysis. The permanent plots were estabished 2 years prior to the decision to use water fluctuation; therefore, plot selection could not have biased results of this study. This is important because the transect method described in this paper is a new method, and additional data from the permanent plot study corroborate findings of the transect study, even though different parameters were estimated.

It is impossible to assess effects of the 2.4 m drawdown independently of the high water levels experienced immediately following drawdown. However, other area lakes that experienced high water levels similar to those of Black Lake (but which were not drawn down) were still heavily infested with a variety of aquatic plants in 1973.

Excellent control of egeria and *Cabomba* was achieved even at depths where water was not completely removed. This suggests that additional stress was placed on these plants that could not be attributed to drawdown alone. High water levels and cold temperatures during spring months could have been responsible for some of the additional stress placed upon these species. A plankton bloom during late May and early June 1973 could also have placed additional stress on these species. All of these factors would have affected the date of initiation of regrowth by remaining individuals.

As others have pointed out,[6,7] proper timing is of great significance in determining success of water level fluctuation. It appears that a fall-winter drawdown is the most effective type of drawdown to use in this area. Degree of success is also related to

TABLE 11.7

**Average Frequency (%) of Aquatic Plants in Twenty-four Permanent 1-Acre Plots During a 3-Year Period**

| | 0—1.8 m | | | 1.8—2.4 m | | | 2.4—3.1 m | | | 3.1—3.7 m | | |
|---|---|---|---|---|---|---|---|---|---|---|---|---|
| Species | 1971 | 1972 | 1973 | 1971 | 1972 | 1973 | 1971 | 1972 | 1973 | 1971 | 1972 | 1973 |
| Floating bladderwort | 49.7[a] | 37.5 | 63.7 | 36.7 | 11.4 | 2.5 | 0.0 | 18.3 | 1.7 | 0.0 | 0.3 | 0.0 |
| Egeria | 42.0 | 71.1 | 1.7 | 22.2 | 58.3 | 0.0 | 14.7 | 25.3 | 0.0 | 2.5 | 10.3 | 0.0 |
| Cabomba | 16.7 | 16.3 | 8.9 | 31.6 | 32.6 | 0.0 | 49.7 | 42.8 | 0.0 | 40.6 | 43.8 | 0.0 |
| Coontail | 0.8 | 0.0 | 27.5 | 0.6 | 4.4 | 0.8 | 1.7 | 2.5 | 1.7 | 0.3 | 10.0 | 0.0 |
| Stonewort | 0.0 | 0.0 | 36.1 | 0.0 | 0.0 | 0.0 | 0.0 | 0.0 | 0.0 | 0.0 | 0.0 | 0.0 |
| Southern naiad | 0.0 | 0.0 | 4.2 | 0.0 | 0.0 | 0.0 | 0.0 | 0.0 | 0.0 | 0.0 | 0.0 | 0.0 |

[a] Individual values are based on 60 samples in each of six plots.

weather conditions present during and immediately following the drawdown. Conditions outlined above appear to be ideal in placing additional stress on existing plant populations. Obviously, it is impossible to predict the degree of control a particular drawdown will yield. However, further study should make possible more reliable predictions concerning expected degree of control under a given set of environmental conditions.

There is definitely a place in aquatic plant management for water fluctuation. This study has demonstrated that it is possible to control heavy infestations of egeria and *Cabomba* utilizing water fluctuation. Untested is the combination of water fluctuation with other methods of control. It is likely, for example, that management plans in the future will combine water level fluctuation with biological control agents such as the white amur. Under such a plan, water level fluctuation would be used to greatly reduce standing crop, and subsequent introduction of the white amur would be utilized to control the population size of aquatic plants. However, this is in marked contrast to the lack of control with floating plants such as alligator weed and water hyacinth, which were not controlled by drawdown (Figure 11.2).

## SUMMARY OF EFFECTS

Water removal (drawdown) produced a decrease in values for temperature, dissolved oxygen, and pH. Effects of drawdown on other physicochemical factors were nonexistent or negligible. Periods of high inflow produced increases in values of ammonia, nitrates, apparent color, orthophosphates, and turbidity, while specific conductance decreased with periods of high inflow. Other characteristics were unaffected by inflow rates or effects were inconclusive as to whether directly related to high inflow. Generally, no changes in water quality that would produce drastic changes in population levels of desirable aquatic organisms were observed. These data suggest that water level fluctuation, while producing temporary minor changes in water quality, is an ecologically sound method for use in management of a submersed aquatic plant such as *Cabomba* in Louisiana reservoirs.

(A)

(B)

FIGURE 11.2. Drawdown operations on Lake Cheniere, Louisiana (1966 to 1969). Alligator weed (A) and water hyacinth (B) were not controlled by drawdown operations.

# REFERENCES


1. **Barton, L. V.,** *Seed Preservation and Longevity,* Interscience, New York, 1961.
2. **Devlin, R. M.,** *Plant Physiology,* Van Nostrand Reinhold, New York, 1969.
3. **Fassett, N. C.,** A monograph of *Cabomba, Castanea,* 18, 116, 1953.
4. **Fernald, M. L., Ed.,** *Gray's Manual of Botany,* 8th ed., American Book Company, New York, 1950.
5. **Gleason, H. A.,** *The New Britton and Brown Illustrated Flora of the Northeastern United States and Adjacent Canada,* Vol. 2, The New York Botanical Garden, New York, 1952.
6. **Lantz, K. E., Davis, J. T., Hughes, J. S., and Schafer, H. E., Jr.,** Water level fluctuation — its effects on vegetation control and fish population management, *Proc. Southeast. Assoc. Game Fish Commissioners,* 18, 1, 1964.
7. **Pierce, P. C., Frey, J. E., and Yawn, H. M.,** An evaluation of fishery management techniques utilizing winter drawdowns, *Proc. Southeast. Assoc. Game Fish Commissioners,* 17, 1, 1963.
8. **Riemer, D. N. and Illnicki, R. D.,** Reproduction and overwintering of *Cabomba* in New Jersey, *Weed Sci.,* 16, 101, 1968.
9. **Wood, C. E. J.,** The genera of the Nymphaeceae and Ceratophyllaceae of the Southeastern United States, *J. Arnold Arbor. Har. Univ.,* 40, 94, 1959.
10. **Arber, A.,** *Water Plants — A Study of Aquatic Angiosperms,* Hafner, New York, 1963.
11. **Beal, W. J.,** Notes on *Cabomba caroliniana* Gray, *Torrey Bot. Club Bull.,* 27, 86, 1900.
12. **Cook, M. T.,** Notes on the embryology of Nymphaceae, *Bot. Gaz. (Chicago),* 48, 56, 1909.
13. **Gwynne-Vaughan, D. T.,** On some points in the morphology and anatomy of the Nymphaceae, *Linnaeun Soc. Bot. Trans.,* 5, 287, 1897.
14. **Kassman, F.,** Die entwicklung der Chondriosomen und Chloroplasten von *Cabomba aquatica* and *Cabomba caroliniana,* auf grund von dauerbeobachtungen an lebeden zellen, *Planta Arch. Wiss. Bot.,* 1, 624, 1926.
15. **Salisbury, R. A.,** Description of the natural order of Nymphaceae, *Ann. Bot. (London),* 2, 69, 1806.
16. **Smith, E. L.,** The induction period in photosynthesis, *J. Gen. Physiol.,* 21, 151, 1937.
17. **Woods, R. D. and Hargraves, P. E.,** Comparative benthic plant ecology by SCUBA-monitored quadrats, *Hydrobiologia,* 33, 561, 1969.
18. **Sanders, D. R.,** The Life History and Ecology of *Cabmba Caroliniana,* Final Rep., Northwestern State University of Louisiana, Natchitoches, 1974.

# *Part III*
# *Research and Development Studies for the Control of Hydrilla with Herbivorous Fish*

Chapter 12

# HERBIVOROUS FISH FOR AQUATIC PLANT CONTROL*

## INTRODUCTION

During the last 2 decades, research to discover means of controlling obnoxious aquatic plant growth has included evaluation of various herbivorous fish species. The white amur or grass carp has shown sufficient efficacy promising for aquatic weed control in habitats occupied by established game fish populations. The Israeli carp was effective only in controlling filamentous algae and is useless for control of rooted aquatic plants. On the other hand, the white amur fish has shown great promise in controlling submersed and emersed plant types. The purpose of this chapter is to evaluate the efficiency of the white amur fish (*Ctenopharyngodon idella*) as a biological control agent for aquatic weed populations in natural habitats, and to evaluate the effects of space and plant nutrients resulting from weed destruction in the aquatic ecosystem. This research information establishes knowledgeable control and operational procedures for use of white amur fish in aquatic plant control.

## REVIEW OF RESEARCH

In August 1971, meetings were held at the Southeastern Fish Control Laboratory in Warm Springs, Georgia to review research on herbivorous fish for aquatic weed control. Meetings convened at Callaway Gardens Holiday Inn near Pine Mountain. Various state and federal research workers attended and participated in the discussions emphasizing past, present, and future research programs involving the white amur.

Three major categories of studies conducted to make feasible the wide-scale use of white amur are: control methods — toxicants, sex determination, sterilization, hybridization, and harvest methods; production methods and facilities — hatcheries (which, and where), diseases, nutrition, culture methods, and economics; management techniques — population dynamics, manipulation of populations, shaping public and scientific opinion and attitudes, and efficiency in large reservoirs.

Objectives of the original research on herbivorous fish are in various stages of completion. The objectives dealing with the effect amur will have on the fisheries environment as related to its competition with common sport fishes of the southeastern U.S. and the determination of the predator relationships among white amur, bass, bluegills, and catfish have been studied in small ponds. The majority of the research effort has been devoted to measuring the type and species of aquatic weeds eaten by the white amur, measuring the growth rate of the white amur as related to plant consumption, establishing stocking ratios necessary to control various species of aquatic weeds, and discerning the effects of the white amur on water quality.

The white amur prefers certain aquatic plants when several species are offered, but will eat many different ones when given only one species at a time. When commercial prepared feeds and aquatic plants were fed to the white amur, the commercial feeds were preferred; however, the fish grew at a faster rate when eating aquatic plants. Growth rate of the white amur depends on the size of the fish, species of aquatic plants, temperature, and other factors.

Results of initial research indicate that the white amur will control aquatic plant

* From Technical Report No. 4, Aquatic Plant Control Program, U.S. Army Corps of Engineers Waterways Experimental Station, Vicksburg, Miss., June 1973.

growth in enclosed areas. This fish will eat many different species of aquatic plants, but prefers certain ones. The effect of the fish on water quality is related to those fish associated with removal of the plants. Based on present work, drastic changes under field conditions will probably not occur. The white amur appears to be the most promising biological control agent for many aquatic weed problems. However, management of this fish will be essential to obtain maximum benefit of its weed control potential. Research under field conditions will help determine the role of the white amur in an effective aquatic weed control program.

**Research in Arkansas**

The results of amur stocking in various lakes by the Arkansas Game and Fish Commission since 1969 were reviewed. Lake Greenlee in Arkansas was cleared of a heavy coontail infestation in the summer of 1970 by stocking adult fish (12 to 14 lb) at the rate of seven fish per acre. The lake was drawn down; some amur were removed by trammel net, and by June 1970, the water level had been restored by pumping. Additional stocking of 0.5- to 1-lb white amur had no adverse effect on the many species of native fish present after 1 year. The fish population before the addition of amur was about 500 lb/A and in August 1971, it was 650 lb/A (including 175 lb/A of amur). The combination of biological weed control and reservoir drawdown resulted in reduction in the number of bluegills present and an increased hatch of largemouth bass. After evaluation of the Lake Greenlee experiment results, Lake Des Arc (Arkansas, 385 A) was stocked. Plans were made to stock about 10,000 A of water in various areas of the state including lakes from approximately 200 to 6000 A in size.

Slides of the amur spawning operation, hatchery troughs, eggs and fry at various developmental stages, and hatchery ponds used for the white amur project were shown. Some interesting facts covering spawning efforts in 1971 were presented:

1. Broodstock were collected best with a 3-in. trammel net, but injuries to the fish were far too common. In water 18 in. deep or less, 1½ in. trammel nets were the best method to obtain broodstock without injuries.
2. Broodstock were anesthetized with 7- to 12-ppm quinaldine concentrations, affecting males first.
3. Fish in breeding condition are easily sexed by tubercles identifiable on the pectoral fins of males, and lack of tubercles and by abdomen distention in females.
4. Both sexes of fish received hormone injections to ensure good egg and milt production. Males received only one injection 12 hr prior to spawning.
5. Hatching occurs at temperatures of 70 to 80°F (optimum hatching at 75°F).
6. Fry from 0 to 20 hr of age swim vertically and rest on the bottom from 20 to 36 hr of age (approximate ages depending on temperature). Horizontal swimming is not the rule until an age of 36 hr.
7. Natural spawning of the white amur has not been observed in Arkansas.

**Research in Florida**

Research efforts on mechanical removal, biological controls and chemical application techniques are being stressed in Florida to achieve better aquatic weed control. Mechanical harvesting research is being developed to utilize the product as well as to develop more effective and efficient machinery. Florida's research program focuses on aquatic insects, diseases, and the white amur. Improved techniques for underwater chemical application to submerged aquatics is also being studied as the state strives to develop an integrated approach to the aquatic weed problem.

Slides of tabular data showing some results of white amur tests done in plastic pools and aquaria (1970 to 1971 at Fort Lauderdale, Florida) were presented. One fish (average weight of 0.5 to 0.75 lb) was stocked in each pool in most tests. This represents 125 fish per acre. Growth rates of 0.25 to 1.0 lb/month were observed in the pools. Fish grew better on aquatic weeds than on catfish chow or trout chow. The chow proportion eaten when the combination of chow and weed was offered depended on the fish species involved. Germinating *Hydrilla* was preferred over the chow. Amur weighing about 0.75 lb exhibited the following descending order of preference for various plants: *Chara*, southern naiad, *Hydrilla*, and pondweed.

Amur presence in aquaria caused an elevation of $PO_4$ and nitrate readings. In the pools, dissolved oxygen and pH were depressed, but turbidity apparently was not affected by the white amur. Fecal pellets in the pools decayed slowly. Future research will be devoted to water quality measurements, food preference, and the relationships between temperature, food consumption, and growth rates. Feeding rates and white amur growth are proportional to temperature; therefore, different stocking rates will be needed in different areas of the country. Also, the growth of the white amur varies with the species of aquatic plants eaten — another factor that must be considered when determining effective stocking rates for aquatic plant control.

**Research in Georgia**

Important facts gleaned during review of research papers on the white amur were discussed. The River Tone in Japan, where a breeding population of amur has been established, presents an interesting situation:

1. The natural range of the white amur is from latitude 50° N to 23° N. When transported to the North American continent, this range includes all of the contiguous 48 states, about half of Mexico, and that part of Canada lying between the U.S. border and the latitude running just south of Hudson Bay.
2. Presently, the feeder streams of the Ah Kung Tian reservoir in Formosa (Taiwan) and the Tone River in Japan are the only places reporting successful spawning of the white amur outside its native habitat. The Ah Kung Tian reservoir covers 889 A during the rainy season and has annual fluctuations of as much as 24 ft. The surface temperature ranges from 12 to 34°C, and the water is slightly turbid the year around. Spawning grounds in the feeder streams of the reservoir are 400 to 500 yd long, 15 to 40 yd wide, and 3 to 6 ft deep; water velocity is 1.5 to 4.5 fps and the temperature is 73 to 75°F (22 to 25°C). Spawning also occurs in the Tone River in Japan. The section of the river where spawning occurs is 200 to 400 yd wide and 7.5 ft deep; the pH is 6 to 9, and the riverbed consists of sand or coarse gravel. Spawning occurs with a water rise of 2 to 4 ft and with a flow of 1.5 to 3 fps.
3. The literature yielded much material on the selectivity of plant material, the breeding of the amur, and enzymatic reactions in the gut. Proteases are found in the gut of actively feeding fish but not in fasting fish.

Literature review findings concerning the habitats where introduced amur populations are said to have spawned (if accounts reviewed are dependable) indicate that comparatively little water movement is necessary for successful development and hatching of eggs provided that eggs are kept in suspension for a 40-min period following fertilization. During this time, the eggs absorb enough water to become buoyant.

Results of a taste test conducted by ten people at the Southeastern Fish Control Laboratory in Warm Springs, Georgia were reviewed. Seven species of fish including white amur, red snapper, channel catfish, rainbow trout, lake whiting, largemouth bass, and croaker were rated on the basis of flavor, firmness, oiliness, and boniness. The amur ranked second, fourth, third, and seventh in the respective categories. Fillets from 3-lb amur contained a number of bones, but these were large enough to cause no problems in eating.

A research proposal to produce nonbreeding populations of white amur was discussed. Large numbers of exclusively female fish can be produced by various methods: gynogenesis achieved through egg fertilization with sperm treated carefully with X-rays to ensure complete destruction of sperm DNA; hormonal treatment of genetic females producing sex-reversed males; or subjecting fertilized eggs to temperature shock in the meiotic division period. Possibilities of producing sterile fish by hybridizing white amur with other herbivorous fish species should also be investigated, as should the possibility of selective breeding to reduce the number of fine bones, a species characteristic.

**Research in Alabama**

At Auburn University, attempts to spawn white amur were quite successful in 1967 and 1968. However, no young have been produced since 1969. It was thought that the adult fish used in 1969 were too large (approximately 15 kg). One of the current experiments involves feeding methyl testosterone to small amur to see what histological changes occur. Auburn makes a practice of stocking amur in fertilized ponds with fingerling channel catfish and in bass-bluegill ponds to control filamentous algae (*Hydrodictyon, Oedogonium* and others). By fertilization only, 30 to 40 kg/ha of amur have been produced annually per acre in bass, bluegill, and channel catfish ponds regardless of stocking rates. Five state-owned, public, fishing lakes stocked with 10 to 20 amur per acre in 1966 have had no weed problems. The amur remaining in these lakes are of trophy size and create considerable interest among anglers. Amur are routinely used to control weeds in production ponds at the State Fish Hatchery at Eastaboga, Alabama. It was found that too many fish per acre left too late in the year will damage levees planted with bermuda grass. White amur probably could remedy 90 to 98% of the weed problems in most Alabama ponds.

Work done with amur at the Marion National Fish Hatchery in Marion, Alabama was discussed. A chronic weed problem in a 17-A pond was eliminated during four successive years by stocking varying numbers of amur in early summer. The pond is used to spawn and rear bluegill fingerlings. The production cycle gets under way in late winter or early spring.

1. Because of shallow water depth averaging less than 2 ft, a dense growth of filamentous algae and submerged rooted weeds developed by late spring. The following trials were conducted using amur to consume the weed growths:

White Amur Stocked

| Date | Rate (per acre) | Size |
|---|---|---|
| Early July 1968 | 50 | 3-in. fingerlings |
| Late June 1969 | 10 | 3-lb yearlings |
| Late June 1970 | 7 | 10-lb adults |
| Late June 1971 | 50 | 3-in. fingerlings |

2. All trials resulted in effective control by late August or mid-September. An oxygen depletion occurred in September 1970, causing a heavy loss of amur.
3. Amur were spawned in 1971 in a very limited trial involving three females and six males. Females were injected on May 26 with an initial dose of two fresh pituitary glands from recently slaughtered amur males and 100 IU of human chorionic gonadothropin (HCG) per pound of body weight. The females weighed 16.5, 16.0, and 19.0 lb, respectively. Ten hours later, a second dose consisting of three fresh pituitaries plus IU HCG was injected into the body cavity.

4. Only one of the treated females stripped easily 9½ hr later, although eggs could be obtained from the other two. Eggs were taken from the running ripe fish. The others, along with two males, were placed in a 0.1-A pond to see if natural spawning would occur. Results were negative.
5. The female stripped yielded approximately 300,000 eggs that were fertilized with sperm from three males. Based upon an examination of a small sample of 269 eggs about 12 hr after fertilization, 62% showed developing embryos. Because of hatching equipment limitations, only about 10% of the eggs hatched. Most of the fry expired during the yolk sac absorption period.
6. This limited experience suggests that amur are easier to spawn by stripping than goldfish or common carp. It should be mentioned that the brood amur were fed commercial pelleted fish feed over winter. Terrestrial grass and clover was supplied three times weekly prior to spawning at a rate of approximately 0.5% of their body weight per day.
7. Perhaps the only real limiting factor for white amur use in the U.S. is that it is an exotic fish and its production in a river system must be considered. This problem might be solved by use of monosex culture.

**Weed Management in California**

A resumé of various weed control studies underway in California was given. Biological control by means of pigmyweed (*Tillaea aquatica*) and slender spikerush (*Eleocharis acicularis*) appears promising. Plantings of the latter species in areas inhabited by ducks must be protected with wire screens. Herbivorous fishes including *Tilapia* and silver dollar fish have shown some promise, but their susceptibility to cold weather renders limited use. They are now being released in California for aquatic weed control. The California Department of Fish and Game (Davis, California) will allow research on the white amur, providing the fish are carefully regulated. California habitat conditions are probably nearer those in the native range of amur than conditions in any other area of the U.S.

## PROBLEMS IN RESEARCH AND DEVELOPMENT

Opposition of the Exotic Fish Committee of the American Fisheries Society (Neptune City, New Jersey) to white amur use was discussed in some detail. The committee recognized the amur's real potential as a biological weed control agent. Furthermore, it is a good sport and food fish. However, to the committee advocates much experimentation must be done prior to amur release to allay fears that native fish populations would suffer from food competition, reduction in cover and spawning sites as vegetation stands are consumed by amur, turbidity produced by feeding activity, and accelerated eutrophication through fecal releases.

Research on the species in canals, waterways, and river systems in addition to ponds and lakes is needed. Riverine studies made in the U.S. and other parts of the world by other agencies have not shown a wild population following introduction of the species.[1]

An adult white amur population exists in Arkansas' White River which meets most of the amur's known spawning requirements. To determine definitely whether spawning had occurred, the Arkansas Game and Fish Commission (Little Rock, Arkansas) conducted population sampling in areas inhabited by white amur during the second summer after the population had attained sexual maturity in the White River. Other samples and tests showed that the amur is very susceptible to rotenone, the most commonly used fish toxicant. Two hundred and fifty pounds of rotenone was applied to flowing water to obtain a very large sample necessary for spawn determination. Ap-

FIGURE 12.1. Typical infestation of *Hydrilla* in Lake Ochlawaka, Florida July 31, 1974. (Courtesy of the U.S. Army Corps of Engineers.)

proximately 4000 kg of fish were killed over 8.5 km of river. A crew of 15 biologists collected,counted, weighed, and recorded all fish killed. The collection was made over 35 km of river for 3 days assuring that all drifting moribund fish would be in the sample. No amur spawn was found! One 5-year-old female amur was collected. All those examined were in the same age class as the original escape fish. The amur never spawned in Arkansas lakes or ponds where experiments were being conducted.

## SUMMARY

This review of current research effort suggests that the white amur does have a real potential for control of submersed aquatic plants. It is particularly applicable for the control of *Hydrilla* (Figure 12.1), and every effort should be made to exploit the characteristics of the white amur to control this plant.

## REFERENCE

1. **Bailey, W. M. and Boyd, R. L.,**A preliminary report on spawning and rearing of grass carp (*Ctenopharygodon idella*) in Arkansas, presented at the 24th Annu. Conf. Southwestern Assoc. Game and Fish Commission, Atlanta, 1970.

Chapter 13

# COMPARATIVE EVALUATION OF THE WHITE AMUR AND COMMON CARP UPON AQUATIC ECOSYSTEMS*

## INTRODUCTION

The use of white amur (*Ctenopharyngodon idella*) as a weed control agent in surface waters has been strongly protested or banned in many sections throughout the U.S. The chief criticism has been that white amur, like common carp (*Cyprinus carpio*), is a member of the minnow family, and might become as noxious in many aquatic habitats as the common carp. In the Southeastern U.S., common carp are present in most streams and impoundments, but their population density has seldom reached a noxious level. While common carp have limited sport or commercial appeal in this section of the country, their presence in many areas is considered beneficial since their bottom feeding habit decreases accumulation of organic materials.

This chapter's study was designed to evaluate effects that moderately heavy stockings (16 ha) of common carp have on aquatic ecosystems in ponds and to compare effects that equal stockings of white amur have on aquatic ecosystems in ponds with similar habitats. Primary objectives of this study were as follows:

1. To determine differences in water quality (particularly turbidity) and plankton production in ponds with and without common carp or white amur.
2. To estimate any degree of interference with growth or reproduction of other fishes caused by presence of common carp or white amur.
3. To determine if presence of common carp or white amur exerts any noticeable effects on the aquatic ecosystem in commercial catfish production ponds.

## RESEARCH PROCEDURES

The experimental design used to produce water quality and biological situations needed to obtain data on the objectives of this study are described below.

Studies concerned with effects of white amur or common carp stocked in combination with bluegills (*Lepomis macrochirus*) and largemouth bass (*Micropterus salmoides*) were conducted in fourteen 0.10-ha earthen ponds. These ponds were similar in shape, and depth ranged from 6 m at the shallow end to 2 m at the drain. Prior to commencing this study, each pond was drained, partially refilled with water, treated with 5 ppm $KMnO_4$, and then filled completely 1 week prior to stocking with bluegills.

The species and rates of stocking for each of these 0.10-ha ponds are given in Table 13.1. The bluegills used in stocking these ponds were in the 5 cm group and weight of the total number (375) stocked per pond ranged from 1.4 to 3.5 kg. These bluegills were stocked into the 0.10-ha ponds on March 8, 1973. The largemouth bass were in the 2.5 cm group and were stocked on May 10, 1973.

The white amur stocked in the 0.10-ha ponds were in the 25 cm group, and the total weight of the 10 fish added per pond ranged from 1.0 to 1.4 kg. The white amur were

* From Lawrence, J. M. and Forester, T. S., Technical Report 4, Aquatic Plant Control Program, U.S. Army Engineers Waterways Experimental Station, Vicksburg, Mississippi, June 1973.

TABLE 13.1

**Species and Stocking Rates Used in 0.10-ha Ponds**[a]

| Pond number F series | Species[b] | Rate/ha |
|---|---|---|
| 4, 15, 21 | Bluegill | 607 |
| | Largemouth bass | 40 |
| | White amur | 16 |
| 16, 19, 22 | Bluegill | 607 |
| | Largemouth bass | 40 |
| | Common carp | 16 |
| 17, 20, 23 | Bluegill | 607 |
| | Largemouth bass | 40 |
| 24, 26 | White amur | 16 |
| 25, 27 | Common carp | 16 |
| 18 | White amur | 16 |
| | Common carp | 16 |

[a] 1973.

[b] Bluegill added March 8, 1973. Largemouth bass added May 10, 1973. White amur added March 9, 1973. Common carp added March 28, 1973.

stocked into the ponds on March 9, 1973. The common carp stocked in the 0.10-ha ponds were in the 28 cm group and the total weight of the 10 fish added per pond ranged from 1.6 to 25 kg. The common carp were stocked into the ponds on March 28, 1973. All of the 0.10-ha ponds received periodic applications of an 8-8-0 fertilizer. The fertilization rate was adequate for maximum bluegill - bass production.

Studies concerned with effects of white amur or common carp stocked in combination with channel catfish (*Icatalurus punctatus*) were conducted in 0.10-ha similarly shaped earthen ponds. The depth of these ponds ranged from 0.66 m at the shallow end to 1.7 m at the drain. Prior to starting this study, these seven 0.05-ha ponds were drained, partially refilled with water, treated with 1 ppm rotenone in December 1972, and then filled with water. The species and rates of stocking for each of these ponds are given in Table 13.2.

These 0.05-ha ponds were stocked with 200 channel catfish in the 7 to 10 cm groups that ranged in total weight from 0.8 to 1.0 kg/pond on February 22, 1973. White amur used in these ponds were in the 25 cm group, and the total weight of the four fish used ranged from 0.9 to 1.1 kg. They were stocked into the ponds on January 15, 1973.

Common carp used were in the 27 cm group and the total weights of the four fish added per pond ranged from 0.7 to 0.9 kg stocked into the ponds on March 28, 1973. All of these catfish ponds received equal quantities of fish feed on a schedule designed to give maximum catfish production.

All ponds included in this experiment were sampled to obtain data on water quality and plankton production. Water from each pond was collected at intervals throughout 1973 and 1974, and the following parameters were made in the laboratory:*

* All analyses conformed with methods included in "Manual of Methods for Chemical Analysis of Water"; EPA 625-16-74-003. Additional sampling of pond waters for phytoplankton and zooplankton were conducted at intervals throughout 1973 and 1974. Procedures used to collect net plankton samples and to identity and quantify organisms followed procedures outlined in "Biological Field and Laboratory Methods for Measuring the Quality of Surface Waters and Effluents"; EPA 670/4-73-001.

TABLE 13.2

**Species and Stocking Rates Used in 0.05-ha Ponds[a]**

| Pond number E series | Species[b] | Rate/ha |
|---|---|---|
| 52, 53 | Channel catfish | 810 |
| | White amur | 16 |
| 55, 56 | Channel catfish | 810 |
| | Common carp | 16 |
| 51, 54 | Channel catfish | 810 |
| 57 | White amur | 16 |
| | Common carp | 16 |

[a] 1973.
[b] Channel catfish added February 22, 1973. White amur added January 15, 1973. Common carp added March 28, 1973.

Turbidity as Jackson turbidity units (JTU)
pH
Conductivity, Ω/cm
N as NH, ppm
N as NO, ppm
N as NO, ppm
P as orthophosphates
Total alkalinity, ppm CaCO
CO, ppm
Ca, ppm
Mg, ppm
Na, ppm
K, ppm
FE, ppm
Mn, ppm
Cu, ppm
Zn, ppm

Sampling of the bottom muds with an Ekman dredge was conducted at intervals throughout the study period to estimate abundance of benthic organisms in various treatments.

All ponds were drained during the first week of October 1974. All fishes within a pond were collected and sorted into species. The individuals within each species were sorted into inch groups, and the total numbers and total weights of each inch group were determined.

All data adaptable to computer storage were placed on cards and analyzed by various statistical techniques to extract as much information as possible from this study.

## RESULTS AND DISCUSSION

The intent of the initial research design was to establish stressed populations of white amur in populations of bluegill plus largemouth bass in ponds. Similarly crowded populations of common carp in bluegill and largemouth bass populations were established for comparative purposes. To achieve these conditions, relatively large numbers 161/ha of advanced amur and carp fingerlings were stocked prior to addition of the bass fry. Further adversity for the white amur was provided by use of weed-free ponds for all test units.

The same stocking rates of white amur and common carp were used in the fed catfish ponds to determine if this amur and carp number would seriously interfere with commercial catfish production. All conditions described above were created to determine if large numbers of white amur in a pond would affect the game or commercial fish population in a manner comparable to effects produced by common carp.

**Water quality** — All ponds provided data on selected water quality parameters that reflected the effects that each fish population was exerting upon its habitat. Statistical comparisons of all measured parameters for all ponds indicated significant differences for only two parameters, CO and turbidity, in those populations which included common carp. The elevated CO in these carp ponds reflected suppressed phytoplankton production caused by excessive suspended colloidal clays. This colloidal clay concentration in pond waters was reflected in elevated turbidity readings. The data indicate that the feeding activity of fairly large numbers of white amur did not roll bottom muds significantly. Information on the other water quality parameters (not significantly different among all fish populations) are summarized in Tables 13.3 and 13.4.

**Plankton** — Studies of plankton composition in ponds with various fish populations were concerned primarily with the relative abundance of phytoplankton and zooplankton forms. One other insight on phytoplankton production was obtained from Chlorophyll *a* measurements. The only significant differences indicated in these studies were in the catfish combinations where concentration of Chlorophyll *a* was greater in ponds whose fish population included either white amur or common carp. There was little evidence from these studies that white amur will consume any great quantities of either phytoplankton or zooplankton.

**Benthic organisms** — Benthos production within the various species combinations was quite varied. Statistical analyses indicated no significant differences in number of insects found in any combinations, but examination indicates their existence was less in populations containing common carp. There was a significant reduction in benthic worm populations in those ponds containing common carp. Data indicated that carp were very efficient bottom feeders and were competing directly with bluegills for food. White amur, on the other hand, did not exhibit any great preference for either insects or worms in their diet.

## BLUEGILL - LARGEMOUTH BASS PRODUCTION

Throughout the course of these pond experiments, periodic seining of bluegill - largemouth bass combinations occurred. From the information obtained by this sampling technique, it was evident that satisfactory bluegill reproduction occurred in all combination ponds during their first summer (1973) of existence. It was also found that common carp reproduced the first season they were in the experiment, but late fall seining in the bluegill - bass-carp ponds indicated limited survival of young-of-the-year carp.

Seine sampling of the combination ponds in spring of the second year (1974) indicated that largemouth bass reproduction was limited. Early summer seining indicated that bluegill reproduction had occurred in bluegill-bass only ponds, but was extremely limited or absent in the other combination ponds in this experiment. No common carp reproduction was detected in any combination ponds during the second spring (1974).

No reproduction of white amur occurred in any of the ponds where they were stocked in either 1973 or 1974.

In the past, bluegill-largemouth bass combinations have yielded variable production

TABLE 13.3

**Average pH, Chlorophyll *a* Content, and Nutrient Concentrations in Waters from Ponds with Different Species of Fish**

| Treatment (species) | pH | Chlorophyll (ppb) | P *a* (ppm) | *N* as $NH_4$ (ppm) | *N* as $NO_2$ (ppm) | *N* as $NO_3$ (ppm) | $CO_2$ (ppm) | Total alkalinity as $CaCO_3$ (ppm) |
|---|---|---|---|---|---|---|---|---|
| Bluegill + largemouth bass | 7.49 | 2.71 | 0.48 | 0.26 | 0.008 | 0.083 | 0.65 | 17.11 |
| Bluegill + bass + amur | 7.64 | 2.74 | 0.33 | 0.25 | 0.006 | 0.011 | 0.97 | 15.39 |
| Bluegill + bass + carp | 7.58 | 3.06 | 0.41 | 0.25 | 0.03 | 0.18 | 1.21 | 13.83 |
| White amur | 8.03 | 3.56 | 0.38 | 0.25 | 0.005 | 0.20 | 1.10 | 15.71 |
| Common carp | 7.67 | 3.17 | 0.38 | 0.49 | 0.034 | 0.19 | 2.37 | 18.08 |
| Amur + carp | 7.93 | 6.75 | 0.30 | 0.24 | 0.004 | 0.11 | 0.83 | 12.46 |
| Channel catfish | 7.33 | 1.35 | 0.15 | 0.67 | 0.007 | 0.16 | 1.23 | 14.42 |
| Catfish + amur | 7.33 | 3.66 | 0.17 | 0.41 | 0.005 | 0.14 | 1.21 | 15.58 |
| Catfish + carp | 7.56 | 4.04 | 0.13 | 0.30 | 0.004 | 0.18 | 1.21 | 13.50 |

TABLE 13.4

**Averaged Concentrations in Filtered Water and Suspended Matter from Ponds with Different Species of Fish**

| Treatment (fish stacking) | Water sampling | Concentration (ppm) | | | | | | | |
|---|---|---|---|---|---|---|---|---|---|
| | | Ca | Mg | Na | K | Fe | Mn | Co | Zn |
| Bluegill + large- | a | 3.57 | 1.65 | 3.64 | 2.03 | 0.22 | 0.028 | 0.024 | 0.028 |
| mouth bass | b | 0.051 | 0.036 | 0.074 | 0.08 | 0.26 | 0.15 | 0.001 | 0.016 |
| Bluegill + bass + | a | 3.63 | 1.47 | 3.58 | 1.76 | 0.29 | 0.01 | 0.034 | 0.037 |
| amur | b | 0.034 | 0.024 | 0.065 | 0.51 | 0.04 | 0.04 | 0.004 | 0.019 |
| Bluegill + bass + | a | 2.37 | 1.23 | 5.02 | 1.73 | 0.62 | 0.02 | 0.04 | 0.032 |
| carp | b | 0.046 | 0.049 | 0.060 | 0.11 | 1.77 | 0.05 | 0.002 | 0.017 |
| White amur | a | 4.0 | 1.80 | 3.22 | 1.66 | 0.079 | 0.03 | 0.025 | 0.009 |
| | b | 0.05 | 0.024 | 0.08 | 0.045 | 0.271 | 0.03 | 0.004 | 0.009 |
| Common carp | a | 2.11 | 1.38 | 3.70 | 1.68 | 0.48 | 0.013 | 0.029 | 0.019 |
| | b | 0.054 | 0.074 | 0.084 | 0.092 | 2.04 | 0.11 | 0.002 | 0.022 |
| Amur + carp | a | 2.76 | 1.01 | 3.69 | 1.71 | 0.15 | 0.02 | 0.039 | 0.02 |
| | b | 0.053 | 0.057 | 0.065 | 0.144 | 1.57 | 0.09 | 0.004 | 0.014 |
| Channel catfish | a | 3.29 | 1.41 | 4.13 | 3.0 | 0.19 | 0.019 | 0.05 | 0.02 |
| | b | 0.04 | 0.016 | 0.054 | 0.04 | 0.47 | 0.03 | 0.003 | 0.02 |
| Catfish + amur[b] | a | 3.83 | 1.51 | 3.78 | 2.67 | 0.42 | 0.019 | 0.073 | 0.05 |
| | b | 0.064 | 0.077 | 0.066 | 0.16 | 2.55 | 0.20 | 0.005 | 0.021 |
| Catfish + carp | a | 3.9 | 1.05 | 3.5 | 2.72 | 0.18 | 0.012 | 0.04 | 0.028 |
| | b | 0.054 | 0.039 | 0.045 | 0.072 | 1.49 | 0.034 | 0.004 | 0.02 |

[a] Average concentration in filtered water.
[b] Average concentration in suspended matter.

in small earthern ponds. Generally the cause of variability has been poor survival of stocked bass or lack of reproduction by 1-year-old or older bass. Such conditions have produced populations of overcrowded bluegills, i.e., so many small bluegills survived that there was insufficient food for them to grow to catchable (15 cm) size. An examination of the individual draining records of the 0.10-ha combination ponds indicates the variability of bass survival occurring in all treatments in this study.

These pond populations are summarized by methods proposed by Swingle[1] describing the relationships and dynamics of balanced and unbalanced fish populations. A brief summary describing the meaning of terms used in this methodology of data evaluation is given below.

Balanced populations are defined as "those capable of producing satisfactory annual crops of harvestable fish. They were characterized as having (1) a definite range in ratio of the weights of forage and piscivorous species, (2) a narrow range in the ratio of weights of small forage fishes to those of piscivorous groups, and (3) more than 33% of the total population weight in the form of fishes of harvestable size."

The "C" value is the weight in pounds of "C" class species and the "F" value is the weight in pounds of the "F" class species. The range in F/C ratios in balanced fish populations was from 1.4 to 10.0. Populations with F/C ratios from 1.4 to 2.0 were overcrowded with "C" species. Balanced populations with F/C ratios below 3 were inefficient due to the overcrowding of "C" species. This condition was found to reduce the total population weight.

The F/C ratio was a relatively stable value (in ponds), remaining almost constant despite variations in rates of fishing for "F" and "C" species. This ratio is useful for comparison and determination of the condition of fish populations.

The "Y" value in a population is the total weight in pounds of all fishes in the "F"

TABLE 13.5

**Fish Population Relationships in Bluegill-Largemouth Bass Ponds**

| Factor | Pond F-17 | Pond F-20 | Pond F-23 |
|---|---|---|---|
| E | 75.3 | 84.6 | 93.1 |
| E | 24.7 | 15.4 | 6.9 |
| F/C | 3.17 | 5.51 | 13.55 |
| T/C | 1.89 | 3.00 | 8.05 |
| A | 53.9 | 52.2 | 44.7 |
| A | 33.4 | 44.4 | 40.1 |
| I | 7.0 | 1.2 | 0.6 |
| S | 59.6 | 54.4 | 59.4 |
| Kilograms largemouth bass per hectare | 43.5 | 29.6 | 19.4 |
| Total kilograms per hectare | 200.4 | 192.6 | 281.7 |

class small enough to be readily gulped by the average-size adult in the "C" class. The Y/C ratio is an expression of the food available to the "C" class. The most desirable populations were in the range Y/C = 1.0 to 3.0.

The "A" value is the percentage of total weight of a population composed of fish of a harvestable size. In balanced ponds, the range was from 33 to 90. The most desirable populations had "A" values between 60 and 85.

The "E" value of a species is the percentage of weight of a population composed of that species. The "F" class and also "F" species were subdivided into groups of large (fishes of harvestable size), intermediate (those too large to be eaten by the "C" species and too small for harvest) and small (fishes small enough to be eaten by the average-sized individual in the large group of "C" species in the population).

The "A" value is the percentage of the total weight of the "F" class composed of large fish. The "I" and "S" values are percentages of the total weight of the "F" class composed respectively of intermediate and small fishes.

An "A" = 35 appeared to be the minimum value found in desirable populations and apparently expressed the maximum allowable depletion of the adult "F" species if satisfactory production is to be maintained. The most desirable populations were in the range "A" – 60 to 80. Satisfactory populations occurred in the "S" value range from 15 to 40.

The "A", "I", and "S" values were found to be dynamic values shifting with changes due to harvest, predation, and natural mortality.

Pond studies indicated that harvest of adult "F" species increased the pounds of "C" species per acre and that failure to harvest the former groups caused a decrease in the pounds of "C" species in the population.

In the bluegill-bass only series, one pond had poor survival (28%) of stocked bass, but there was slight reproduction by the 1-year-old bass. This condition resulted in overcrowding of 5 to 7 cm group bluegills with an absence of individuals in the 2.5 cm group at draining time. In the other two ponds in this series, the bass survival was 88 and 68% respectively. The 1-year-old bass reproduced, and there were bluegills in all groups. Bluegill survival was 76, 78, and 98% respectively for the ponds listed below.

Fish popuation relationships in ponds stocked with bluegills and largemouth bass only are given in Table 13.5.

In the bluegill - largemouth bass - common carp series of ponds (F-16, F-19, F-22),

TABLE 13.6

**Bluegill-Largemouth Bass Population Relationships in Common Carp Ponds**

| Factor | Pond | | |
|---|---|---|---|
| | F-16 | F-19 | F-22 |
| E | 70.1 | 70.2 | 72.5 |
| E | 29.9 | 29.8 | 27.5 |
| F/C | 2.37 | 2.37 | 2.64 |
| Y/C | 1.17 | 0.5 | 0.2 |
| A | 79.8 | 90.3 | 96.8 |
| A | 27.0 | 58.2 | 63.2 |
| I | 23.6 | 20.1 | 29.1 |
| S | 49.5 | 21.7 | 7.7 |
| Kilograms largemouth bass per hectare | 30.0 | 28.3 | 22.5 |
| Kilograms carp per hectare | 72.3 | 66.7 | 70.8 |
| Total (kilograms per hectare) | 173.0 | 162.0 | 152.6 |

TABLE 13.7

**Bluegill-Largemouth Bass Population Relationships in White Amur Ponds**

| Factor | Pond | | |
|---|---|---|---|
| | F-4 | F-15 | F-21 |
| E | 71.9 | 64.0 | 87.1 |
| E | 28.6 | 36.0 | 12.9 |
| F/C | 2.54 | 2.27 | 6.74 |
| Y/C | 1.69 | 1.12 | 0.5 |
| A | 62.5 | 71.1 | 94.9 |
| A | 21.6 | 20.7 | 86.5 |
| I | 11.8 | 30.1 | 6.3 |
| S | 66.6 | 49.2 | 7.4 |
| Kilograms largemouth bass per hectare | 37.0 | 51.9 | 15.1 |
| Kilograms amur per hectare | 37.1 | 32.3 | 36.3 |
| Total kilograms per hectare | 168.2 | 168.0 | 134.8 |

stocked bass survival was 88, 68, and 84% respectively, but reproduction by 1-year-old bass occurred in only one pond. Stocked bluegill survival was 48, 63, and 79% respectively. Bluegill reproduction occurred in all three ponds in 1973, but was practically eliminated in 1974.

Bluegill - largemouth bass population relationships in ponds stocked with common carp are given in Table 13.6.

In the bluegill-largemouth bass-white amur series (F-4, F-15, and F-21), stocked bass survival was 76, 82, and 64% respectively, and bass reproduction occurred in only two of the three ponds. The stocked bluegill survival was 41, 77, and 89% respectively. Bluegill reproduction occurred in all three ponds in 1973, but was virtually eliminated throughout 1974.

Bluegill-largemouth bass population relationships in ponds stocked with white amur are given in Table 13.7. At draining time, largemouth bass production in the bluegill-bass only ponds averaged 32 kg/ha, in the bluegill-bass-white amur ponds the bass production was 30.9 kg/ha, and in the bluegill-bass-common carp combinations the bass production was 26.8 kg/ha. Statistical analyses of these production data did not indicate any significant difference in bass production among treatments.

At draining time, average bluegill production in bluegill-bass only ponds averaged 191 kg/ha, in bluegill-bass-white amur ponds 93 kg/ha, and in bluegill-bass-common carp ponds 65 kg/ha. There was a significant decrease in bluegill production in both common carp and white amur ponds when compared with production in bluegill-bass only ponds. This overall decrease apparently was correlated with significant reduction in weight of 5 to 7 cm group bluegills in these same ponds.

An analysis, including comparisons with previous bluegill-bass combinations in these 0.10 ha ponds, of the bluegills reproductive success in the presence of white amur or common carp indicated that reproductive interference probably started during the latter part of the first growing season (late summer, 1973). At this time, the common carp and white amur were of sufficient size (average weight 0.45 kg per individual) to present a physical hazard to bluegill bedding. In common carp ponds, the food for bluegills was limited due to competition and occlusion of light by turbidity. All of these factors interacted in the respective ponds where they were present and resulted in drastic reduction of small bluegills. Interference with bluegill reproductive success by both white amur and common carp during the second growing season almost eliminated production of small fish.

The general conclusion drawn from these production data would be that in a weed-free pond, a population of 45 white amur/ha in combination with bluegills and largemouth bass, would interfere with bluegill reproduction. This would eventually result in decreased largemouth bass production since their food supply (small bluegills) would be significantly reduced.

Common carp, at a rate of 45 per hectare in combination with bluegills and largemouth bass would also interfere with bluegill reproduction. This situation would eventually decrease bass production.

In large impoundments, the largemouth bass depend mainly on gizzard and threadfin shad for food. The presence of large numbers of either white amur or common carp in these habitats would probably interfere with sunfish reproduction, but this would have limited effects upon bass production.

The exact mechanism whereby white amur and common carp interfered with bluegill (and other sunfish) reproduction is unknown. Based on knowledge of the behavioral characteristics of the species involved in these pond studies, the following situations might have existed.

White amur and common carp do not feed extensively until water temperatures reach or exceed 20°C. Largemouth bass spawn at approximately 20°C; therefore, their nesting activities might have been subject to limited interference by amur or carp.

Bluegills start bedding at approximately 26°C at which time both amur and carp are actively feeding. Both species travel in schools and were continually patrolling the shallow pond edges. Thus, their consistent periodic physical presence over bedding areas might have repelled the sunfish. It is known that excessive disturbances in sunfish bedding areas will cause sunfish to abandon their nests. The extensive feeding by common carp during those periods when bluegill were spawning could have produced silt sufficient to interfere with hatchability of bluegill eggs. No evidence was obtained that either amur or carp consumed sunfish eggs or fry, but this does not eliminate it as a possible factor for reducing sunfish production in these ponds.

## COMMERCIAL CATFISH PRODUCTION

In ponds containing only channel catfish, the average production was 1175 kg/ha. Survival of the stocked catfish ranges from 97 to 100%.

In the fed channel catfish-white amur ponds, production of channel catfish averaged 1131 kg/ha. The survival of stocked catfish ranged from 94.5 to 97.5%. The averaged production of the white amur in these catfish ponds was 135 kg/ha. Total catfish plus white amur production averaged 1265 kg/ha.

In the fed channel catfish-common carp experiment, the entire fish population in one pond was killed by oxygen depletion in June 1974. Total catfish production to that time was 888 kg/ha, and for common carp it was 164 kg/ha. Survival of catfish in this pond was 96.5% prior to kill. In the remaining channel catfish-common carp pond, a total of 1235 kg/ha of catfish was recovered by draining in October 1974. Survival of catfish in this pond was 99.5%. The common carp production in this pond was 77 kg/ha. The common carp in each of the catfish ponds reproduced in spring of 1973, but there was no carp reproduction in 1974.

Based on data obtained in these limited tests, there was no evidence that white amur or common carp caused any significant decrease in catfish production.

## WHITE AMUR-COMMON CARP PRODUCTION

White amur (16/ha) plus common carp (16/ha) were stocked in a 0.10-ha pond and in a 0.5-ha pond. The 0.10-ha pond was fertilized, and production of common carp and white amur was 147 and 19 kg/ha, respectively. The 0.05-ha pond was fed with commercial catfish feed, and production of common carp and white amur was 291 and 95 kg/ha, respectively.

Common carp (161 per hectare) were stocked alone in two 0.10-ha ponds. These ponds received inorganic fertilization. Production in these ponds ranged from 49 to 170 kg/ha. In the pond with lowest production, a population of green sunfish existed. These fish apparently prevented reproduction of the population of common carp. In the other pond, the carp reproduced in 1973, but did not spawn in 1974.

White amur (161 per hectare) were stocked alone in two other 0.10-ha ponds. These ponds received inorganic fertilization. Production of white amur in these ponds ranged from 60 to 69 kg/ha. In the pond with lowest production, there was a population (30 kg/ha) of green sunfish. However, these sunfish did not appear to significantly interfere with white amur growth.

## SUMMARY

In bluegill-largemouth bass combinations in weed-free ponds, the addition of 16 white amur or common carp per hectare significantly reduced (52 and 66% respectively) production of bluegill. This reduction apparently resulted from interference with bluegill reproduction since in both combinations, there was a significant decrease in bluegills.

The exact mechanisms of reproductive interference are unknown, but it is postulated that the chief cause was physical, i.e., the constant presence of cruising white amur or common carp over bluegill beds frightened the bluegills and caused them to leave the area.

In fed channel catfish ponds, the addition of 16 white amur per hectare caused no significant decrease in catfish production. White amur presence increased total pond production approximately 91 kg/ha over that obtained by channel catfish alone. The addition of 161/ha common carp into fed channel catfish ponds resulted in a slightly higher production of catfish and an overall gain in pond production of approximately 136 kg/ha.

In these studies, the bluegill-largemouth bass populations were subjected to stressed conditions by high densities (161/ha) of white amur or common carp. This number of either the amur or carp reduced bluegill production in weedless ponds. It is still common practice to use 2, 5, or as many as 10 white amur per hectare for weed control in ponds. The effects these stocking rates may have on bluegill production in weedy ponds are unknown.

## REFERENCE

1. **Swingle, H. S.,** Relationships and dynamics of balanced and unbalanced fish populations, *Auburn Univ. Exp. Stn. Bull.*, 274, 21, 1950.

Chapter 14

# WEED CONTROL OF THE WHITE AMUR*

## INTRODUCTION

Studies undertaken at Auburn University during the past 10 years to determine the value of herbivorous fish as biological control agents indicate that the white amur (*Ctenopharyngodon idella* Val.) has greater potential for that purpose than other species studied so far. In these tests, species of *Chara, Potomogeton, Eleocharis,* and *Rhizoclonium* have been well controlled. None of the ponds tested became muddy or developed phytoplankton blooms as a result of the feeding activities of the white amur.[1]

Although the white amur has been studied in various research programs in the U.S., it has not been widely accepted for aquatic weed control, seemingly because of fear that it might become naturalized and cause a problem in freshwater fishing resources. Though some potential that this fish could become naturalized exists, there is no particular evidence from observations in the U.S. or other countries that this would be the case.

The environmental benefits derived from use of a biological control agent like the white amur suggest that more definitive and extended research be done to find means to utilize this potential in solving aquatic weed problems that have reached major proportions in some situations.

Previous experiments with white amur as a biological weed control agent showed that the stocking rate of the white amur is an important factor in determining the degree of aquatic weed control. It has been shown that white amur will eat a wide variety of aquatic plants and will effectively control many. However, most experiments neglected to measure effects of the white amur on other fish species. This must be done if the white amur is to be considered for aquatic weed control in the U.S. This experiment attempted to determine effects of white amur at various stocking rates on (1) channel catfish and striped bass, and (2) aquatic plants: water hyacinth, *Spirogyra,* and *Hydrodictyon.* An attempt was also made to determine effects of aquatic plants on these fish species.

## MATERIALS AND METHODS

Twelve E series ponds were used in the experiment. The experimental design was completely randomized to minimize effects due to location and previous use of each pond. Treatments, including stocking rates of fishes and aquatic plants, are listed in Table 14.1.

By April 6, 1970 full ponds were prepared by treating them with 10 lb of HTH-brand® compound, a chemical-containing chlorine used to disinfect swimming pools. Each pond was fertilized monthly with the equivalent of 8-8-2 inorganic fertilizer (112 kg/ha). All fish stocked were measured (total length [TL], mm) and weighed (g) individually.

Taking seine samples in several ponds on June 10 revealed that common carp, golden

* By Lawrence, J. M., Agricultural Experiment Station, Auburn University, Auburn, Ala. From Technical Report 4, Aquatic Plant Control Program, U.S. Army Engineers Waterways Experimental Station, Vicksburg, Miss., June 1973.

TABLE 14.1

**Treatments in Aquatic Weed Control Experiments with White Amur[a]**

| Treatment number | E pond number | Number of fish per hectare: Channel catfish[b] | White amur[c] | Largemouth bass[d] | Weight of aquatic plants per hectare (kg): Water hyacinth[e] | Spirogyra[f] | Hydrodictyon[f] |
|---|---|---|---|---|---|---|---|
| 1 | 2, 11, 12 | 560 | 0 | 56 | 0 | 0 | 0 |
| 2 | 5,[g] 7, 10 | 560 | 0 | 56 | 560 | 56 | 56 |
| 3 | 4, 6,[g] 8 | 560 | 45 | 56 | 560 | 56 | 56 |
| 4 | 1, 3,[g] 9 | 560 | 90 | 56 | 560 | 56 | 56 |

[a] E ponds, 1970.
[b] Stocked May 1, 1970.
[c] Stocked May 21, 1970.
[d] Stocked April 30, 1970.
[e] Weight obtained after drying out on the ground for 72 hr, although 1 in. of rainfall occurred in 60 hr.
[f] Wet weight.
[g] Drained October 26, 1970. (All others were drained October 27, 1970.)

shiners, and threadfin shad had been introduced into the ponds as eggs attached to water hyacinths. To regain some degree of uniformity among ponds within treatments, 190 g of Fintrol® (antimycin) were added to each pond. This concentration does not kill catfish; however, the other fish were killed. The dead fish were removed, identified, measured, weighed, and counted for a 3-day period (June 16 to 19). White amur were restocked on June 29. For unknown reasons, pond E1 was not treated with Fintrol®. This was not discovered until the ponds were drained, and pond E1 was stocked with white amur twice. It was thus considered as a fifth treatment, with a stocking rate of 180 white amur per hectare, channel catfish, largemouth bass, and striped bass. Striped bass fingerlings were added to all ponds when the white amur were restocked. Largemouth bass were not restocked. For each pond, the extent of surface area covered by water hyacinth was recorded weekly when possible.

## DISCUSSION OF RESULTS

The treatments in E ponds for aquatic weed control experiments with the white amur are listed in Table 14.1. For each treatment, various data pertaining to fishes stocked and their growth are summarized in Table 14.2.

Apparently, white amur in ponds containing water hyacinths and channel catfish did not interfere with catfish growth, yet controlled the plants to the extent that catfish actually showed a slight increase in standing crop over those catfish in control ponds. Obviously, the most important limiting factor on growth and standing catfish crops in this experiment was not white amur presence, but the presence and size of water hyacinths in the ponds. This was also true c. striped bass. The white amur ate the water hyacinth roots (actual observations) and thus controlled the size of the plants.

Aside from the single replicate of white amur stocked at the rate of 180 per hectare, water hyacinths were not completely controlled over a period of 146 days. However, the stocking rate of 45 per hectare seemed most efficient in controlling water hyacinths. Perhaps this rate (45 per hectare) over a period of 2 or more years, with the aid of either partial or total winter kill or chemical plant control, might eliminate water hyacinths.

Lowest catfish survival (73%) occurred in ponds containing no white amur but having aquatic plants. This indicates that perhaps the largest size of water hyacinths (275 g) interfered with the ability of the catfish to seek and capture food organisms. The average catfish weight at draining was only slightly greater in ponds containing white amur, but was greatest (120 g) in control ponds without plants and white amur. The second greatest average weight of catfish (103 g) occurred in the single replicate of Treatment 5, where there were no water hyacinths after August 28. This occurred in spite of a high standing common carp crop in the pond. Catfish standing crops were in the same order. For white amur, average weight gains and actual weights at draining were greater in ponds stocked at the rate of 45 per hectare. Survival rates for striped bass were similar in all ponds except Treatment 5. Average weights were greatest in Treatments 1 and 5.

Aquatic plants, *Spirogyra* and *Hydrodictyon,* never became firmly established in the ponds. Consequently, they have been disregarded in the experiment. Water hyacinths rapidly became a nuisance in all ponds where stocked. No clear-cut distinctions were discernible except in the single pond in Treatment 5 when water hyacinths were completely eliminated within 100 days.

Standing crops of water hyacinths in control ponds (no white amur) and those stocked at the rate of 90 white amur per hectare were approximately equal, although the average size of each water hyacinth plant was 62% greater in the control ponds.

TABLE 14.2

**Summary of Fish Data Obtained in Aquatic Weed Control Experiments with White Amur[a]**

| | Channel catfish treatment number | | | | | White amur treatment number | | | | | Striped bass treatment number | | | | | Large-mouth bass treatment number | Total treatment number | | | | |
|---|---|---|---|---|---|---|---|---|---|---|---|---|---|---|---|---|---|---|---|---|---|
| | 1 | 2 | 3 | 4 | 5 | 1 | 2 | 3 | 4 | 5 | 1 | 2 | 3 | 4 | 5 | 5 | 1 | 2 | 3 | 4 | 5 |
| Number stocked (per ha) | 56 | 56 | 56 | 56 | 56 | 0 | 0 | 5 (5) | 9 (9) | 18 | 12 | 12 | 12 | 12 | 12 | 6 | 67 | 67 | 71 (76) | 76 (685) | 91 |
| Number recovered | 55 | 41 | 55 | 55 | 56 | — | — | 3 (3) | 9 (7) | 18 | 8 | 7 | 6 | 8 | 3 | 3 | 63 | 48 | 63 (70) | 71 (77) | 80 |
| Survival (%) | 98 | 73 | 98 | 96 | 100 | — | — | 100 (67) | 100 (69) | 100 | 70 | 57 | 50 | 70 | 30 | 60 | 93 | 72 | 91 (90) | 93 (91) | 89 |
| Average weight stocked (g) | 25 | 25 | 26 | 26 | 25 | — | — | 183 (144) | 228 (150) | 162 | 1.7 | 1.7 | 1.7 | 1.7 | 1.7 | (fry) | — | — | — | — | — |
| Average weight at draining (g) | 120 | 78 | 87 | 81 | 103 | — | — | 625 | 568 | 190 | 26 | 8.3 | 13.1 | 9 | 45 | 127 | — | — | — | — | — |
| Weight increase (g) | 95 | 53 | 62 | 56 | 78 | — | — | 442 (208) | 340 (146) | 28 | 24.3 | 6.6 | 11.4 | 7.3 | 43.3 | 127 | — | — | — | — | — |
| Weight increase (%) | 380 | 212 | 239 | 216 | 436 | — | — | 242 (145) | 149 (97) | 17 | 1440 | 388 | 670 | 430 | 2550 | 12,700 | — | — | — | — | — |
| Total weight stocked (kg/ha) | 29 | 31 | 33 | 32 | 30 | — | — | 18 (14) | 45 | 64 (30) | 4 | 4 | 4 | 4 | 4 | T[b] | 30 | 31 | 66 | 107 | 95 |
| Standing crop (kg/ha) | 144 | 70 | 106 | 96 | 128 | — | — | 61 (24) | 112 (40) | 75 | 5 | 1 | 2 | 1 | 3 | 9 | 182 | 154 | 309 | 312 | 383 |
| Days in pond | 179 | 179 | 179 | 178 | 179 | — | — | 120 (26) | 120 (26) | 134 | 119 | 119 | 118 | 118 | 119 | 178 | — | — | — | — | — |
| Average gain per fish (g/day) | 0.53 | 0.30 | 0.35 | 0.31 | 0.43 | — | — | 3.7 (8.0) | 2.8 (0.6) | 0.0 | 0.20 | 0.05 | 0.10 | 0.06 | 0.36 | 0.71 | — | — | — | — | — |

*Note:* Numbers in parentheses (Treatments 3 and 4) pertain to white amur stocked on May 21 and harvested (dead) by applying Fintrol® on June 16, 1970.

[a] 1970, E ponds (includes mean values, by treatment).
[b] Trace, less than 0.01 kg/day.

From Sutton, D. L. and Blackburn, R. D., Feasibility of the Amur (*Ctenopharyngodon idella* Val.) as a Biocontrol of Aquatic Weeds in Technical Report No. 4, Herbivorous Fish for Aquatic Plant Control, U.S. Army Engineer Waterways Experiment Station, Vicksburg, Miss., 1973.

The lowest standing crop (excluding Treatment 5) occurred in ponds stocked at the rate of 45 white amur per hectare.

## REFERENCE

1. **Avault, J. W., Smitherman, R. D., and Shell, E. V.,** Evaluation of eight species of fish for aquatic weed control, in Proc. World Symp. Warm Water Pond Fish Culture, Food and Agriculture Organization, Geneva, 1968, 109.

Chapter 15

# FOOD HABITS OF THE WHITE AMUR (*CTENOPHARYNGODON IDELLA*)*

## INTRODUCTION

It becomes increasingly important that biological methods be used for controlling undesirable aquatic plants in artificial impoundments and natural waters since many of our existing chemical control methods are potentially hazardous to the ecological balance of ponds, lakes, rivers, and to man himself. A number of organisms exist that can effectively control or eliminate certain aquatic plants. Several fish species have shown promise as weed control agents. One of these is the white amur, *Ctenopharyngodon idella* Valenciennes, also known as grass carp.[1] This fish has been in the U.S. for a decade, but it is an exotic species, indigenous to large rivers in China. One major reason that white amur are not widely used in this country is the fear that they may become established in our natural waters, as did the common carp, and compete for food organisms with native fish. Hora and Pillay[2] reported that white amur do not spawn in ponds but in large rivers. If this is so, and it seems to be, they can possibly be used in ponds not connected to natural waters.

Hora and Pillay[2] classified white amur as omnivorous and stated that it has a distinct preference for vegetable food, such as grass, leaves, weeds, and all kinds of aquatic plants. Young fry feed on zooplankton. Stevenson[3] found that small (80 to 120 mm) white amur preferred *Daphnia,* chopped earthworms, and chironomid larvae to canned spinach, *Spirogyra,* or *Eleocharis* when the fish were kept in aquaria. When transferred to earthen ponds, the fish frequently ate cut blades of Bermuda grass floating on the pond surface. It was also reported[2] that in 60 days, 24 white amur per acre reduced, but did not completely eliminate, *Chara, Najas, Eleocharis,* and *Polygonum.* Thirty white amur per acre eliminated *Chara* and *Najas,* but not *Anacharis,* within 90 days. However, in both cases, the fish were given commercial fish feed.

Avault[4] reported that *Chara, Eleocharis,* and *Potamogeton* were eliminated from 0.1-acre earthen ponds within 90 days by 40 white amur (12 to 16 in.) per acre. Cross[5] compiled a list of aquatic plants eaten by white amur, in order of preference. He found that 9-in. white amur ate *Daphnia,* tubifex worms, and *Asellus,* an isopod crustacean, as well as vegetation, when in aquaria.

Avault et al.[1] found that white amur (250 to 400 mm) stocked at rates of 20 to 40 per acre controlled a wide variety of aquatic weeds. They recommended use of the lowest stocking rate at which weeds are controlled. Cross[5] reported that under certain conditions, white amur will eat more than their own weight of plant material in a day. Stott and Orr[6] estimated that the conversion rate of aquatic weeds (wet weight) to fish flesh by white amur is about 224:1.

In the present study, white amur were tested in two types of stocking — alone in fertilized ponds with aquatic plants added and stocked with other fish species in a fertilized pond with little natural macrophyte vegetation. Their effectiveness in aquatic weed control and their food habits compared with other fishes in the population were observed.

* By Kilgen, R. H. and Smitherman, R. O., Department of Fisheries and Allied Agricultures, Auburn University, Auburn, Ala. From Technical Report 4, Aquatic Plant Control Program, U.S. Army Engineers Waterways Experimental Station, Vicksburg, Miss., June 1973.

TABLE 15.1

**Stocking Rates, Average Weights and Lengths, and Standing Crops of White Amur, by Treatment[a]**

| Treatment number | Number fish per acre | Average total length (mm) | Average weight (g) | Average standing crop (lb/acre) |
|---|---|---|---|---|
| 1[b] | 0 | — | — | 0 |
| 2 | 20 | 175 | 64 | 3 |
| 3 | 40 | 176 | 64 | 6 |
| 4 | 80 | 184 | 76 | 13 |
| 5 | 160 | 186 | 78 | 27 |

[a] July 16, 1969.
[b] Control ponds, no fish.

## MATERIALS AND METHODS

### White Amur Alone

Ten 0.1-acre earthen ponds at the Fisheries Research Unit, Auburn University Agricultural Experiment Station, Auburn, Ala. were used for this experiment. Aquatic plants were obtained from Lake Seminole, Ga. on June 11, 1969, and stocked in ponds the following day. Eurasian milfoil (*Myriophyllum spicatum*), collected from a relatively pure stand of plants, was stocked into each pond at the rate of 900 lb/acre. Two other species, alligator weed (*Alternanthera philoxeroides*) and water hyacinth (*Eichhornia crassipes*), were entangled and inseparable. A combined weight of 800 lb/acre was stocked into each pond. Two other plant species, *Potamogeton diversifolius* and *Chara* spp., were added to the ponds on June 30 at the rates of 400 and 100 lb/acre, respectively. The plants were obtained from other ponds on the station. Ponds were fertilized on June 17, June 30, and July 24 with $NH_4NO_3$ at the rate of 75 lb (8-0-0) per acre.

White amur (age I+, size 7 in.) were stocked on July 16 about 1 month after the first plants were stocked. Although some ponds had poor plant growth, most ponds had an adequate amount of plants for testing white amur. Stocking rates, average weights and lengths, and standing crops are listed by treatment in Table 15.1

Records were kept on the amount of plants in each pond to determine efficacy of different stocking rates of white amur. The areal percent composition covered by each species was estimated.

On August 18, one white amur was removed from each pond for analysis of its stomach contents. Each fish was replaced by another white amur somewhat smaller in size (average weight, 97 g). When ponds were drained, stomachs of one half the number of white amur in each pond were removed and preserved for analysis. Since it was difficult to distinguish different intestine parts, the entire gut was considered as the stomach.

### White Amur in Combination with Other Species

A 5.2-acre earthen pond (S15) was stocked with a combination of 500 channel catfish, *Ictalurus punctatus* (5 in.); 50 largemouth bass, *Micropterus salmoides* (1 in.); 25 redeye bass, *M. coosae* (3 to 6 in.); 25 spotted bass, *M. punctulatus* (3 to 6 in.); 50 Israeli carp, a variety of the mirror strain of common carp, *Cyprinus carpio* (6 to 11 in.); 1000 fathead minnows, *Pimephales promelas* (1 to 3 in.); and 50 white amur (5 to 7 in.) per acre during February to May 1968. This experiment was designed to evaluate competition and growth of these species when stocked together.

The pond received superphosphate fertilization during 1968 and 1969. In addition, experiments with cage culture of channel catfish and also of *Tilapia aurea* were conducted in the pond in 1969. Wastes from the feed and feces supplied organic fertilization. Fish were fed 1.25 lb of a trout ration daily from June 1 to June 25, 1969.

All species were collected from May to September 1969 by angling or seining. Stomachs were immediately removed and preserved for content analysis. Food habits of fathead minnows were not studied.

## RESULTS

### White Amur Alone

All ponds were drained on October 23, 1969, 133 days after the first plants were stocked and 99 days after the fish were stocked. An analysis of average length and weight variance of the white amur at stocking time showed that the sizes of the smallest and largest fish were not significantly different (5% level). There were significant differences in sizes of fish recovered at draining. Their average lengths, weights, survival, and standing crops are listed by treatment in Table 15.2 which also includes standing crops of water hyacinth and alligator weed and presence or absence of other plant species.

Water hyacinth and alligator weed were weighed the day after draining the ponds (after allowing them to dry out for the same amount of time as prior to stocking — 24 hr). Other plant species could not be easily removed, and the effect of white amur on these plants could only be evaluated by their presence or absence.

Figure 15.1 shows the stomach contents of fish taken on August 18 and October 23. No attempt was made to identify insects found in the stomachs, since only pieces of mature insects were found. No insect larvae were found in any of the stomachs.

### White Amur in Combination with Other Species

Pond S15 was drained on December 10, 1969. The fish recovered, percent survival, standing crops, and ranges in length and weight are listed in Table 15.3. Figure 15.2 shows the stomach contents of adult fish in the pond for May through September 1969.

## DISCUSSION

### White Amur Alone

The average weight (at draining) of white amur stocked at the rate of 20/acre was significantly higher (5% level) than those stocked at rates of 40, 80, and 160/acre. They grew from 64 g at stocking to 552 g at draining (99 days).

Stomach analyses revealed white amur will resort to eating insects when plants are no longer available. This was more evident when bottom vegetation was eliminated and floating plants were unavailable due to their growth in shallow water. Highest standing crops of water hyacinths were produced in control ponds. Lowest standing crops of hyacinths were present in ponds containing 40 white amur per acre, but higher standing crops were present in ponds with 80 and 160/acre. This was due to plants being rooted in shallow water around pond edges and thus being unavailable to the fish.

Alligator weed standing crops were similar in all ponds, regardless of white amur stocking rates. Eurasian milfoil, *Chara,* and *Potamogeton* were eliminated in 99 days or less in ponds stocked with 40 or more white amur per acre. Rates of white amur stocking which will effectively control floating weeds could not be determined due to shallow water around pond edges.

TABLE 15.2

**Average Lengths, Weights, Percent Survival, and Standing Crops of White Amur, Standing Crops of Water Hyacinth and Alligator Weed, and Presence or Absence of Other Plant Species (by Treatment)[a]**

| Treatment number | Number fish per acre | Average total length (mm) | Average weight (g) | Survival (%) | Average standing crop (lb/acre) | | | Other Plants[b] | | |
|---|---|---|---|---|---|---|---|---|---|---|
| | | | | | White amur | Water hyacinth | Alligator weed | Milfoil | *Chara* | *Potamogeton* |
| 1 | 0 | — | — | — | 0 | 7320 | 100 | P | P | P |
| 2 | 20 | 356 | 552 | 100 | 24 | 4090 | 50 | P | P | P |
| 3 | 40 | 281 | 283 | 100 | 25 | 440 | 50 | A | A | A |
| 4 | 80 | 290 | 297 | 88 | 46 | 1080 | 50 | A | A | A |
| 5 | 160 | 238 | 154 | 100 | 54 | 3440 | 80 | A | A | A |

[a] October 31, 1969.
[b] P denotes presence of plants at draining; A denotes absence of plants at draining

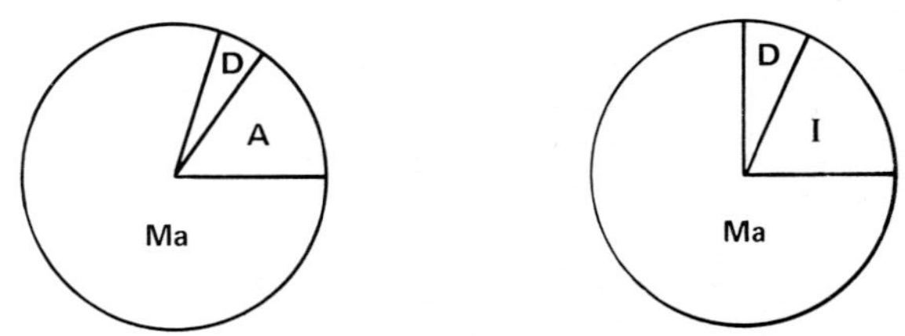


August 18
(6)

October 23
(29)

FIGURE 15.1. Stomach contents of white amur stocked alone in ponds. (Insects were all mature forms.) Code to food categories: A — algae, D — debris, I — insects, Ma — macrophytes. Number of fish is shown in parentheses.

TABLE 15.3

**Fish Recovered, Percent Survival, Standing Crops, and Ranges of Lengths and Weights in Pond S15, Including Fish Taken by Angling and at Draining**

| Species | Survival (%) | Standing crop (lb/acre) | Range in length (in.) | Range in weight (lb) |
|---|---|---|---|---|
| Channel catfish | 79 | 311[a] | 6.20[a] | 0.05—2.8[a] |
| Largemouth bass | 38 | 21[a] | 5—14[a] | 0.06—1.5[a] |
| Redeye bass | 61 | 3[a] | 4—9[a] | 0.03—0.3[a] |
| Spotted bass | 52 | 6 | 11—13 | 0.55—0.8 |
| Israeli carp | 89 | 224 | 19—23 | 4.4—10.0 |
| Fathead minnow | — | 0.3 | — | — |
| White amur | 81 | 96 | 16.22 | 1.3—5.1 |
| Total | | 661.3 | | |

[a] Includes reproduction.

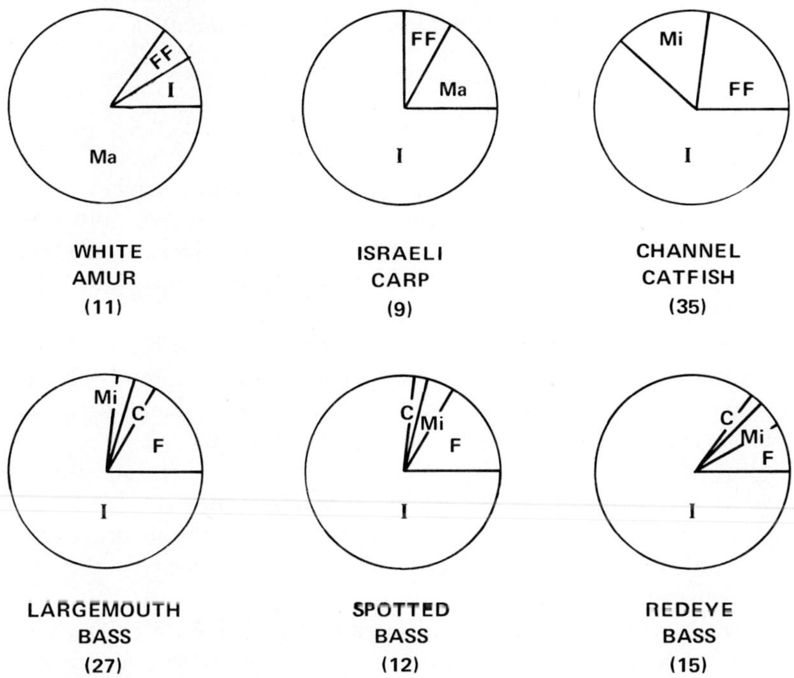


FIGURE 15.2. Stomach contents of fish taken by angling from a 5.2-acre pond during May through December 1969. Code to food categories: C — crustacea, F — fish, FF — fish feed, I — insects, Ma — macrophytes, Mi — miscellaneous. Number of fish shown in parentheses.

## White Amur in Combination with Other Species

White amur in pond S15 grew from an average of 60 g (0.13 lb) to 1470 g (3.24 lb) in 617 days (20.5 months). Although they made up 7% of the total number of fish stocked (including fathead minnows), they comprised 14.5% (by weight) of the total standing crop. Their food (Figure 15.2) consisted mainly of macrophytes (84% by volume) with only a relatively small amount of insects (9%), mainly chironomid larvae and, rarely, *Chaoborus* larvae and chironomid adults. The plant material in white amur stomachs appeared as shredded blades of grass. It is very likely that this material was centipedegrass (*Eremochloa ophiuroides*), a lawn grass whose roots and stolons occasionally trail into the water at the pond edge. On several occasions, white amur were observed with their heads out of water, nibbling at overhanging grasses. No aquatic macrophytes were available to the fish at any time during the experiment. The pond bottom was completely clear of vegetation at draining, but many chironomid larvae were seen.

Israeli carp (Figure 15.2) preferred insects (75%) — mainly *Chaoborus* larvae and chironomid larvae and adults — to macrophytes (16%). In general, a more insectivorous habit was indicated by Israeli carp.

The largemouth and spotted bass (Figure 15.2) had similar stomach contents (77% insects and 17% fish). Redeye bass (Figure 15.2) ate fewer fish (9%) and more insects (85%). The stomach contents of channel catfish (Figure 15.2) consisted mainly of insects (62%), but supplemental feed made up a significant amount (23%).

In general, it seems that food habits of white amur are not at all similar to the other fish species in this experiment. Although white amur did eat aquatic insects, this food item comprised only a very small amount by volume of stomach contents (9%). Similar to the case where they were stocked alone, white amur resorted to eating insects only when absolutely necessary. This was evident in this experiment in which virtually no plant material was available.

## CONCLUSIONS

White amur, when stocked alone at rates of 45 (or more) per hectare, effectively eliminated *Chara* spp., *Potamogeton diversifolius,* and Eurasian milfoil in less than 99 days and decreased the amount of water hyacinths in 0.1-acre earthen ponds. Under these conditions, their diet consisted mainly of macrophytes and algae (75 to 95% by volume) and only a small amount of mature insects (0 to 18%).

When stocked with channel catfish, Israeli carp, and three basses — largemouth, redeye, and spotted — the white amur diet consisted of 84% macrophytes and only 9% insects (larvae). In this case, with virtually no aquatic plants in the pond, white amur resorted to nibbling grass roots and stolons at the pond edge.

These findings demonstrate benefits of using white amur in situations where biological methods of weed control are preferable to chemical methods. They also indicate that white amur do not pose a threat as a competitor for food organisms eaten by game fish. They also provide an excellent food source. As game fish, they can be taken by angling using worms as bait.

The pharyngeal teeth (Figure 15.3) and the horny pad (Figure 15.4) are particularly adapted for masticating aquatic vegetation. A white amur approximately 900 g in weight is shown in Figure 15.5.

FIGURE 15.3. Pharyngeal teeth of a large white amur. (From Gangstad, E. O., Technical Report 4, Aquatic Plant Control Program, U.S. Army Engineers Waterway Experimental Station, Vicksburg, Miss., June 1973. With permission.)

FIGURE 15.4. Pharyngeal teeth with the horny pad reversed to show its hard surface. (From Gangstad, E. O., Technical Report 4, Aquatic Plant Control Program, U.S. Army Engineers Waterway Experimental Station, Vicksburg, Miss., June 1973. With permission.)

FIGURE 15.5. A white amur approximately 900 g in weight. (From Gangstad, E. O., Technical Report 4, Aquatic Plant Control Program, U.S. Army Engineers Waterway Experimental Station, Vicksburg, Miss., June 1973. With permission.)

## REFERENCES

1. **Avault, J. W., Jr., Smitherman, R. O., and Shell, E. W.,** Evaluation of eight species of fish for aquatic weed control, in *Proc. World Symp. Warm-Water Pond Fish Culture,* FAO Fish. Rep., 44, VII/E-3, Vol. 5, Food and Agriculture Organization, Rome, 1968, 109.
2. **Hora, S. L. and Pillay, T. V. R.,** Handbook of fish culture in the Indo-Pacific region, *FAO Fish. Tech. Pap.,* 14, 1962.
3. **Stevenson, J. H.,** Observations on grass carp in Arkansas, *Prog. Fish Cult.,* 27(4), 203, 1965.
4. **Avault, J. R., Jr.,** Preliminary studies with grass carp for aquatic weed control, *Prog. Fish Cult.,* 27(4), 207, 1965.
5. **Cross, D. G.,** Aquatic weed control using grass carp, *J. Fish Biol.,* 1(1), 27, 1969.
6. **Stott, B. and Orr, L. D.,** Estimating the amount of aquatic weed consumed by grass carp, *Prog. Fish Cult.,* 32(1), 51, 1970.

Chapter 16

# ENVIRONMENTAL ASPECTS OF WHITE AMUR CULTURE

## INTRODUCTION

Common names of the Chinese grass carp are Wuan Ue, Ts'ou yu, Ts'yu, Waan yu, ca Cham, Chow hu, Sogyo, and white amur. The scientific name is *Ctenopharyngodon idella* (Val). The original habitat of the white amur includes the rivers draining to the Pacific between the Amur River, bordering northern China, approximate latitude 50 N, and the West River, or SiKiang, approximately latitude 23 N, and in all the rivers between. The habitat area also includes the middle and lower Amur River (as far upstream as Blagouschensk), Sungari, Ussuri, Lake Khanka, China southward to Canton and is found in rivers, lakes, and ponds of Siberia (Amur region), Manchuria, North China, and South China.[7,29,41]

The white amur has spawned naturally in certain rivers of Taiwan, Japan, and the Ukraine and has become naturalized in the River Tone in Japan. It is presently found in Malaysia, Taiwan, Japan, the U.S., eastern Europe, Holland, Germany, and south, central, and north China. It is grown for food in Thailand, Malaysia, Japan, Ceylon, and the U.S.S.R.[17] It has recently been introduced into Israel and Europe.[19-23] In autumn of 1963, 6000 white amur and silver carp fingerlings were taken from a farm near Moscow and reared in ponds in Turkmen SSR (Charezlon region, Krasnodar Territory, Moscow region, Leningrad region, Byelorussia).[25,26] Over 1500 white amur and silver carp were released into the Volga in November 1955 to 1956 and were found to survive. White amur (35 cm and 61 cm) were taken in the Kutum River in 1957. A 4- to 5-kg white amur was caught in the Ural delta in 1959.[30]

The progeny of white amur brought from the Amur River were first bred in the U.S.S.R. in 1961 in the Turkmen Republic. In 1964 to 1965, the experimental fish hatchery "Goryachy Kluch" (Krasnodar Territory) exported approximately 90 million larvae of white amur, silver carp, and bighead to various regions of the U.S.S.R., Poland, Czechoslovakia, Hungary, and Bulgaria.[40] Currently the white amur is widely used for food and weed control in Russia and China.

## PHYSIOLOGY AND ADAPTATION

White amur have survived 1½ to 2 months in 9 and 10 g/ℓ NaCl of the Sea of Azov and the Aral Sea, at 17 and 20.5°C, and a 40.8-g white amur survived up to 24 days in 30% seawater (10.5 g/ℓ NaCl).[15] High pH value and abundant calcium and bicarbonate in solution are extremely healthy for fish growth and reproduction.[37] Natural spawning of amur was discovered in 1962 in Wu-Shan-Tou Reservoir in Taiwan. In Eurasia, the northern boundary of possible appearance of herbivorous fish is the isotherm for the average annual temperature of +5°C. Above the isotherm, farming of herbivorous fishes would be unprofitable. Exceptions are hot water basins and areas with warm water supplies from power plant cooling systems. Plant-clearing activities by this fish begin to be more effective in regions having an average annual temperature of 7 to 8°C. Oxygen needs for amur are less difficult to fulfill than for ordinary carp in water 10°C or less. The white amur is much more sensitive to warm water than are ordinary carp.[27-29]

## DESCRIPTION AND MORPHOLOGY

The upper jaw of the white amur is a little longer than the lower and has thick lips. No teeth are present on the jaws, vomer, palatines, or tongue. Buccal cavity has many low, longitudinal folds on the roof of the mouth and a few transverse folds on the front part of it. A labial fold runs along the upper lip. The tongue is not free of the floor of the mouth, and a mucous membrane covers the tongue which has fine transverse folds on the surface. Gill rakers are short, broad, and 19 to 25 in number. The peritoneal cavity is dark with melanin chromatophores scattered in the peritoneum. It has a short, stout esophagus which extends to the "stomach-like part" by a narrow, compressed strip. It does not have a true stomach, but a "stomach-like part" which has thick walls corresponding to the stomach of other fishes. The mucous membrane of the esophagus consists of high longitudinal folds which turn to complicated, ramulose, high folds at the junction of the esophagus and "stomach-like part". The wall of the intestine has transverse folds but changes to longitudinal folds near the anus. The liver is irregular and is spread over the right side of the digestive mass tract. The gall bladder is large, it consists of semitransparent thin membranes with melanin chromatophores scattered on the surface, and it is located on the right side of the "stomach-like part". The pancreas is irregular in form and is situated on the left side of the "stomach-like part". The bile duct opens to the anterior of the "stomach-like part". The air bladder is large, dumb-bell shaped, and covers all of the dorsal side of the peritoneal cavity, consisting of two chambers. The large anterior chamber is formed by an opaque membrane; the posterior chamber is formed by a semitransparent membrane. The pneumatic or duct originates from the lower part of the anterior end of the posterior chamber and communicates to the beginning of the "stomach-like part".[33-35]

Pharyngeal teeth are mounted in sockets on the lower pharyngeal bones, which are produced forward as anterior processes which curve toward each other, and are joined in the middle line by a ligamentous symphysis. The upper row of teeth consists of two teeth on either side; a lower row of four teeth on the right side and five on the left side are strong comb or file teeth. An ascending process on each pharyngeal bone is attached to strong muscles which pass upward into the skull. Small fish (3 cm or less) have serrated cutting surfaces on the lower pharyngeal teeth, while in larger fish, the teeth are thicker and tend to have doubled and flattened cutting and rasping surfaces. Broadening and flattening of the teeth in larger fish is not associated with wear since the replacement teeth look the same. The change in teeth is associated with age and food selection; young flourish on the soft vegetation (thus, the cutting teeth), while the large fish can also manage tough, fiberous plants (cutting and rasping surfaces).[31,32]

The teeth of the white amur are 2 to 3 mm apart in 30- to 40-cm fish, and 3 to 5 mm apart in 60- to 80-cm fish. There are three to six serrations per millimeter on the cusps, with spacing wider toward the inner side where serrations tend to be hooked. The two rows of pharyngeal teeth can interlock for about one third of their length. The horny pad or masticatory plate has zigzag ridges on its lower surfaces. The pharyngeal teeth can be pressed against the plate and can move towards and away from each other, with some interlocking giving a transverse cutting action, and in a fore and aft direction with a rasping, grinding action.[1,16,18-20]

The body of the white amur is elongated and moderately compressed. The head is broad with a short, round snout; the upper jaw is slightly longer than the lower. The forehead of the white amur is very wide. The mouth is subterminal. The gill rakers are short, widely spaced, and approximately 12 in number. The short dorsal fin is inserted slightly in advance of the ventrals and is not provided with serrated osseous spines as in common carp. It is composed of three simple and seven branched rays. The anal fin commences behind the dorsal, is of moderate length, and is composed of three simple and eight branched rays. The scales are large and cycloid. The one below the dorsal fin is about eye diameter and has 24 to 27 apical striae. The circuli are fine. There are no scales on the head, and those on the abdomen are small. The lateral line is complete and slightly decurved, extending along the middle of the depth of the tail. The tube of the lateral line is slender, simple, and a little decurved.

Coloration of the white amur is olive brown on the upper side and silvery on the lower side. The sides of the body are lighter and have a slightly golden sheen. All fins are dark. The base of each scale is dark brown. The male can be distinguished from the female externally by the structure and relative size of the pectoral fins (usually associated with the approach of maturity). The upper surface of the pectoral fin is rough to the touch due to deciduous, finely serrated ridges along each ray of the pectoral fin, a thickening of the first ray of the pectoral fin, or the presence of "pearl organs". There are no corresponding external features distinguishable in the female.[4,13,16,19,20,28,41]

## FEEDING HABITS

At the hatching stage, white amur feed on plankton and are almost exclusively vegetarian at lengths greater than 30 mm. Fry eat rotifers and crustaceans with occasional larvae of chironomids and algae. At 17 to 18 mm, macrophyte material is first eaten. The importance of rotifers declines and that of chironomids increases. Macrophyte material is the main staple at 27 mm. From 30 mm on, the fish is a vegetarian. Small fish (smaller than approximately 2 in.) will eat tubifex, daphnia, and other small invertebrates, but later they become herbivorous and eat higher plants. It is said that macrophytes are first eaten when the fish is approximately 1 in. long and that the fish is wholly vegetarian at 1¼ in. Adults prefer vegetable food — grasses, leaves, weeds, and water plants.[2,4,8,17] From 5 to 30 days after hatching, the bulk of the food eaten is animal material. After 20 days, the role of algae increases, and at 1 to 1.5 months, filamentous algae and higher plants are eaten almost exclusively. At 1 month (approximately 1 in.), they eat duckweed, *Spirodela,* or any other tender plants.[2,38,41] Plants with high palatability are typically succulent, with negligible fiber such as *Hydrilla, Anacharis,* and *Elodea.* The other extreme is represented by fibrous and woody plants, emergent reeds, sedges, and rushes. The latter can be controlled by cutting. Then the grass carp will browse on the new, young shoots.

The pharyngeal teeth of the white amur are specialized to tear and triturate vegetable material. The plants with little vascular material are easily broken into very small fragments. Monocotyledon leaves, grasses with parallel veination, are transversely cut into segments the length of which corresponds to the distance between the pharyngeal teeth. The fore-and-aft rasping action splits leaves down the lines of the fibrous bundles and strips away parenchyma between the bundles. In land dicots with reticulate veination, the tears are irregular and follow the lines of the principal veins. Plants with a high percentage of dry matter are mixed with water during mastication.[3] Movement of the pharyngeal teeth against each other and against a horny pad on the roof of the pharynx grinds the food. Plants are gripped between the pharyngeal teeth and horny pad and are torn off by violent lateral movement. With food held against the horny pad, the pharyngeal teeth move to and from each other to provide a transverse cutting action, or they move fore and aft to provide a grinding action. Both movements are used when the fish is eating plants.[13,24]

Amur prefer to feed in calm weather at dawn and dusk, but may feed during the night. They show little discrimination of food when feeding intensely. Discrimination increases as the temperature drops. At higher temperatures, the fish is a voracious feeder, and daily consumption can equal body weight. Feeding is most intense when the temperature is above 16°C. At 10°C, feeding may be intermittent. Below 10°C, the fish will only eat the most tender vegetation, but will become less selective as the temperature rises. They feed only above 15°C and in the middle water layers.[13] There appears to be a minimum plant length with which the amur can deal. If food is chopped too fine, the fish become choked and constipated; deaths have been reported for fish which were fed minced plants.[24] In small ponds, stocked at the usual rate for commercial production, the growth rate slowed down after 3 months, is best at 70°F and is dictated by climate.[10-12,14]

## GROWTH AND REPRODUCTION

Amur River fish are 9 to 10 years old before they mature, and the female may spawn as many as three times in a year.[8,9] It takes between 9 and 12 years to attain sexual maturity, depending on the length of the growing season.[13] In the basins of the Yangtze, the amur matures at 5 to 6 years. In the Amur River, it matures even later.[5-7] Time to reach maturity seems affected by latitude: 15 months for males, 20 months for females at the Malacca Experimental Station, Malay Peninsula.[38] Females reach maturity at the age of 3 to 4 years in the Turkmen Republic, 5 years in the Krasnodar Territory, 8 to 9 years in the Ukraine (the Kiev region), and 10 years in the Moscow region. As a rule, the males mature a year earlier than the females. Only at the end of the 1950s was it found that females can reach maturity in the usual carp-rearing ponds and can be used for breeding purposes.[40] Attempts at imitating natural conditions for breeding to obtain sexual products have been unsuccessful. Organs are elongated, paired, and intimately associated with the kidney. Ovaries are greenish, granular in texture, and lie below and beside the air bladder. At the end of spring (March) or the beginning of summer (April), when the temperature rises to about 20°C, ovaries become enlarged and nearly fill the entire body cavity. Eggs are plainly visible. Testes are located in the same position as the ovaries. They are smaller, paler, and look creamy rather than granular. Sexual maturity in the amur is reached in 6 years in Hungary.[43]

In mating, the female is followed by more than two males. They swim against the current in the center of the stream where the current is strongest. After swimming and chasing about for some time and distance, the male will occasionally push his head against the belly of the female and then lean closely to one side. It is probably at this moment that the ova and milt are shed. It takes several spawnings to complete shedding of all ova and milt.[4]

The mating season extends from April to about mid-August; the earliest that fry have appeared on the West River (in China) is April 24. Most spawn from the end of May to the middle of June. Spawning is rare after June 22. Eggs are shed through a short oviduct, then through the genital aperture to the exterior. When ripe, eggs and milt continue to drop as the fish is held. Eggs near the oviduct are ripe. Those near the front are still immature. Not all eggs and milt are extruded at one time. It takes several occasions of sudden water rises to complete the spawning in any one season.[41]

Japanese estimated that a 15½-lb white amur contained 458,000 eggs.[8] A 7-year-old female weighing 7.4 kg (16.3 lb) and measuring 76 cm (30 in.) had 816,000 eggs.[28] At Malacca, 37,000 eggs per kilogram of fish were derived from an injected female.[38] For every 7- to 9-year-old mother fish, it is possible to press out 500,000 eggs. The Soviets expect to obtain 250,000 young fish and one year later have 100,000 survivors from the original 500,000 eggs. As little as 1 kg of dry spawn from the amur contains approximately 1 million eggs (700,000 to 1.3 million).[43] Under captive conditions, the male will develop mature sperm, but the female will not ovulate eggs. The first induced spawning by pituitary injection occurred at Harbin, in Manchuria, in July 1958. The donor fish for pituitary injection is usually the common carp, *Cyprinus carpio,* but other cyprinids may be used. The only successful injections have been composed of pituitaries of cyprinids.[7]

## SUMMARY AND CONCLUSION

A survey of the world literature on the white amur indicates that it has been widely used in Asia and Europe for food and for control of heavy growths of aquatic weeds. There is reason to believe that utilization in the U.S. is both possible and very desirable from the standpoint of efficiency and economy. However, proper utilization will require better management than has been given in the past.

## REFERENCES


1. **Inaba, D. and Nomura, M.**, On the digestive system and feeding habits of young Chinese carps collected in the River Tone, *J. Tokyo Univ. Fish.*, 42(1), 17, 1956.
2. **Prowse, G. A.**, Experimental criteria for the study of grass carp feeding in relation to weed control, *Prog. Fish Cult.*, 33(3), 128, 1971.
3. **Hickling, C. F.**, On the feeding process in the white amur, *Ctenopharyngodon idella, J. Zool.*, 148, 408, 1966.
4. **Michewicz, J. E., Sutton, D. L., and Blackburn, R. D.**, The white amur for aquatic weed control, *Weed Sci.*, 20, 106, 1972.
5. **Kuronuma, K.**, Do Chinese carps spawn in Japanese waters?, *Indo-Pac. Fish. Coun. Proc.*, Sections II and III, 5, 126, 1954.
6. **Singh, S. B., Sukumarah, K. K., Pillai, K. K., and Chakrabarti, P. C.**, Observations of efficacy of grass carp, *Ctenopharyngodon idella* (Val.), in controlling and utilizing aquatic weeds in ponds in Asia, *Indo-Pac. Fish. Coun. Proc.*, Section II, 11, 220, 1966.
7. **Hickling, C. F.**, The artificial inducement of spawning in the grass carp, *Ctenopharyngodon idella* Val. a review, *Indo-Pac. Fish. Coun. Proc.*, Section II, 12, 236, 1967.
8. **Stott, B.**, Aquatic weed control by grass carp (*Ctenopharyngodon idella* Val.), Proceedings of the Third British Coarse Fish Conference, 62, 1967.
9. **Tang, Y.-A.**, Reproduction of the Chinese carps *Ctenopharyngodon idellus* and *Hypophthalmichthys molitrix* in a reservoir in Taiwan, *Jpn. J. Icthyol.*, 8, 1, 1960.
10. **Opszynski, K.**, Carp polyculture with plant feeding fish: grass carp (*Ctenopharyngodon idella* Val.) and silver carp (*Hypophthalmichthys molitrix* Val.), *Bull. Acad. Pol. Sci., Cl. 2,* 16, 677, 1968.
11. **Stott, B. and Robson, T. O.**, Efficiency of grass carp (*Ctenopharyngodon idella* Val.) in controlling submerged water weeds, *Nature (London),* 226 (5242), 870, 1970.
12. **Avault, J. W., Jr., Smitherman, R. O., and Shell, E. W.**, Evaluation of eight species of fish for aquatic weed control, *FAO Fish. Rep. 44,* 5, V 11/E-3, 109, 1968.
13. **Cross, D. G.**, Aquatic weed control using grass carp, *J. Fish Biol.*, 1(1), 27, 1969.
14. **Tan, Y. T.**, Composition and nutritive value of some grasses, plants and aquatic weeds tested as diets, *J. Fish Biol.*, 2(3), 253, 1970.
15. **Cross, D. C.**, The tolerance of grass carp, *Ctenopharyngodon idella* (Val.), to seawater, *J. Fish Biol.*, 2(3), 231, 1970.
16. **Hora, S. L. and Pillay, T. V. R.**, *FAO Fish. Tech. Pap.*, 14, 27, 1962.
17. **Bardach, J. E., Riglher, J. H., and Mclarney, W. D.**, *The Status and Potential of Aquaculture, Particularly Fish Culture,* Vol. 2, Wiley-Interscience, New York, 1973.
18. **Akhmerov, A. K.**, Parasites and diseases of the amur fishes during acclimatization in pond fisheries of the R.S.F.S.R., in *Proceedings of the Conference of Fish Diseases,* Akad. Nauk. SSSR, Moscow, 1957, 108.

19. **Pentelow, F. T. K. and Stott, B.**, Grass carp for weed control, *Prog. Fish Cult.*, 27(4), 210, 1965.
20. **Avault, J. W., Jr.**, Preliminary studies with grass carp for aquatic weed control, *Prog. Fish Cult.*, 27(4), 207, 1965.
21. **Stevenson, J. H.**, Observations on grass carp in Arkansas, *Prog. Fish Cult.*, 27(4), 203, 1965.
22. **Sills, J. B.**, A review of herbivorous fish for weed control, *Prog. Fish Cult.*, 32(3), 158, 1970.
23. **Prowse, G.A.**, The role of cultured pond fish in the control of eutrophication in lakes and dams, *Verh. Internat. Verein. Limnol.*, 17, 714, 1969.
24. **Prowse, G. A.**, Letter to The Editor, *Prog. Fish Cult.*, 28(2), 119, 1966.
25. **Lemenko, E. P. and Lyakhnovich, V. P.**, Comparative morpho-physiological peculiarities of white amur, silver carp and carp reared in Byelorussian ponds, in *Ecological Physiological Peculiarities of the Blood of Fish*, Akad. Nauk. SSSR, Moscow, 1968, 28.
26. **Konradt, A. G. and Faktorovich, K. A.**, A case of dropsy in grass carp, in Selective Breeding of Carp and Intensification of Fish Breeding in Ponds, Bull. State Sci. Res. Inst. Lake and River Fisheries, 61, 216, 1966.
27. **Drews, R. A.**, Raising fish for food in southeast Asia, in *Fish as Food*, Vol. 1, Borgstrom, G., Ed., Academic Press, New York, 1961, 122.
28. **Berg, L. S.**, *Freshwater Fishes of the USSR and Adjacent Countries*, Vol. 2, Izdatel'stvo Akademii Nauk. SSSR, Moscow-Leningrad, 1949, 466.
29. **Doroshev, S. I.**, Survival of young grass carp (*Ctenopharyngodon idella*) and silver carp (*Hypophthalmichthys molitrix*) in waters of various salinities from the Azov and Aral Sea, in *Problems of the Fisheries Exploitation on Phytophagous Fishes in Waters of the USSR*, Akad. Nauk. Turkmensk. SSR, Ashkhobad, 1963, 144.
30. **Pavlov, A. V. and Nelovkin, P. D.**, Information on the Results of Transplanting Grass Carp and Silver Carp into the Lower Volga Basin. *Ref. Zh. Biol.*, 1964(41116), 169, 1961.
31. **Prikhod'ko, V. A.**, Production and breeding of grass carp and silver carp in Ukraine, in *Phytophagous Fish*, Landa, N. G., Ed., Nauka, Moscow, 1966, 40.
32. **Bobrova, Y. P.**, On development of the gonads and the process of fertilization in the white amur, in *Genetics, Selection and Hybridization of Fish*, Cherfas, B. I., Ed., Akad. Nauk. SSSR, Moscow, 1969, 123.
33. **Stroganov, N. S.**, Food Preferences of the Grass Carp. *Ref. Zh. Biol.*, 1964.
34. **Kilgen, R. H. and Smitherman, R. O.**, Food habits of the white amur (*Ctenopharyngodon idella*) stocked in ponds alone and in combination with other species, *Prog. Fish Cult.*, 33, 123, 1971.
35. **Okoniewska, Z.**, Nutritional and organoleptic properties of herbivorous fish flesh — white amur and white tolpyga, *Prezemysl Spozywczy*, 22(7), 304, 1968.
36. **Konradt, A. G.**, Methods of breeding the grass carp, *Ctenopharyngodon idella*, and the silver carp, *Hypophthalmichthys molitrix*, State Research Institute of Lake and River Fisheries, Leningrad, USSR, *FAO Fish.Rep.*, 44, 195, 1965.
37. **Lin, S. Y.**, Induced Spawning of Chinese Carps by Pituitary Injection in Taiwan, *Chin.-Am. J. Comm. Rural Reconstr. Taiwan Sp. Bull.*, 5, 31, 1965.
38. **Gangstad, E. O., Raynes, J. J., and Burress, R. M.**, Aquatic Plant Control Program; Herbivorous Fish for Aquatic Plant Control, Tech. Rep. 4, U.S. Army Engineers Waterways Experimental Station, CE, Vicksburg, Miss., June 1973.
39. **Crowder, J. P. and Snow, J. R.**, Use of grass carp for weed control in ponds, *FAO Fish. Bull.*, 2(1), 6, 1969.
40. **Vinogradov, V. K.**, Techniques of rearing phytophagus fishes, All-Union Institute of Pond Fisheries, Moscow, USSR, *FAO Fish. Rep.*, 44, 227, 1968.
41. **Lin, S. Y.**, Life history of waan ue, *Ctenopharyngodon idellus* (Cuv. and Val.), *Lignan Sci. J.*, 14(1), 129, 1935.
42. **Bailey, W. M. and Boyd, R. L.**, A Preliminary Report on Spawning and Rearing of Grass Carp (*Ctenopharyngodon idella*) in Arkansas, paper presented at 24th Ann. Conf. Southeastern Assoc. Game Fish Commissioners, Atlanta, Ga., 1970.
43. **Antalfi, A. and Tolg, L.**, *Herbivorous Fish*, Agricultural Publishing House.

Chapter 17

# ASSESSMENT OF EFFECTS OF THE WHITE AMUR ON THE AQUATIC ECOSYSTEM

## INTRODUCTION

The herbivorous fish *Ctenopharyngodon idella* (Val.), commonly called the white amur or Chinese grass carp, appears to be the most promising organism presently available for biological control of many submersed aquatic plant problems. The white amur was introduced to the U.S. in 1957.[1] Since that time, the Arkansas Game and Freshwater Fish Commission has released this fish in the natural aquatic ecosystem in an effort to control many of their aquatic plant problems. They have reported no detrimental effect on the natural system by these fish and are confident that the white amur will become a part of their aquatic vegetation management program.[2]

The introduction of any exotic organism involves risk that the organism may meet little resistance and become a problem rather than correct the problem for which it was introduced. On the other hand, indigenous populations of organisms may severely limit the ability of the exotic organism to establish itself. This information needs to be assessed according to the relationship which may develop between the white amur and the established natural populations in association with the ability of this herbivorous fish to control aquatic vegetation.

## BACKGROUND INFORMATION

The white amur has been cultured in China for centuries, primarily for its ability to convert vegetation into fish protein. The Chinese and several other countries use a polyculture of fish in their aquaculture in order to maximize utilization of food available in the water. Most available information indicates that white amur begin to feed on zooplankton soon after hatching. As development proceeds, however, their diet changes to a herbivorous one. Higher vegetation, rather than algae and phytoplankton, constitutes a major portion of their diet. Some studies indicate that fish production in general may increase when white amur fish are present in the aquatic system. Natural reproduction of the white amur has been reported in only a few isolated cases outside native habitat of the fish. Until artificial spawning techniques were recently developed, young white amur fish could only be obtained by catching the fry soon after hatching from rivers in northern China. Certain water quality parameters are required before natural spawning will occur. The reproductive organs of the fish develop, but fail to mature when these specific water quality factors are absent.[3-6]

## RELEVANCY OF THE PROJECT

Information has been generated about aquatic plants preferred by the white amur and the ability of these fish to control aquatic plant growth. Also, a review of literature was completed and published. At the Research Center in Ft. Lauderdale, Fla., considerable expertise has developed in the care and handling of white amur fish. Initial results indicate that the white amur does have potential for controlling nuisance growth of some aquatic plants.[7-10]

The first objective of the current study is to draft a statement regarding the possible impact of white amur fish on the natural aquatic ecosystem. The effect of the white amur on game fish populations should be considered before these fish are released. The high stocking rate of white amur will simulate a condition in which this might become extremely important — for example, if natural spawning is interfered with. Likewise, information on natural reproduction is the key element in establishing an effective aquatic plant management program with the white amur. Establishment of an effective stocking rate will depend on maintenance of sufficient numbers of herbivorous fish to control problem plants without the fish itself becoming a problem.[1,2,10-12]

The predator-prey relationship has been developed by stocking white amur and game fish in large containers under controlled conditions. In some cases, the white amur had an abundance of vegetation. In other cases, no vegetation was available. Various stocking rates and combinations of game fish and white amur have been developed. If possible, a relationship must be established between game fish and small white amur. Development and maturation of the reproductive organs of the white amur must be evaluated by subjecting the fish to various water quality factors, especially waters that vary in alkalinity. These ecological conditions for widespread applications within the U.S. should be studied further and are part of the current research program.[14]

## ECOSYSTEM MODEL

There are several components that most lake ecosystems have in common. The detailed composition and operation of these components, plus the effect of external physical forces acting on them, distinguish lakes from one another. Therefore, the model that is used to initiate an effort like this one is usually general and is then refined according to specific characteristics of each ecosystem.[13,14]

Four general kinds of plant communities are distinguished: (1) native submersed species, such as pondweed (*Potamogeton* sp.), predominant around the edges of the lake; (2) hydrilla, capable of growing over and shading out the submersed species; (3) phytoplankton, concentrated within the top layers of the water and having a shading effect on the submersed species; and (4) algal periphyton, which grows on the surfaces of hydrilla and other broad-leaved plants. Productivity in the native submersed, hydrilla, and phytoplankton communities is regulated partially by solar radiation and partially by availability of phosphorus. Although other elements are certainly critical to plant growth, phosphorus is believed to be one of the main limiting nutrients in aquatic ecosystems.

The periphyton population is considered unique because it is just as dependent on the presence of a substrate (hydrilla and native submersed plants) as it is on phosphorus and sunlight. In the flow of energy, this kind of interaction is distinguished as "open", indicating an effect rather than an actual flow. Phosphorus and sunlight are actually used by the periphyton population, but its productivity is limited by the amount of substrate. Therefore, productivity of periphyton is included as a drain on the phosphorus, but phosphorus does not have an influence in the periphyton.

Fish use the periphyton and native plants for food. When the white amur is introduced, it is assumed to feed solely on the hydrilla. Sediments form from dying plants, consumers, and waste products and accumulate in a detrital pool from which phosphorus is mineralized.

After the model was conceptualized, differential equations describing these interactions were formulated. These equations were then translated into difference equations, and the model was simulated on a digital computer using the simulation language DYNAMO.

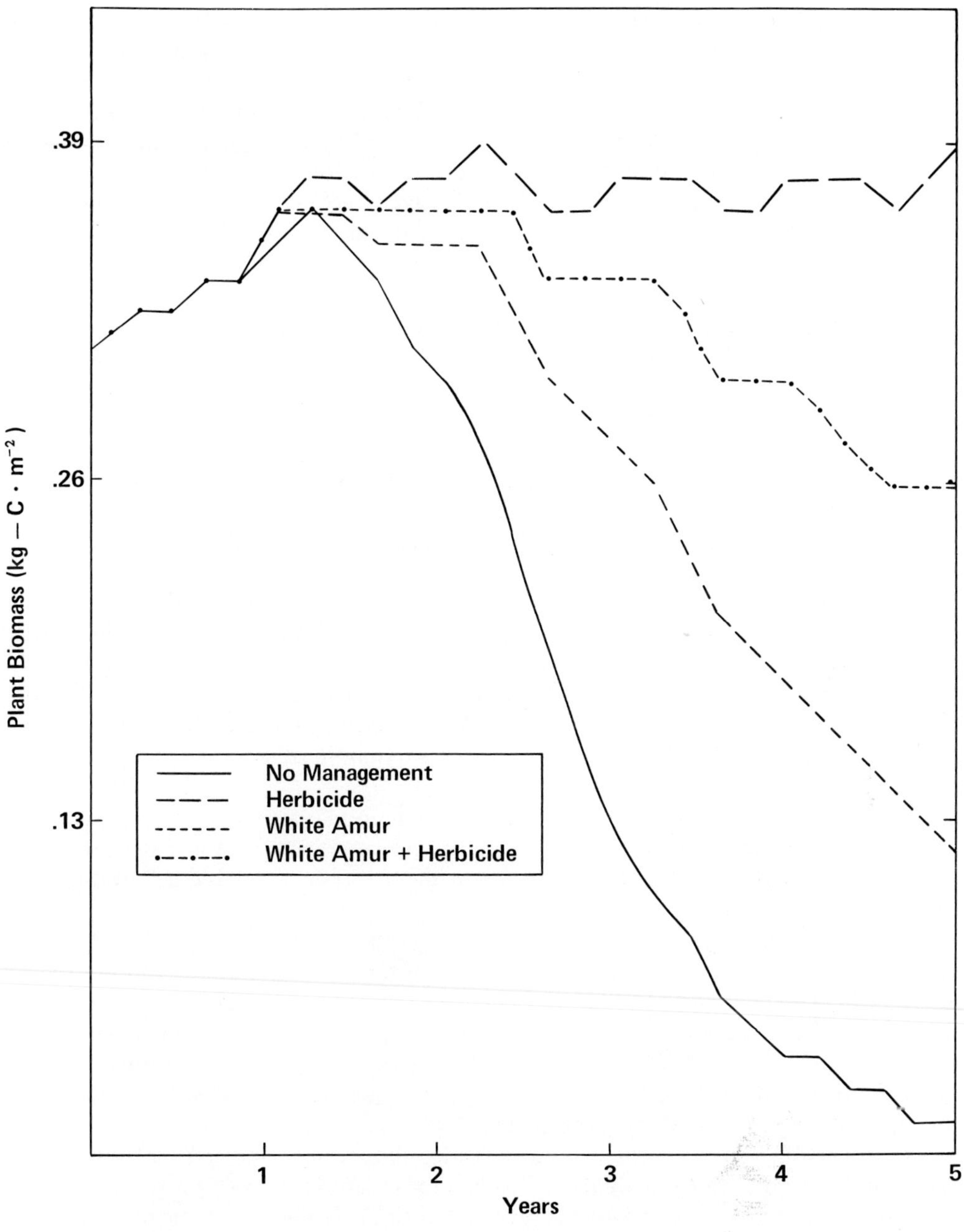


FIGURE 17.1. Biomass imulations of a computer model from available data. (From Ewel, K. C. and Fontaine, T. D., Possible effects of the introduction of the white amur into Lake Conway, Florida, misc. pap. A76-1, U.S. Army Engineers Waterways Experimental Station, Vicksburg, Miss., 1975, 64. With permission.)

## SIMULATIONS

Four simulations were run (Figure 17.1). In one case, hydrilla was allowed to grow unrestrained. The results from this simulation are entitled ''No Management''; application of herbicide (''Herbicide'') was modeled in the second simulation; introduction of the white amur (''White Amur'') was modeled in the third; and use of both agents (''White Amur and Herbicide'') was modeled in the fourth.

Addressing first the fate of hydrilla under various management schemes, the simulations indicate that highest standing stocks will occur under "No Management" conditions. Standing stocks declined in magnitude when herbicides or the white amur were used. A combination of the two controls ("Herbicides and White Amur") resulted in the lowest standing stock (7 times less than that found under undisturbed conditions); the presence of the white amur alone was more successful than herbicide alone (5 and 2.5 times less than "No Management", respectively). None of the management programs completely eliminated the hydrilla, and in all cases, hydrilla achieved a dynamic steady state.

Tubers and turions were always manufactured by the plant, the rate depending on the productivity of hydrilla. When hydrilla was under the most stress, these reproductive structures served as growth initiators for hydrilla, thus allowing it to achieve a dynamic steady state. When no control was used, tubers and turions accumulated rapidly during the time frame used.

Phytoplankton increased when hydrilla was most heavily controlled, reaching more than twice the concentration found under undisturbed conditions. Increased availability of phosphorus seemed to be responsible for this increase. When the white amur or herbicides were used, the concentrations were equal to or slightly less than twice the normal amount, respectively.

The native submersed plant population showed interesting responses under different management strategies. Since one of the assumptions of the model was that hydrilla could shade out native species, the native submersed plants were able to compete with the uncontrolled hydrilla for only slightly more than 1 year and then declined in biomass to about one tenth their original state. When herbicide was applied, native species showed their best growth and seemed to attain a dynamic steady state. Other management schemes proved less favorable.

Algal periphyton showed best development when no management was supplied, with hydrilla acting as the main substrate. As management efforts were initiated, decreases in algal periphyton resulted, primarily from decreases in available substrate (either hydrilla or native species).

Native fish also showed best growth when no management was supplied. In this model, the supply of periphyton seems to control the fish population, and periphyton is primarily dependent on hydrilla as a substrate. The least amount of growth resulted when herbicide was applied. Intermediate values were achieved for other management schemes. Since all native fish (herbivores, carnivores, and omnivores) were considered one unit in this model, the results obtained depict general trends only. Further work toward a reliable fish submodel is necessary.

When hydrilla growth was most successfully suppressed, values of phosphorus in the total water column were highest. This cannot be ascribed to faster recycling, since the recycling rate from sediments was set as a constant. Higher total phosphorus values probably resulted instead from a diminished uptake rate caused by a decline in the standing stocks of vegetation. Even when hydrilla is controlled, there is enough of the weed present to exert a shading effect on the submersed vegetation. Only the phytoplankton thrives, but its turnover time is so rapid that its effect on the phosphorus level is only slight. The larger biomass and slower turnover time of the hydrilla population are sufficient to reduce the level considerably when growth is not controlled.

The accumulation rate of the shallow component of detritus was greatest under "No Management" conditions; it was least when white amur and white amur plus herbicides were used for weed control. These low values occur because much of the biomass becomes vested in the white amur population.

When white amur was used alone as a management strategy, it achieved a higher standing stock than under conditions of management by herbicide and white amur combined because more food was available for the white amur to eat. As with the native fish component, more work is needed to refine the biomass number, standing stock, and turnover time relationships of this component. The preliminary model suggests that control of hydrilla by the white amur will result in a phytoplankton bloom. However, the population of native fish will be enhanced and the accumulation rate of organic detritus on the lake floor will be decreased.

## REFERENCES


1. **Swingle, H. S.,** Control of pondweeds by use of herbivorous fishes, *Proc. South. Weed Conf.*, 10, 11, 1957.
2. **Stevenson, J. H.,** Observations on grass carp in Arkansas, *Prog. Fish Cult.*, 27, 203, 1965.
3. **Cross, D. G.,** Aquatic weed control using grass carp, *J. Fish Biol.*, 1, 27, 1969.
4. **Lin, S. Y.,** Induced spawning of Chinese carps by pituitary injection in Taiwan, *Chin.-Am. Jt. Comm. Rural Reconstr. Taiwan Spec. Bull.*, 5, 31, 1965.
5. **Dorosher, S. I.,** The survival of the white amur and tolstobick fry in the Sea of Azov and Aral Sea water of varying salinity, in Symp. Prob. Fish. Exploit. of Plant Eating Fishes in the Water Bodies of the USSR, Ashkhabad Academy of Science, Turkmen, U.S.S.R., 1963, 144.
6. **Cross, D. G.,** The tolerance of grass carp *Ctenopharyngodon idella* (Val.) to seawater, *J. Fish Biol.*, 2, 231, 1970.
7. **Yeh, Y. T.,** A preliminary report on the oxygen consumption, energy requirements, asphyxiation point and respiratory quotient of fish, *Acta Zool. Sin.*, 11(2), 117, 1959.
8. **Anon.,** Report of the subcommittee of fisheries, in Proc. 10th Pacific Sci. Congr., Honolulu, 1961, 277.
9. **Tang, Y. A., Hwang, Y. W., and Lin, C. K.,** Reproduction of the Chinese carps, *Ctenopharyngodon idellus* and *Hypophthalmichtys molitrix,* in a reservoir in Taiwan, *Jpn. J. Ichthyol.*, 13(1/2), 1, 1960.
10. **Sutton, D. L.,** Utilization of hydrilla by the white amur, *Hyacinth Control J.*, 12, 66, 1974.
11. **Hickling, C. F.,** Biological control of aquatic vegetation, *PANS,* 11, 237, 1965.
12. **Sills, J. B.,** A review of herbivorous fish for weed control, *Prog. Fish Cult.*, 32, 158, 1970.
13. **Odum, H. T.,** *Environment, Power and Society,* Wiley-Interscience, New York, 1971, 330.
14. **Ewel, K. C. and Fontaine, T. D.,** Possible effects of the introduction of the white amur into Lake Conway, Florida, misc. pap. A76-1, U.S. Army Engineers Waterways Experimental Station, Vicksburg, Miss., 1975, 64.

# Part IV
# Utilization of Herbivorous Fish in China

Chapter 18

# DISTRIBUTION AND BIOLOGY OF ECONOMICALLY IMPORTANT FISHES*

## INTRODUCTION

According to the characteristics of geographical groupings of fishes in different parts of the country, the distribution of freshwater fishes in China can be divided into five major zones, namely, the Amur River Zone, the Yangtze-Yellow River Plain Zone, the South China Zone, the Northwest Zone, and the Tsangpu-Lu River Zone. This division depends upon the Nanking Mountains as the principal dividing line. North of the mountain chain the land is considered to be part of the Ancient North Belt; this Belt is subdivided into the Siberia Belt which includes the Yangtze-Yellow River Plain Zone. South of the mountain chain is the East Oceanic Belt, the Chinese part of which is the Sino-Indian Belt, which contains the South China Zone. The Northwest Zone and Tsangpu-Lu River Zone are located in the vast lands of western China.[1,2]

Included in the Amur River Zone is that part of the Amur basin that is within the Chinese territory. The principal fish are typically northern species plus some cyprinids introduced from the Yangtze-Yellow River Plain Zone.

The Yangtze-Yellow River Plain Zone includes the lower section of the Yangtze, the Yellow River basin, and northern China. Cyprinids are the main fishery resource. The four domestic carps are especially abundant. A few typically northern or southern species are also mixed in, such as *Monopterus albus* and *Anguilla japonica* of the south and salmonids, cottids, and sticklebacks of the north. Completely absent, however, are the purely northern species of ling, pike, and arctic grayling and the purely southern species of *Megalops cyprinoides* and *Claris pulicaria.*

Included in the South China Zone are Kwangtung, Kwangsi, Kweichow, Yunnan (southeastern part), Fukien, Taiwan, and Hainan Island. Fishes in this zone include the typically tropical fauna of many catfishes and species belonging to Homalopteridae and some cyprinids that are similar to India-Burma varieties. With the exception of the anadromous Taiwan salmon, the northern types of lamprey, sturgeon, trout, and sticklebacks are entirely absent.

The Northwest Zone embodies Sinkiang, northern Tibet, western Inner Mongolia, Tsinghai, Kansu, and Shensi. Most fishes are high plateau types. The mainstay comprises about 20 species of the subfamily *Schizothoracini* and over a dozen kinds of catfish-like species. Quite a few of these species have economic value, but they have yet to be utilized.

The Tsangpu-Lu River Zone covers the river basins of the Tsangpu, Lu, Kinsha, and Lantsang Rivers in southern Tibet, Sikang, and the western parts of Szechwan and Yunnan. Fish fauna is distinctly a mixture of South China and Northwest types.

Among the more than 500 species of freshwater fishes in China, about half have economic value as food fish. The more important ones are listed below:

* From *Culture of Freshwater Fish in China,* Science Publishers, Peking, China, 1966, translated by Koo, T. Y., Dept. of Fisheries, U.S.D.I., Solomons, Md., Walker, C., Ed., Senior Research Scientist, U.S. Department of the Interior, Washington, D.C.

| | | | |
|---|---|---|---|
| 1. | *Huso dauricus* | 17. | *Barbus* spp. |
| 2. | *Ascipenser sinensis* | 18. | *Elopichthys bambusa* |
| 3. | *Oncorhynchus keta* | 19. | *Xenocypris* spp. |
| 4. | *Salanx* spp. | 20. | *Hemibarbus maculatus* |
| 5. | *Cyprinus carpio* | 21. | *Spinibarbichthys sinensis* |
| 6. | *Carassius auratus* | 22. | *Onchostoma laticeps* |
| 7. | *Mylopharyngodon piceus* | 23. | *Parasilurus asotus* |
| 8. | *Ctenopharyngodon idellus* | 24. | *Pseudobargrus* spp. |
| 9. | *Hypophthalmichthys molitrix* | 25. | *Esox reicherti* |
| 10. | *Aristichthys nobilis* | 26. | *Coilia* spp. |
| 11. | *Squaliobarbus curriculus* | 27. | *Ophiocephalus* spp. |
| 12. | *Parabramis pekinensis* | 28. | *Monopterus albus* |
| 13. | *Megalobrama* spp. | 29. | *Anuilla japonica* |
| 14. | *Erythroculter* sp. | 30. | *Hilsa reevesii* |
| 15. | *E. ilishaeformis* | 31. | *Cirrhina molitorella* |
| 16. | *Hemiculter* spp. | 32. | *Mugil* spp. |

Although China has rich resources in freshwater fisheries, the yield in 1949 amounted to only 140,000 metric tons. Since then, restoration and development measures have brought considerable increase each and every year. In the 9 years from 1950 to 1958, the production was increased from 330,000 to 1,500,000 metric tons, an increase of 4.5 times. Moreover, the relative weight of freshwater fisheries in the total landings of all fisheries was very high. The total harvest of all fisheries in that year was 1,500,000 metric tons, of which 550,000 or 35.9% came from freshwater fisheries. In 1958, total harvest reached 3,600,000 metric tons, and freshwater fishery increased its share to 39.9%. Such figures are not even approached by other nations with highly developed fisheries.

The high ratio of contribution by the freshwater fisheries to the total fisheries production in China is due to the combined effect of rich natural resource and the highly developed culturing business. In 1954, production from fish culture accounted for 31% of all freshwater fisheries harvest. Fishing for wild species was therefore an important factor. The production from that source climbed to 60% of total freshwater fisheries harvest. It is clear, therefore, that China's freshwater fish resource provides an amazing potential.

As discussed above, China not only has a large number of species of freshwater fishes, but also many economically important species. The four domestic carps are China's freshwater fish special. They are tasty, fast growing, and widely distributed in all large river basins. Their production is especially rich in the Yangtze valley. In 1958, the combined production in freshwater fisheries from the three provinces of Kiangsu, Anhwei, and Hupeh accounted for 50% of the nation's total and a major part of this production was due to the four domestic species of carps. Aside from this, the fingerling production of these four carps is also enormous, and it has been on the increase in recent years. By 1958, the production from the Yangtze and West Rivers already reached 52.1 billion fish. This continuing increase in fingerling production assures the further advancement of fish culture in China.

The important role which the common carp and goldfish play in Chinese freshwater fishery should not be overlooked. These two fishes are not only abundant but also widely distributed. They are usually the largest yielding species in many areas, and they generally contribute as much as 20% of the total production in a water area. They are the principal natural species that fishermen seek. At different localities, both carp and goldfish have their distinct races of subspecies, each possessing distinctive characteristics. Some varieties, for instance, tend to fatten easily; others grow fast. Carefully planned selection and breeding will open the door to further development of freshwater fishery.

Among the many highly valued species, *Parabramis pekinensis, Magalobrama terminalis, Brythroculter ilishaefermis, Blopichthys bambusa,* and *Siniperca chautsi* stand out as the important ones. These fishes are important because their quality and quantity are good and their distribution is widespread. If we can discover the secrets of their habitats and requirements, we should be able to increase their abundance in their natural habitat and eventually bring them under culture.

Along the lower valley of the principal rivers in China, there are a number of migratory species, such as the shad, chum salmon, eel, and mullet, and some other forms that are gradually becoming river residents, such as *Acipenser, Huso, Salanx, Coilia,* etc. All these are high class food fishes. The catch of the chum salmon from the Amur River reached a high of 2.1 million kg in 1951. The average catch of *Salanx* in Tai Lake per boat per year may reach 15,000 kg. The per boat catch of the Yangtze shad averages 1500 kg per season. The catch of sturgeons fluctuates from year to year but the maximum may attain tens of thousands of kilograms. Their roe is especially esteemed.

Species such as *Xenocypris argentea, Scualiobarbus curriculus, Hemiculter leucisculus, Ophiocephalus argus, Monopterus albus,* and the Mongolian and redfin minnows play a certain part in freshwater fisheries due to their common occurrence and great abundance although they are only medium sized. Among them, *Hemiculter leucisculus* is the most common. On one fishing ground in Talai Lake in Inner Mongolia, the daily catch of this fish may reach 2000 kg. After processing, they offer a fine fishery product.

What has been discussed above merely represents the major food fishes in China. Actually, fish used as food year round number over 200 species. Because the catches of many miscellaneous species are made on a small scale and are often scattered, their statistics are difficult to collect. In any event, it can be conservatively estimated that such catches may amount to 5 to 10% of all freshwater fisheries.

## HABITAT AND GROWTH

An understanding of the habitat of the fishes and of their biological rhythms is essential to the development of fisheries, either for natural fishing or for artificial rearing. Knowledge of growth rate, feeding habit, reproduction, and migration of fishes is basic to improvement of conservation and production. These four areas are also the basic subjects of fishery biology studies.

Each species of fish has its own habitat, and each is closely adapted to its environmental conditions. Generally, closely related species live in habitats that are similar to each other; on the other hand, divergent species live in distinctly different habitats.

The growth rate of fishes varies greatly according to species. Fast-growing species may attain several kilograms up to 100 kg in a few years. Slow-growing species only put on a few hundred grams in a year. Sizes of adult fishes also vary tremendously. Giant types may weigh up to 100 kg; tiny forms may weigh only a few grams. Generally speaking, growth rate of a fish changes with life stages. Some fish grow the fastest during their first year; others do not pick up weight appreciably until the second or third year. Many long-living species will maintain relatively high rate of growth even after a considerable period of growth. Some short-lived species almost stop their growth within a relatively short period of time. In quite a few species the growth rate in the two sexes is different, although it is more marked in some than in others.

The growth of fishes results from feeding upon various kinds of food from the environment. These foods are various aquatic animals and plants of all sizes, consisting of phytoplankton, zooplankton, benthic animals, vascular plants, and bacteria. However, not all these aquatic organisms can be directly utilized by fish, nor are all needed by some fish. On the contrary, most fish are quite selective as to their food. Actually, aquatic organisms serving as fish food are but one of the factors in the environment in which fish live. Consequently, fish are well adapted to environmental factors in the aquatic habitat. Through long periods of evolution, the body shape and structure (particularly the digestive system) of the fish acquired various forms in adaptation to the convenience of feed, digestion, and assimilation.

Fish do not feed directly on every kind of aquatic life, but all aquatic organisms must depend, for their growth, ultimately upon the microorganisms (such as algae), which, in turn, depend upon organic matter and nutrients in the water. In this way a food chain is formed. Quite obviously, the productivity of a water body is determined by the amount of organic material and nutrients present in the water.

Generally speaking, the food chain among aquatic organisms starts with phytoplankton (algae, including bacteria). These form the food material for invertebrates (zooplankton and benthic animals) which become an important food item for many fishes; these, in turn, are fed upon by many predaceous fishes. Of course, there are some species of fish that feed directly on phytoplankton or vascular plants, thus skipping one or two intermediate food stages.

Feeding habits of fishes may change in different seasons, in different growing stages, and in different localities. This is because growth and reproduction of aquatic organisms are closely related to seasonal and climatic changes. Accordingly, fish may feed temporarily upon some nonessential food or reduce their feeding activity for the time being. For example, during the early spring when the weather is cold, fish that normally feed upon vascular plants may sometimes feed on animals. The change of food habits as the fish grows is only natural, since the feeding organ of juvenile fish is not fully developed and the young fish must take whatever is easily caught as food. For fishes in different localities, the selection of food may often depend upon what is available. For instance, in waters rich in prawns, fish-eating predaceous fishes often contain prawns as the major portion of their stomach contents. According to their primary food habits, fishes can be simply classified into three categories: carnivores, herbivores, and omnivores.

## BREEDING HABITS

Breeding habits of fishes also vary a great deal. Most fishes spawn during spring or early summer; however, there are quite a few species that spawn in early spring or in the fall. Eggs of some species possess oil globules and, therefore, are freely buoyant and planktonic; however, some demersal eggs can also become suspended in the water column after being stirred up by currents. Some fish eggs are adhesive; they become adherent to vegetation or other submersed objects immediately after being laid. In still other species, the parent fish builds a nest before laying eggs inside and guards the nest during the incubation period. There are some species that grow an ovipositor during the spawning season and deposit their eggs inside the body of other animals (such as the mussel) in order to prevent predation by enemies. From the above, it is evident that some fish require flowing water; others need stagnant water for spawning.

The fecundity of fishes differs from species to species. Some female fish have very high fecundity, having several hundred thousand or even more eggs; others possess only a few hundred or a few thousand to a maximum of tens of thousands of eggs.

Even within the same species, the fecundity is different in different ages. During each spawning season, the gonads in some species mature all at one time, and all eggs are deposited in one spawning; in other species, they mature in stages and consequently spawn in several batches.

The breeding habits of fishes are not only determined by the internal conditions in the fishes themselves but are also affected by external conditions. In China the climatic conditions vary from one place to the other, and the four seasons do not fall in the same months at all localities. Therefore, even within the same species, the breeding habit may vary from one locality to another. The latitude, topography, and other major environmental factors also impose certain limitations to the breeding of fishes. This accounts for the existence of some special species of fish in some geographical areas.

A species of fish usually selects one definite place as a habitat. For instance, some live in the ocean, some in freshwater; some choose running water as their habitat while others prefer a quiet water body. Nevertheless, there are cyclic patterns in the life of some fishes, in which the demand of life changes during the life span or within the time periods of the year.

## MIGRATION HABITS

The migration of fishes is usually for one of three reasons: feeding, breeding, or over-wintering. For feeding migration, fish always seek areas where food is abundant. A lake that drains into a river is often the place where fish fatten themselves. This accounts for lake to river migration. Even within the confines of the same body of water, the distribution of food organisms for different species of fish may vary, which often leads to short distance migration of some fishes. During the breeding season, many species of fish require specific spawning conditions and frequently migrate upstream in search of spawning locations. A few species that normally reside in the ocean must migrate long distances into rivers to spawn. This is the most representative and most typical kind of migration phenomenon of fishes. When winter comes, most species of fish migrate into deeper water or bottom to over-winter where the temperature is higher.

## LIFE HISTORIES OF FISH

The following gives brief accounts of the life history and breeding habits of the major economic species of fish in China. Most of the species described are the objects of freshwater fish culturing species.

### The Grass Carp (White Amur) (*Ctenopharyngodon idellus*)

The grass carp (Figure 18.1) is closely related to *Mylopharyngodon piceus.* Both species are similar in distribution, body size, breeding habits, nursery area of the young, and over-wintering conditions. The young of the grass carp usually winter in relatively deep holes and are not associated with the adults. Growth rate is slightly slower than that of the black Chinese roach, but in the young stage it is about the same. A fish of six winters averages about 60 cm.

The grass carp derives its name from the fact that it feeds on grass. The main food of the adult grass carp is comprised entirely of high aquatic vegetation. The plant species that are eaten depend upon what are available. Generally, *Vallisneria spiralis, Hydrilla verticillata, Najas minor, Spirodela polyrhiza,* and *Wolffie arrhiza* are the favorite food of the grass carp. In actuality, however, almost any aquatic plants are

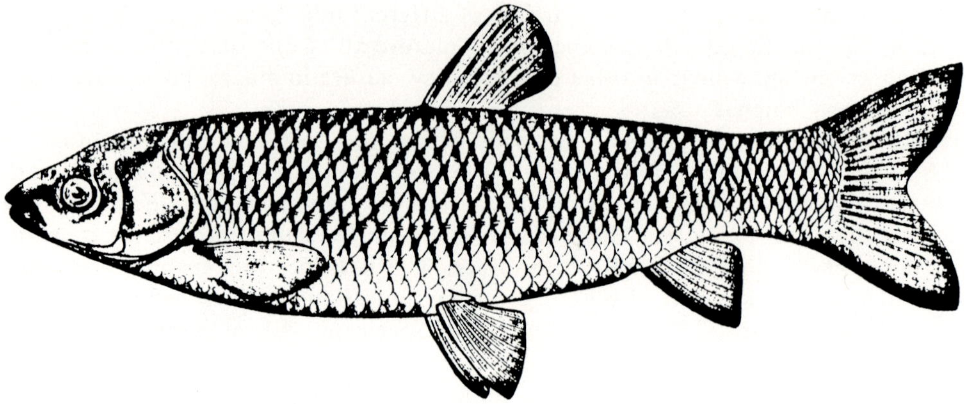

FIGURE 18.1. Grass Carp (White amur) (*Ctenopharyngodon idellus* [Val.]).

eaten by the fish. Only in localities where there is an abundance of aquatic vegetation do the fish select the tender, young kinds of plants. Even some terrestrial plants are also eaten by the grass carp. *Lactuca sativa* which grows near the edge of the water is one of these. During flooding conditions, grassy areas which become immersed in water often provide excellent nursery conditions for the grass carp.

Due to its vegetarian habit, the grass carp has a short intestine. In the adult, the intestine is about 2.5 times the length of the body. Its pharyngeal teeth possess a saw-like surface and are capable of cutting aquatic plants. During the winter, the intestine of the grass carp is often empty or contains only a small amount of plant seeds. It is likely that the grass carp feeds very little or not at all during the winter in order to adapt itself to the absence of aquatic vegetation during that period.

The young grass carp has entirely different food habits. Its main food comes from animal source. Fry of less than 1 cm in length feed principally on planktonic animals and algae, and their intestine is straight, measuring only half as long as the body length. Following the growth of the body, the intestine also increases in length, and the food of the young fish gradually consists of rotifers, small midge larvae, and planktonic mollusks. When the young fish grow to more than 3 cm in length, they become almost completely vegetarian.

### The Black Chinese Roach (Black Amur) (*Mylopharyngodon piceus*)

The black Chinese roach (Figure 18.2) is a large fish of the grass carp subfamily; its maximum weight may reach 60 kg. Individuals weighing 18 to 32 kg are commonly found in the rivers and lakes. They have a wide distribution and a fast growth rate. One-winter fish may grow to 0.284 to 0.567 kg; two-winter fish, 2 to 3 kg; and three-winter fish, 6 kg under good environmental conditions. Larvae grow to 2 cm in 1 month, 4 cm in 2 months, and 8 cm in 3 months. Each year from late April to mid-June (varying somewhat from place to place depending upon the weather), the adult fish migrate upstream to a suitable location to spawn. To date, none of the five domestic cyprinids can be made to spawn in still water. The youngest age to reach maturity is four-winter fish, which are about 100 cm long. The female contains 400,000 to 1 million eggs. The eggs, as well as the newly hatched larvae, float freely downstream. In about 2 weeks, the larvae develop into a stage of active feeding and seek the shore areas or bays in which to feed and grow. The parent fish also return to the tributaries or lakes to feed after spawning. Those adults that are not sexually mature generally

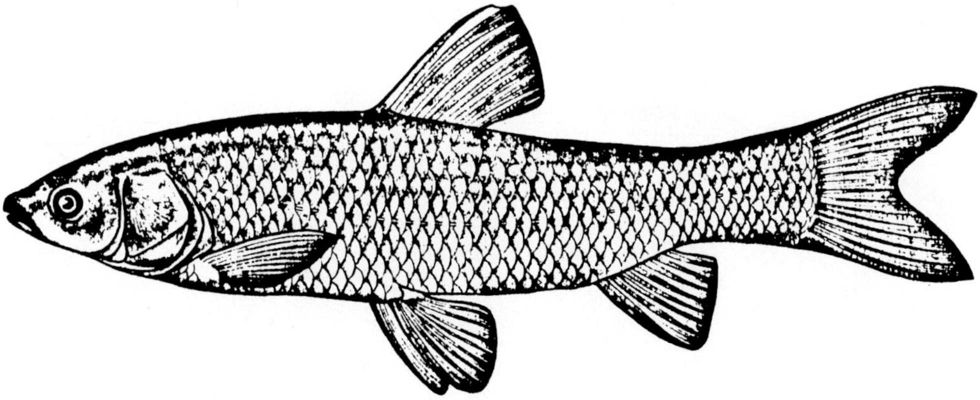

FIGURE 18.2. Black Chinese Roach (Black amur) (*Mylopharyngodon piceus* [Richardson]).

remain in larger tributaries or lakes where there is an abundant supply of snails. During the winter they stay near the riverbed in the deeper tributaries of the main river.

The food habit of the black roach is very simple. The main food is snails. They may also feed on other small mollusks. The smaller black roach also feed on some benthos such as dragonfly nymphs and other insect larvae, gonospores of bryozoa, mollusk larvae, etc. The young black roach feed on planktonic animals. As a whole, the black roach is a carnivore; therefore, its intestine is relatively short, usually about 1.25 times the length of the body. Since mollusks are benthic, the black roach, which depends on mollusks for food, also adopt a bottom-dwelling habit. Under artificial rearing, the black roach fry like to eat bean cake, grain residue from alcohol fermentation, and other similar vegetable food. Black roach have molar-like pharyngeal teeth which are capable of crushing the shells of snails. This kind of structure is evidently an adaptation to its feeding habits.

### The White Bighead (Silver Carp) (*Hypophthalmichthys molitrix*)

The white bighead (silver carp) (Figure 18.3) has a breeding habit similar to the grass carp and the black roach. Maturity is reached at 4 to 5 years, with the body length at 50 to 60 cm and body weight at 4.5 to 6 kg. In the Pearl River, the white bighead reach maturity at both a younger age and smaller size. These may belong to a special geographic race. The food consists of phytoplankton, although zooplankton is the main food during the young stage. The white bighead is also a large fish. It is not uncommon to find individuals weighing 12 to 18 kg. The 2-year-old fish normally grow to 30 cm and weigh 0.59 kg; 3-year-old fish may reach 40 cm and weigh more than 1.1 kg. The eggs and newly hatched fry drift downstream with the current. Juvenile fish enter bays or the tributaries to lakes to feed. In the spring, immature adults move from the main stream into tributaries or lakes where food is abundant. The mature adults migrate upstream to spawn, move into lakes to feed, and return to deep riverbed to over-winter.

The feeding method of the white bighead is still another type. The gill structure is similar to that of the striped bighead, but the gill rakers are much finer. They are so fine as to be able to retain microscopic phytoplankton (algae), which become the main food of the fish. Rotifers and small mollusks are also found in the intestine. During the juvenile stage (body length around 15 cm), some planktonic animals comprise the principal food item.

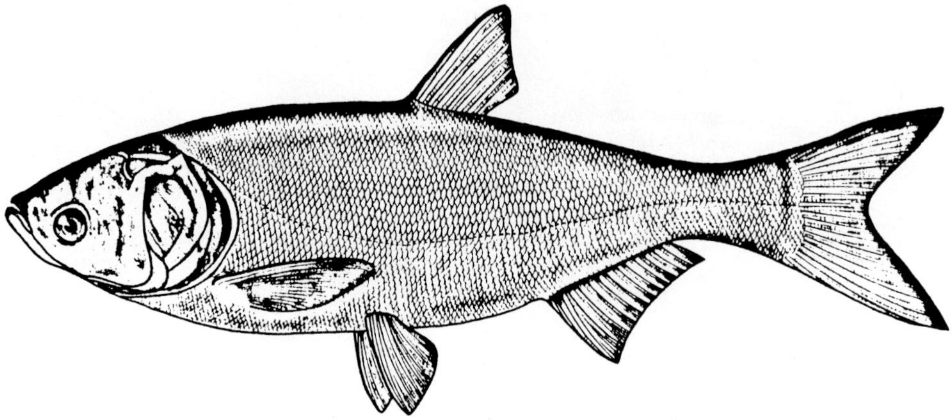

FIGURE 18.3. White Bighead (Silver carp) (*Hypophthalmichthys molitrix* [Val.]).

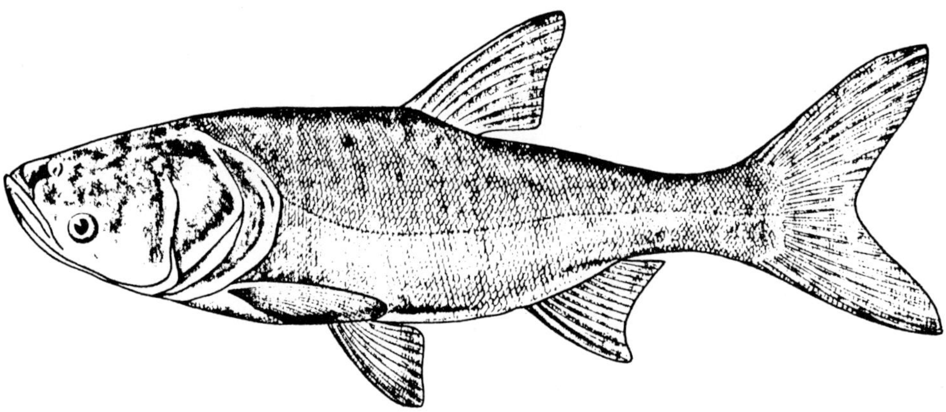

FIGURE 18.4. Striped Bighead (*Aristichthys nobilis* [Richardson]).

The white bighead may be regarded as a typical algae-feeding fish. The intestine is very short — not even as long as the body while feeding on planktonic animals during the young stage. However, as the feeding habit changes, the intestine gradually lengthens, until finally the intestine becomes ten times longer than the body. The intestine is very full during the summer; it contains a large portion of mud when the quantity of phytoplankton becomes small. Although the white bighead also feed in the winter, the amount of feed taken at that time is minimal.

### The Striped Bighead (*Aristichthys nobilis*)

The striped bighead (Figure 18.4) and the white bighead are very closely related. The former are slightly faster growing than the latter and are generally larger. However, they are not as active as the white bighead, which readily leap out of water when disturbed. The food of the striped bighead consists mainly of planktonic animals. The striped bighead generally live in the middle layer of a water column. In breeding habits and migration cycles, they are similar to the white bighead. In various bodies of water, the school of the striped bighead is smaller. This is probably due to the large mortality of the young fish.

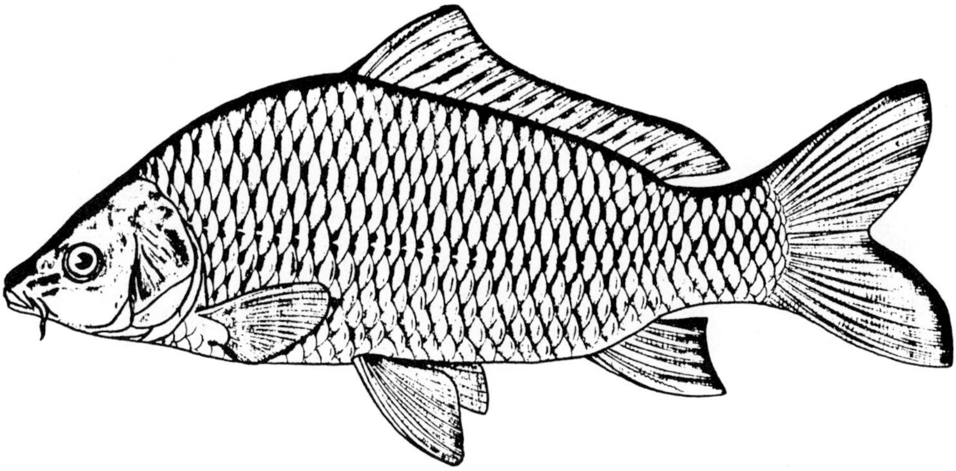

FIGURE 18.5. Common carp (*Cyprinus carpio* [Linn.]).

As with the white bighead, the planktonic animal food is gathered in the densely arranged gill filaments before it is passed on to the digestive tract. The major components of the food are the rotifers, crustaceans, daphnias, and copepods, as well as various species of algae. In numbers, algal cells often outnumber those of animal nature; in volume, animal food far outweighs vegetable food. The feeding behavior of both species of bighead indicates that they feed continuously. As long as the mouth is open, food continues to flow into the oral cavity without interruption. The intestine of the striped bighead is five times the body length.

**The Common Carp (*Cyprinus carpio*)**

The common carp (Figure 18.5) is a large fish, often weighing 13.5 to 36.5 kg. However, fish weighing 3 to 3.5 kg are the most prevalent. The growth rate is slower than that of the grass carp and the black roach. One-winter fish grow to 10 to 17 cm; two-winter fish, 22 cm; and three-winter fish, 30 cm. In general, females are heavier than males. Maturity is reached when the fish is 4 years old when the body is about 35 to 40 cm in length and weighs 1 to 1.5 kg. Some males may mature at 3 years of age when body length reaches 25 to 30 cm. Maturity may be reached even earlier in populations that live in warm waters. The carp will spawn in any type of water body. A female contains 300,000 to 500,000 eggs, which are laid in several batches. The eggs are adhesive; they become attached to aquatic vegetation immediately after being spawned. The egg has a diameter of 1.7 mm. Hatching takes place in 1½ to 2 days in a water temperature of 25°C. Newly hatched larvae usually hang upon surface leaves or stems where oxygen is rich.

The carp is an omnivorous species. Mollusks, higher aquatic plants, benthic animals, and minute algae are all regular food for the carp. There are three rows of pharyngeal teeth of many shapes and forms which can both cut aquatic vegetation and crush molluscan shells. Because the carp are not selective in their food and because they are very adaptive to a variety of environmental conditions, they are a species of high vitality. It is not accidental that the carp has become a widespread species.

The food taking of the carp varies according to seasons, food composition in the water, and size of the fish itself. In summer, carp feed most actively. In winter, feeding virtually stops. Juvenile carp 15 mm long feed primarily on algae and rotifers. Young above 3 cm long feed mainly on crustaceans, daphnias, copepods, and chironomid

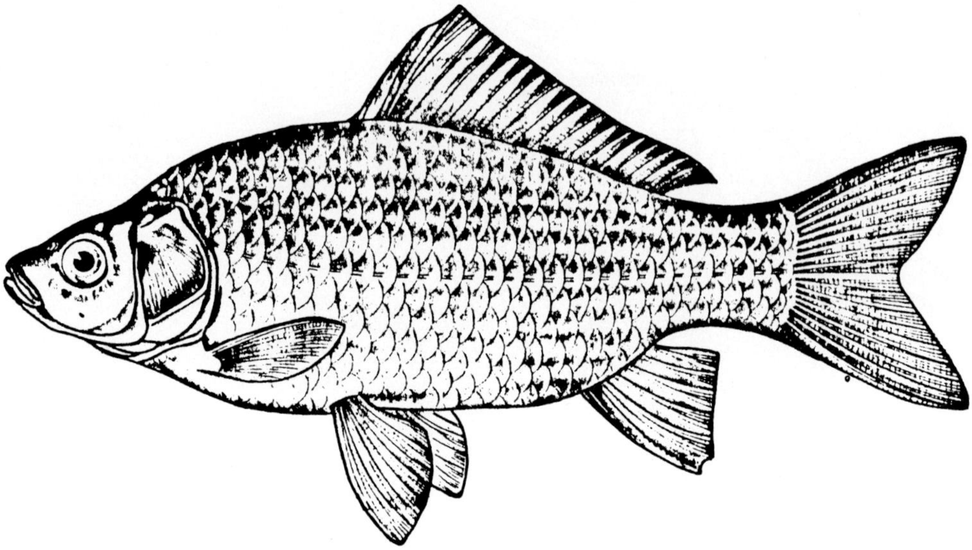

FIGURE 18.6. Goldfish (*Carassius auratus* [Linn.]).

larvae, and other insect larvae. When the fish grow to 20 cm long, larvae of *Tipulicae* become the major food. The 1 to 2-year olds depend principally upon higher aquatic plants and snails for food. Large amounts of algae and bits of detritis are also present in the intestine.

**The Goldfish (*Carassius auratus*)**

This is a medium-sized fish, but because of its excellent flavor, wide distribution, large production, and strong adaptive ability (Figure 18.6), it is economically very significant. The most commonly seen goldfish weigh about 0.34 kg; the larger ones may reach more than 1.1 kg. Even larger ones are relatively rare, although a few may weigh as much as 3.5 kg. One-winter fish may measure 5 cm; two-winter fish, 10 to 12 cm; and three-winter fish may measure 18 cm and weigh 0.34 kg. After this point, growth becomes slower. A five-winter fish measures only 22 to 25 cm and weighs 0.59 kg. Goldfish in northeastern China grow faster and are larger.

Goldfish breed in still water. The eggs are adhesive and become attached to aquatic vegetation immediately after being laid. Egg diameter is about 1.6 mm, slightly smaller than that of the carp. Fecundity of the female averages 200,000 eggs. The incubation period and fry habit are similar to those of the carp. The eggs are laid in batches. Spawning period lasts from April to July. Maturity is usually reached at 3 years of age, but a few may mature at 2 years.

In a school, females usually outnumber males; they are also a little larger in size. During the breeding season, tubercles appear on the head and pectoral fins. This is especially prominent in the males. The larger schools are found in quiet water in the lakes and ponds. Goldfish do not migrate into deeper waters of a river in the winter. The young usually gather in the bays of a lake and among thick vegetation in the school area.

Goldfish are also omnivorous in feeding habit. They differ from the carp principally in that they seldom feed on snails. The occasional occurrence of snails in the stomach of goldfish could be due to accidental ingestion with other food. The adult goldfish feed primarily on plants. The intestine may be five times the length of the body. In the intestine there is usually a large amount of fine material which contains various

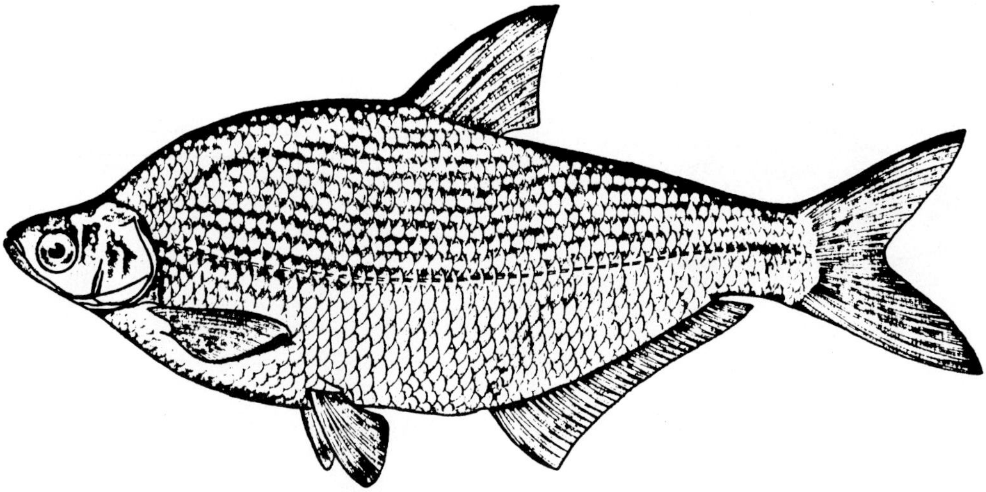

FIGURE 18.7. Peking Bream (*Parabramis pekinensis* [Basilewsky]).

algae, the most common of which are diatoms and filamentous algae. The goldfish frequently peck at algal growth on rocks or other immersed objects. The larger fish feed more on high plants. The food of the young goldfish is almost the same as that of the carp, i.e., algae, rotifers, daphnids, copepods, chironomid larvae, and other insect larvae. They are carnivorous at this stage of life. The fry of less than 15 mm, however, take algae as the principal food. Omnivorous fishes usually are not very selective about their food; for instance shrimps are sometimes found in the intestine of the goldfish.

### The Peking Bream (*Parabramis pekinensis*)

This is a widely distributed species. The maximum weight may reach nearly 1.3 kg; average weight is about 0.68 kg. One-winter fish may grow to 8 cm; two-winter fish, 14 cm; four-winter fish, 22 cm (Figure 18.7). Maturity is not reached until 5 years old, at which time the body is about 25 cm. May to June is the breeding season. Spawning takes place in flowing water in the river. During a freshet, the tributaries to the larger lakes also may be used as the breeding place. Eggs are nonadhesive; they drift downstream with the current. After fry develop free-swimming ability, they move to shoal areas along the shore or in small bays where there is not much water flow and use these areas as their nursery ground. The adult fish also move into small tributaries to feed after spawning, but they migrate into deeper waters to over-winter.

The Peking bream is a herbivore. Its intestine may be three times as long as the body. Principal food consists of *Vallioneria, Potomogeton,* etc. Feeding activity is the strongest between spring and summer. During the cold winter when vegetation is lacking, Peking bream feed on some small fish or on the fruits of some higher aquatic plants. During the young stage, the bream are omnivorous. Fry that are under 2 cm in length depend upon algae as the principal food. Those above 3.5 cm can prey upon small snails, chironomid larvae, and some higher aquatic plants. It is when the fish reach 1 year old that they become principally a plant feeder.

### Mud Carp (*Cirrhina molitorella*)

This fish (Figure 18.8) is a member of the subfamily Xenocyprininae. It occurs mainly in warm waters. It is one of the major cultured species in the Pearl River system

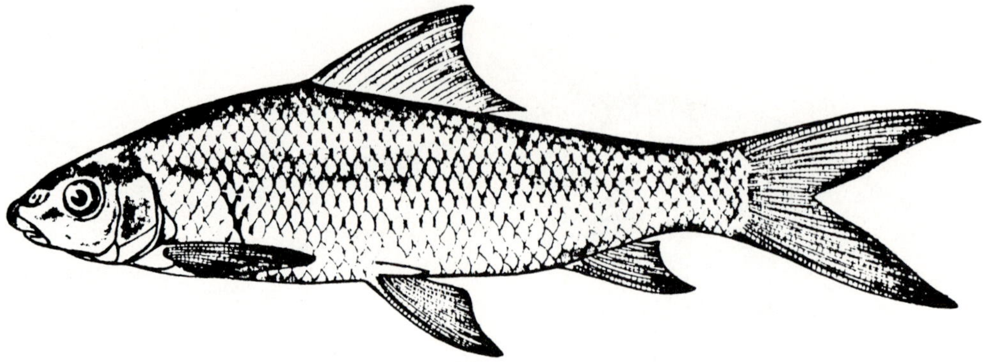

FIGURE 18.8. Mud carp (*Cirrhina molitorella* [Cuv. et Val.]).

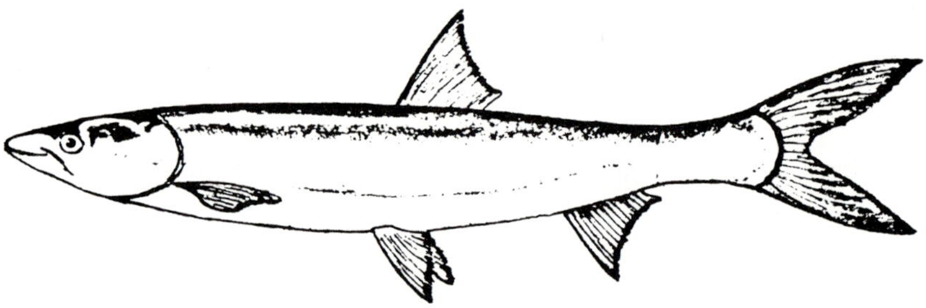

FIGURE 18.9. Carnivorous cyprinid (*Elopichthys bambusa*).

and is produced in large quantities. In natural water bodies, this fish may grow to 2.5 kg or more. The largest number of schools are composed of fish weighing approximately 0.9 kg. Fish may grow to 8 to 9 cm in a year; in 2 years, fish may reach 20 cm in length and 0.23 kg in weight and become edible at this stage. Spawning starts in May and continues until September. The youngest age at maturity is 3 years when the fish weighs a little over 0.56 kg. A few may become mature at a body weight of only 0.28 kg. Spawning grounds are located from the middle to the head water of a river. These fish abound in the tributaries of the West River in Kwangsi Province, such as Hung-sui River, Liu River, Ching River, and Kwei River. This species has a low tolerance to low temperature. They cannot survive in waters below 7°C. They over-winter in the deep riverbed during the cold season.

This fish is a benthic species. It likes to nip attached growth on bottom materials. Its food consists mainly of algae, especially diatoms, but it also feeds on aquatic plants and a variety of bottom materials.

### Carnivorous Cyprinid (*Elopichthys bambusa*)

This fish (Figure 18.9) is a large, carnivorous cyprinid. Maximum size can reach over 60 kg. The most frequently seen weigh from 2 to 6 kg. They have a fast growth rate. One-winter fish may grow to 25 cm; two-winter fish, 35 to 40 cm. They will put on an additional 10 cm every year after that until 6 years of age. After this, length increment slows down, but there is appreciable increase in weight. For instance, four-winter fish measure 55 cm in length and weigh around 1.8 kg; six-winter fish, 70 cm and 3.6 kg; and seven-winter fish, 75 cm and 4.8 kg.

Maturity is not reached until 4 to 5 years old. Breeding habit is similar to that of

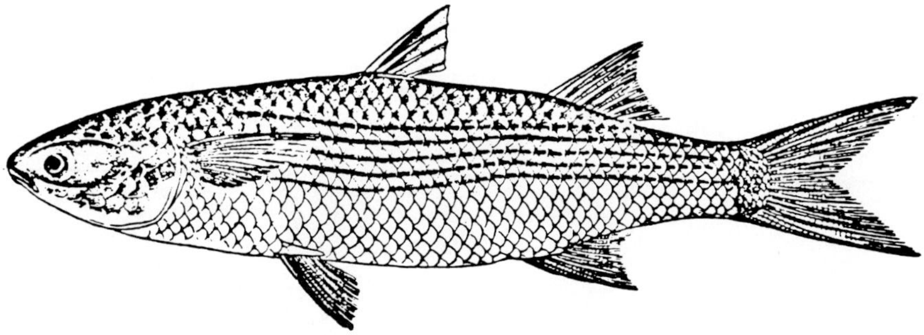

FIGURE 18.10. Chinese mullet (*Mugil cephalus*).

grass carp and the bigheads. Spawning takes place in moving water in rivers. After absorption of water, the egg measures 6 mm in diameter and is larger than most other species that have similar breeding habit. After spawning, *Elopichthys* move into secondary water bodies to feed and return to the main river in winter, staying in deeper parts to over-winter. The juveniles are also often found in lakes and ponds feeding along the shores; they also return to the main river to over-winter.

*Elopichthys* is a notoriously voracious fish. Living in the upper and middle layers of the water column, it preys entirely upon other species of fish and is an active swimmer. Because the mouth is not especially large, prey usually consists of slender-bodied species. Larger, deeper-bodied fish or bottom-dwelling species are seldom seen in the stomach contents. Larvae 2 cm long already start feeding on other fish fry. The intestine of *Elopichthys* is relatively short. In examinations of the food contents of this fish, the intestine is often found empty, attesting to a fast rate of digestion and excretion.

### The Chinese Mullet (*Mugil cephalus*)

This mullet (Figure 18.10) is a relatively abundant species of the family Mugilidae which lives in the estuarine environment. It occurs near the river mouth along the sea coast (Fukien, Chekiang, Kiangsu). It spawns in early spring in the coastal area. The young enter the rivers (Chiang-tang, Ming, and Yangtze Rivers) during the last third of March. They grow fairly fast. One-winter fish may reach 18 cm in length; two-winter fish may reach 35 cm and weigh around 0.59 kg. Five-winter fish may exceed 50 cm.

The mullet is an estuarine species. It feeds on materials off the bottom mud, ingesting few living organisms and little organic matter. It has a rather peculiar digestive system. The intestine is very long. The stomach has a thick wall, performing a grinding function. The young mullet migrate into the river during early spring. At that time, they feed on plankton.

### The Chinese Perch (*Siniperca chautsi*)

The Chinese perch (Figure 18.11) is a well-known predaceous fish. It is especially abundant in lakes and ponds. It has a fast growth rate and may reach a maximum weight of 6 kg. One-winter fish may reach 12 cm; two-winter fish, 20 cm; and three-winter fish, 25 to 30 cm. At this age maturity is reached. Males may mature 1 year earlier. The fish spawn in flowing waters in the main stream, in the larger tributaries, or in good-sized lakes. Eggs are laid in separate installments, generally occurring in May and June. Eggs are small, measuring only 2 mm in diameter. They possess an oil globule, thick membrane, and are only slightly transparent. They are planktonic and

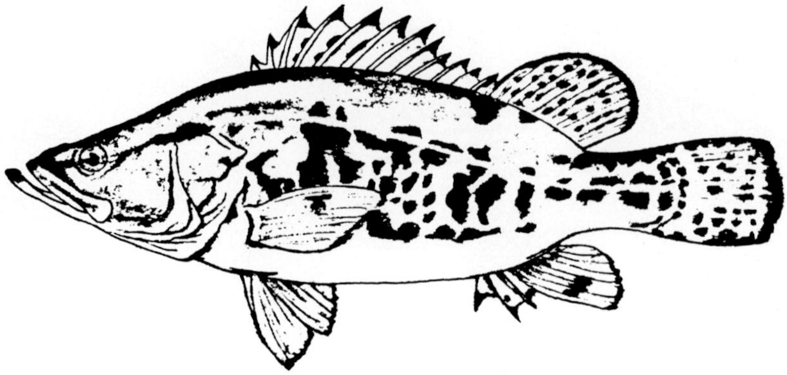

FIGURE 18.11. Chinese perch (*Siniperca chautsi*).

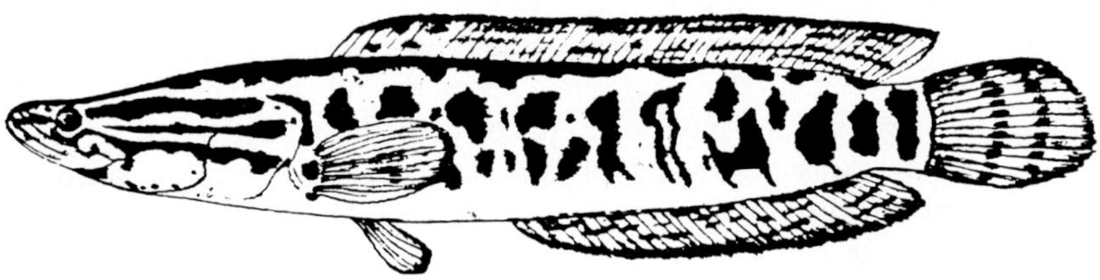

FIGURE 18.12. Snakehead (*Ophiocephalus argus*).

drift freely in the water. Incubation period lasts from 3 to 4 days. While growing to 1 cm long, the young find auxillary water bodies in a nursery area. The adult fish or past-spawning parents often stay in the clearer major tributaries or lakes. Some also live in the main rivers. During the winter, all sizes live near the bottom to over-winter and are rather inactive.

The Chinese perch is a voracious fish. Its digestive system is different from that of predaceous cyprinids. On both the upper and lower jaws, there are sharp teeth, which facilitate prey-catching. The intestine is very short, but there is a thick-walled, very elastic stomach. During periods of active feeding, the stomach is often greatly distended. Food contents often include goldfish, bream, nace, catfish, etc., and even its own young. Some of the prey fish may be longer than half its own body length. In waters rich in shrimps, they too are often part of the food contents.

The habit of preying starts with the juvenile stage in the Chinese perch. The prey is often as long as the perch itself. As the perch grows longer, its prey consists of relatively shorter fish. Mysids are a frequent food item for the young perch.

### The Snakehead (*Ophiocephalus argus*)

The snakehead (Figure 18.12) is a voracious fish which lives among thick vegetation. It grows fairly rapidly, reaching a maximum weight of 6 kg. Males are larger than females. One-winter fish may grow to 25 cm; two-winter fish may reach 35 cm and weigh about 0.59 kg; four-winter fish, 55 cm and more than 2.3 kg. Generally, maturity is reached at two winters of age. The fecundity is about 50,000 to 60,000 eggs, which are laid in separate installments. For spawning, a nest is built with pieces of higher plants near the surface of the still waters of bays or lakes. Egg diameter is about 2 mm; eggs contain a large oil globule and are therefore buoyant. After the eggs are

FIGURE 18.13. Sheat fish (*Parasilurus asotus*).

laid, the male guards the nest. Hatching takes place in 2 days. The yolk sac becomes completely absorbed when the young reaches 8 cm long. The young are usually schooled in swimming. The snakehead lives year-round in lakes and bays and over-winters in the deeper parts. It possesses an epibranchial organ, which is an auxillary breathing organ. Because of this, the snakehead can tolerate relatively low oxygen content in the water.

The snakehead preys upon other fishes and shrimps. Its prey consists of the common species, such as the goldfish, pike, etc. It does not chase after its prey; rather, it hides among the vegetation and waits for the prey to pass by, whereupon it darts out to catch the prey. The young snakehead, aside from feeding on other young fishes, feeds upon insect larvae, crustaceans, and other animal food.

### The Sheat Fish (*Parasilurus asotus*)

The sheat fish (Figure 18.13) is among the most widely distributed, highly productive, and economically important species in China. It has a fast growth rate and grows to 30 to 37 kg. One-winter fish may attain a body length of 15 to 20 cm. The fast growth rate is maintained during the second year, at which time the fish may reach above 30 cm. Thereafter, the increase in length slows down somewhat, but the body increases rapidly in weight, especially in females. Maturity is reached when the fish becomes three winters old. By that time, the body measures about 40 cm long and over 0.40 kg in weight. Fecundity is several tens of thousands of eggs. Spawning takes place during May and June. Eggs are slightly adhesive and are scattered over aquatic plants near the bottom of still water. The egg is greenish, measuring 4.5 mm in diameter. It generally takes 5 to 6 days to hatch. During spring and summer, the sheat fish feeds in lakes and tributaries where aquatic vegetation abounds; during autumn and winter, it enters the main stream or the deep layers of the branch rivers to over-winter.

The sheat fish is also a voracious species. It has a large mouth which is provided with teeth. The main food consists of various fishes (goldfish and other shore-living cyprinids), shrimps, and aquatic insect larvae. Higher aquatic plants which are sometimes found in the intestine are probably ingested with other foods. A sheat fish 5 cm long can swallow several prey fish 2 cm in length. Compared with other carnivorous fishes, the sheat fish starts to prey at a much younger stage.

The sheat fish uses a different method in catching its prey. It adopts the strategy of ambush by hiding in thick vegetation near the bottom and quietly watching the activity of other fishes. When the prey comes within striking distance, the sheat fish opens its wide mouth and sucks in the prey together with water.

### The Chinese Chub (*Erythroculter ilishaeformis*)

This fish (Figure 18.14) is the largest member in the family Cyprinidae. It has a delicate flavor and a high economic value. The largest individual may weigh up to 12

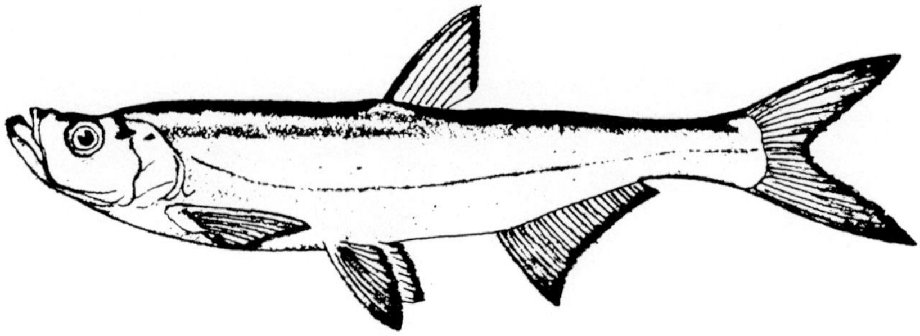

FIGURE 18.14. Chinese chub (*Erythroculter ilishaeformis*).

kg, but it may take more than 10 years to grow to this size. More generally, the average weight is 1.5 to 3 kg. Growth is rapid under 5 years of age. Generally, one-winter fish are about 15 cm long; two-winter fish, 22 to 25 cm; and six-winter fish, 50 cm. Mature individuals weigh about 1.8 kg. Spawning occurs in lakes and rivers. Eggs are somewhat adhesive and attach to water plants. In northeastern areas, the eggs are nonadhesive and drift freely with the current. Spawning takes place during June and July. Fecundity is about 30,000 to 50,000 eggs. After water absorption, eggs measure about 4 mm in diameter. Incubation takes only 2 days. After spawning, the parent fish enters into lakes to feed or may frequent shallow portions of the river for food. Juveniles often feed along the shore areas. When winter comes, all sizes of fish gather in deeper places to over-winter.

The Chinese chub is a voracious fish which feeds on other fishes in upper and middle layers of the water. It differs from *Elopichthys bambusa* in that its young stages (less than 10 cm in length) seldom prey upon fishes. Stomach contents of the adults also contain fishes of the slender such as *Hemiculter leucisculus,* pike, etc. Occasionally, young of its own are also found in the stomach of the adult fish. Prinicipal food items of the young stages consist of aquatic insects, daphnias, shrimps, and even small amounts of higher water plants.

All other related species also feed upon other fishes. However, because of their own small size, plus the fact that they have relatively small mouths, their victims are usually small, miscellaneous species. Also, aquatic insects, shrimps, and other small crustaceans form a significant portion of their diet.

### The Nace (*Xenocypris argentea*)

The nace (Figure 18.15) is the most widely distributed cyprinid in China. The largest individual attains only a little over 0.5 kg. Those that are normally used for food weigh around 0.25 kg. The growth rate is slow: one-winter fish grow only to about 6 cm; annual increment thereafter is only about 3 cm. Five-winter fish may grow to 18 to 20 cm. Females are usually slightly larger than the males. Maturity is reached in the three-winter fish when the body length is about 13 cm. Spawning takes place in May to June in running water. Eggs are transparent and measure about 3.6 to 4.0 mm. They hatch after 2 to 3 days. The larvae are brought by the current to still waters in eddies, where they feed and grow. After autumn they move to deep places to over-winter. After spawning, the parent fish enters into lakes to feed, only to return to the river to over-winter.

The nace is primarily a phytoplankton feeder. Diatoms and green algae are the most frequent items of food. Adult intestines often contain detritus and fine sand and mud particles. The intestine of adult fish is about 5 times the length of the body, but that

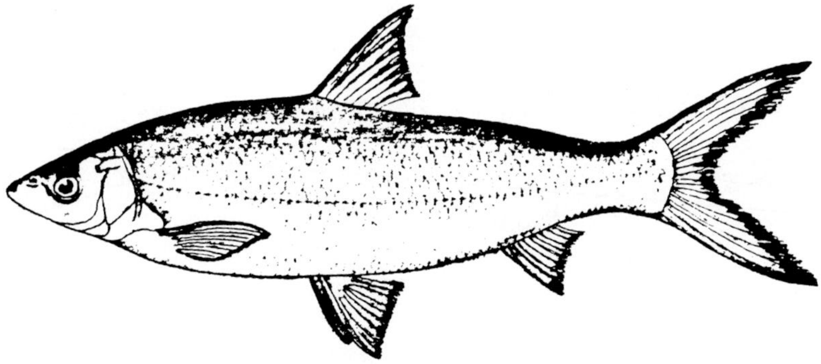

FIGURE 18.15. Nace (*Xenocypris argentea*).

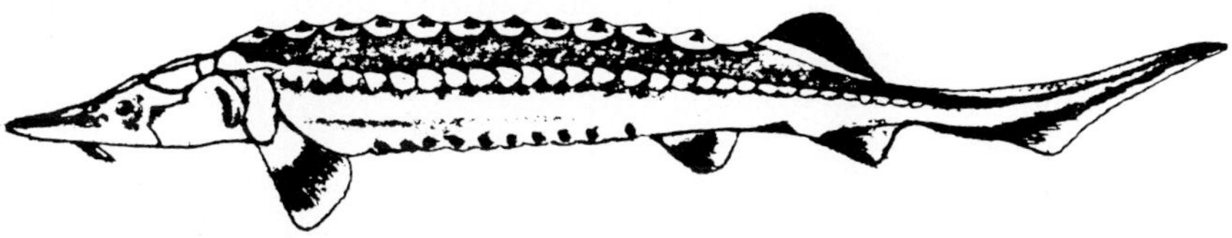

FIGURE 18.16. Chinese sturgeon (*Acipenser sinensis*).

of young fish is only twice as long as the body. Young nace under 1.5 cm feed mainly on animal food, mostly daphnias and rotifers. Thereafter the feeding habit changes entirely to a herbivorous one. In winter, feeding almost completely stops.

The feeding habits of other related species are similar, except that they also take small amounts of higher aquatic plants.

**The Chinese Sturgeons (*Acipenser* sp.)**

Sturgeons are produced in both Heilung and Yangtze Rivers in China. However, each river produces a different species. The sturgeon produced in Heilung River is *Acipenser schrenchi,* which is smaller than the other. The maximum size it may reach is nearly 30 kg at an age of over 20 years. The Yangtze sturgeon (Figure 18.16), *A. sinensis,* may grow to 100 kg.

With the Heilung sturgeon, the hatch of the year may grow to 15 cm; one-winter fish, 30 cm; three-winter fish, 50 to 60 cm with a weight of about 0.55 kg; five-winter fish, 80 cm and weigh 2 kg; eight-winter fish, 100 cm and more than 4.5 kg. Sturgeons are long-lived fish and are late in reaching sexual maturity. Generally they become mature at 10 to 13 years of age, when they are about 1 m in length and weigh about 6.5 kg. The spawning period lasts a fairly long period; they spawn any time from May to September. Fecundity is about 100,000 to 200,000 eggs. Spawning grounds are located in main streams where there are gravel beds and where the current is rapid. The newly hatched sturgeon drift downstream with the current. When they are able to swim, they school in slow-moving streams near the bottom to feed. After autumn they migrate to deeper riverbeds and continue to feed. Post-spawners generally remain in the main stream. The Yangtze sturgeon spawn in the upper reaches of the river in the autumn.

A closely related species, the Amur sturgeon (*Huso dauricus*), is also found in the Heilung River. This is a large sturgeon which may grow to 100 kg. Its spawning season

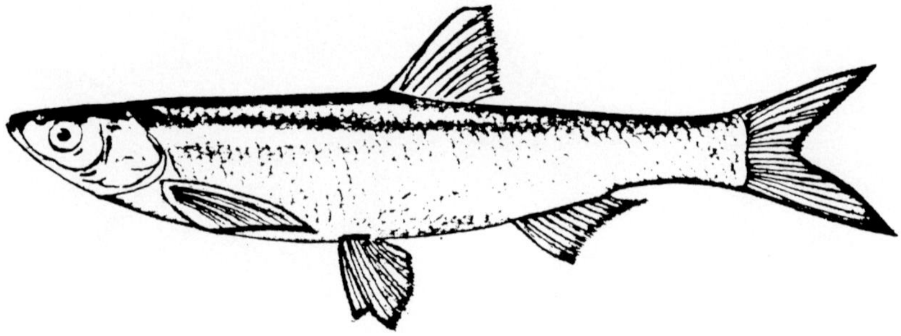

FIGURE 18.17. Hemiculters (*Hemiculter leucisculus*).

is in May to June. All species of sturgeon are highly valued fishes. Their eggs are especially esteemed.

Sturgeons are large fishes. They feed on bottom fishes and benthic animals in the river. Because of their large mouth, gravel is often ingested with the food. Young sturgeon feed on insect larvae. When they grow to about 70 cm, the frequency of food fishes in their stomachs increases noticeably.

**The Hemiculter (*Hemiculter leucisculus*)**

This is a small fish (Figure 18.17). However, because of its wide distribution and large quantity, it could be developed into an economically useful species. There are several species of Hemiculter. The commonest are *H. leucisculus* and *H. peii.* They occur in various bodies of water, but the former are more often found in still water, while the latter occur primarily in the rivers. This difference in habitat is due to the slight difference in spawning habit, for the former has adhesive eggs while the latter has free-drifting eggs.

The largest individuals of *H. leucisculus* may grow only to about 18 cm and weigh 10 oz. One-winter fish may attain a length of 5 to 6 cm; two-winter fish, about 10 cm; three-winter fish, 14 cm. In general, 2- to 3-year-old fish may mature and spawn. Spawning season is in May to June. The male develops a white spot during the spawning season.

*H. peii* is even smaller, weighing 0.14 to 0.17 kg at the maximum. The most commonly seen are 20 to 30 g in weight. One-winter fish measure about 4 to 5 cm; two-winter fish, 8 to 9 cm; and three-winter fish, 11 cm. Two-winter fish mature and spawn. Spawning season is also in May to June. Egg diameter is about 3 mm. Hatching takes place after the first day.

*Hemiculter* live in shallow areas in the rivers and lakes during all life stages. Only in the winter do they move into deeper places to over-winter.

Although there are many species of *Hemiculter,* the food habits are fundamentally the same in all species, i.e., they are omnivorous. Because they are small, their food items are small organisms. The main food items are algae, rotifers, small crustaceans, land and aquatic insects, and some higher aquatic plants. Young fish below 5 cm in length depend primarily on rotifers and algae as food. After that, they can eat many kinds of food. *Hemiculter* is a prolific fish and can spawn in various bodies of water. This is related to the fact that they are nonselective in their foods.

**The Yellow Catfish (*Pseudobagrus fulvidraco*)**

In the catfish family, the yellow catfish (Figure 18.18) is the most widespread and the most productive in China. Individuals are often seen that weigh about 0.23 kg,

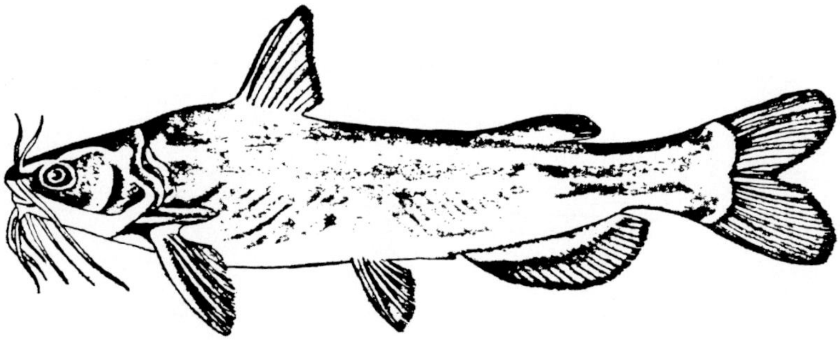

FIGURE 18.18. Yellow catfish (*Pseudobagrus fulvidraco*).

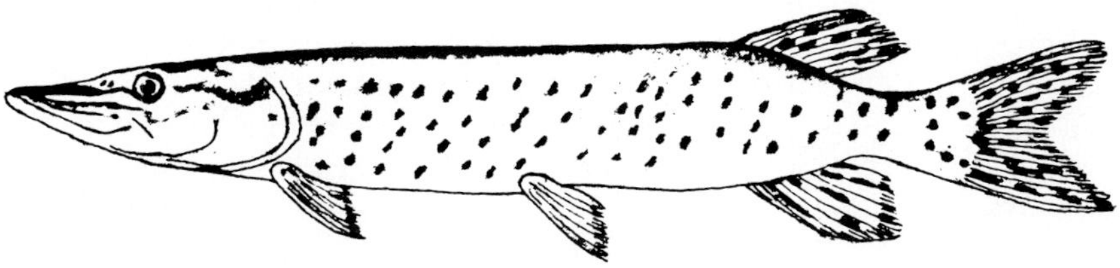

FIGURE 18.19. Chinese pike (*Esox reicherti*).

but larger ones may weigh over 0.5 kg. Males are usually larger than females. One-winter fish may grow to 8 cm in length; two-winter fish, 15 to 18 cm; thereafter the growth slows down, and five-winter fish measure only about 25 cm. Maturity is reached after three winters of growth. Fecundity is about 5000 eggs. Spawning takes place during May to June on gravel bottom in still water near shore. The male catfish digs a nest, which measures about 10 cm in diameter and 10 cm in depth. The male also guards the nest after eggs are laid; often it fights and chases other fishes away. Eggs measure 2.5 mm in diameter and are slightly yellowish. They hatch within 2 days. The young feed along the shore areas where the current is slow and enter into deeper waters during the winter. During the spawning season, the adult fish enters into lakes and branch rivers, and remains there to feed after spawning.

The yellow catfish is a carnivorous species. Its food includes small fishes, shrimps, various terrestrial and aquatic insects (especially mayfly and midge larvae), mysids, small molluscs, and other aquatic invertebrates. Sometimes large quantities of daphnias and higher aquatic plants are also included. In summer time, the yellow catfish may swim on its back in still water and catch falling insects from the air. This phenomenon is frequently observed in evening and night hours. The young yellow catfish depends mainly on benthic animals, chiefly chironomid larvae, for food; this is due to the inferior location of its mouth. When winter comes, small fishes become more important in the food of the yellow catfish, and at the same time, benthic animals diminish in number. During the period of nest guarding, the male yellow catfish does not take in any food.

### The Chinese Pike (*Esox reicherti*)

The Chinese pike is a fast-growing, cold-water fish in northeastern China (Figure 18.19). It has significant economic value. Pike that are often seen on the market weigh about 1 to 2 kg and measure about 45 to 55 cm in length. A large pike may grow to 1

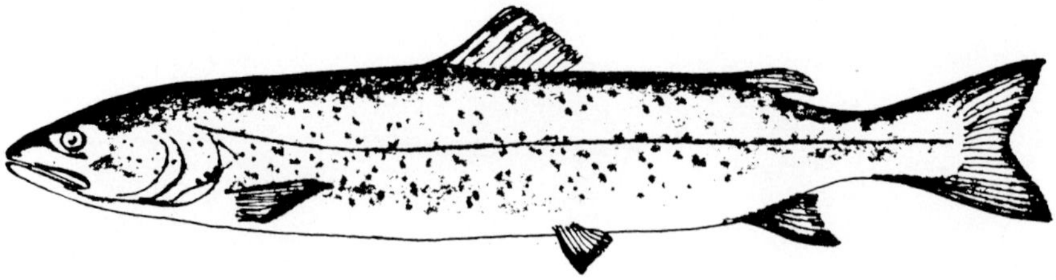

FIGURE 18.20. Chinese Taimen (*Hucho taimen*).

m and weigh 12 kg. One-winter fish measure about 30 cm and weigh 0.40 kg; two-winter fish measure about 40 cm and weigh about 1 kg; three-winter fish, 40 to 50 cm and 1 to 1.5 kg; five-winter fish, 60 cm and more than 2.25 kg; eight-winter fish, 80 to 90 cm and almost 10 kg. Males mature at 2 years of age, females at 3 years.

Pike spawn in early spring, generally in April after ice melts. Spawning occasionally may be delayed until the end of May, as a result of unfavorable conditions, such as low water level. In any event, the spawning period is quite short. Pike that weigh about 2 kg contain 30,000 to 50,000 eggs; large pike weighing more than 5 kg may produce 150,000 eggs. The eggs are large, yellow, and slightly adhesive. They attach to aquatic vegetation on the bottom. Incubation takes about 10 days. Adult pike migrate into accessory waters in early spring and start to spawn as soon as ice breaks up. Larger individuals may return to the rivers. In early winter, they move into deeper riverbeds and continue to be active and feed. The young pike live among thick vegetation near shore.

The pike is a voracious fish. It has a large mouth and sharp teeth. Predation starts during the young stage, when the fish feeds on fry or juveniles of other fish species. After spawning, the pike is a strong and active feeder. At this time, it lives in branch rivers, bays, and small tributaries to the main rivers. It may take a prey half as long as itself. When the weather turns cold, the pike migrates into deeper waters where it feeds upon those fish species that happen to over-winter there.

**The Chinese Taimen (*Hucho taimen*)**

The Chinese Taimen is the largest salmonid in China (Figure 18.20). It occurs in the Heilung River system. Maximum size may reach 22 to 30 kg in weight, 120-130 cm in length, and 10 years of age. The most commonly seen, however, average approximately 5 kg in weight. One-winter fish grow to 10 cm, and annual increment is about 1.2 kg. Four-winter fish are about 45 cm and weigh about 2 lb 10 oz. At this age, maturity is reached, and spawning takes place during May. Spawning fish turn a bright red color. Fecundity is about 20,000 eggs. Spawning takes place on gravel beds where there is rapid water movement. Parent fish remain here to feed after spawning and do not return to the main river until after autumn.

The taimen is also an extremely active, voracious fish. Its food consists of various cold-water fishes. In the stomach of larger individuals, frogs occassionally may be found.

## SUMMARY OF OBSERVATIONS

The fish-producing areas in China can be divided into five major areas; the Amur River basin, the Yangtze-Yellow River basin, the South China area, the Northwest Area, and the Tsangpu-Lu River basin. The majority of the important fish species originated in the Yangtze-Yellow River basin.

The major fish species can be divided into two groups: the herbivores and the carnivores. The carnivores include the Chinese perch, snakehead, sheat fish, sturgeon, yellow catfish, pike, and taimen. The herbivores include the carp species, black roach, the bigheads, Peking bream, and the mullet species. Generally, the herbivores prefer to feed in calm, still waters, while the carnivores feed in either fast or slow waters. The breeding and migration habits vary widely with the individual species. Size and other habitat requirements also vary with each species.

## REFERENCES

1. **Roxby, P. M. and Freeman, T. W.,** *China Proper; Physical Geography, History and Peoples,* Division of Naval Intelligence, Great Britain, London, 1944.
2. **Whitaker, D. P. and Shin, R. S.,** Area Handbook for the Peoples Republic of China, DA Pam 556-60, U.S. Government Printing Office, Washington, D.C., 1972.

Chapter 19

# FISH PRODUCTION IN CHINA*

## FISH FRY PRODUCING AREAS

The major fry producing areas in China are located in the Yangtze and Pearl valleys. In recent years, other producing areas have been discovered, but they are all of minor importance. In five provinces along the Yangtze, production in 1957 was 13 billion fry or 56.3% of the total production in the entire country. The Pearl River is second only to the Yangtze River in importance, but the principal species here is *Cirrhina molitorella*. Its production in 1958 was 10 billion fry, accounting for 42% of the national total (see Figure 19.1).[1-4]

### The Yangtze River System

**The main Yangtze** — Fry are produced from Hupeh to Kiangsu along a 1600-km stretch. In Hupeh, the principal species harvested is the white bighead; next are black Chinese roach and grass carp. In 1957, the production was 4.5 billion, 19.2% of the national total. In 1958, the production increased to 12 billion, taking the leading position in the country. In Anhui Province, the production in 1957 was 4367 million, accounting for 18% of the national total, second only to Hupeh. The white bighead accounts for 50%, the striped bighead 20 to 30%, and grass carp and black Chinese roach combined around 20%. In Kiangsu Province, the production in 1957 was 970 million. The 1958 production was 1.9 billion, consisting of grass carp and roach, 70 to 75%, and the bigheads, 25 to 30%.

**Tributaries of the Yangtze** — Three major tributaries to the Yangtze River also produce large numbers of fry. These are the Han, the Hsaing, and the Kan rivers.

### The Pearl River System

The Pearl River is made up of three main branches: West, North, and East Rivers. The West River is the most important contributor to fry production. The major species produced in this area is *Cirrhina molitorella,* which accounts for 80% of the total. Grass carp is next, accounting for 18%, and the white bighead accounts for 2%.

### Other River Systems

Four other river systems in China also produce fish fry, but they are relatively minor in importance. These river systems are the Hwai, the Amur, the Great Canal, and the Chientang.

## COLLECTION OF FRY

### Fry "Bloom"

Fry "bloom" is a term applied to the phenomenon of mass appearance of fry in a river, and this is the time for fry collection. Fry bloom coincides with the spawning of the fish. The timing and duration of the bloom vary geographically. In the Yangtze River, the fry producing period lasts from mid-April to early July. In the Pearl River, it lasts from mid-April to the end of August.

* From *Culture of Freshwater Fish in China,* Science Publishers, Pekinr, 1966, translated by Koo, T.Y., Dept. of Fisheries, U.S.D.I., Solomons, Md., and Walker, C., U.S.D.I., Washington, D.C.

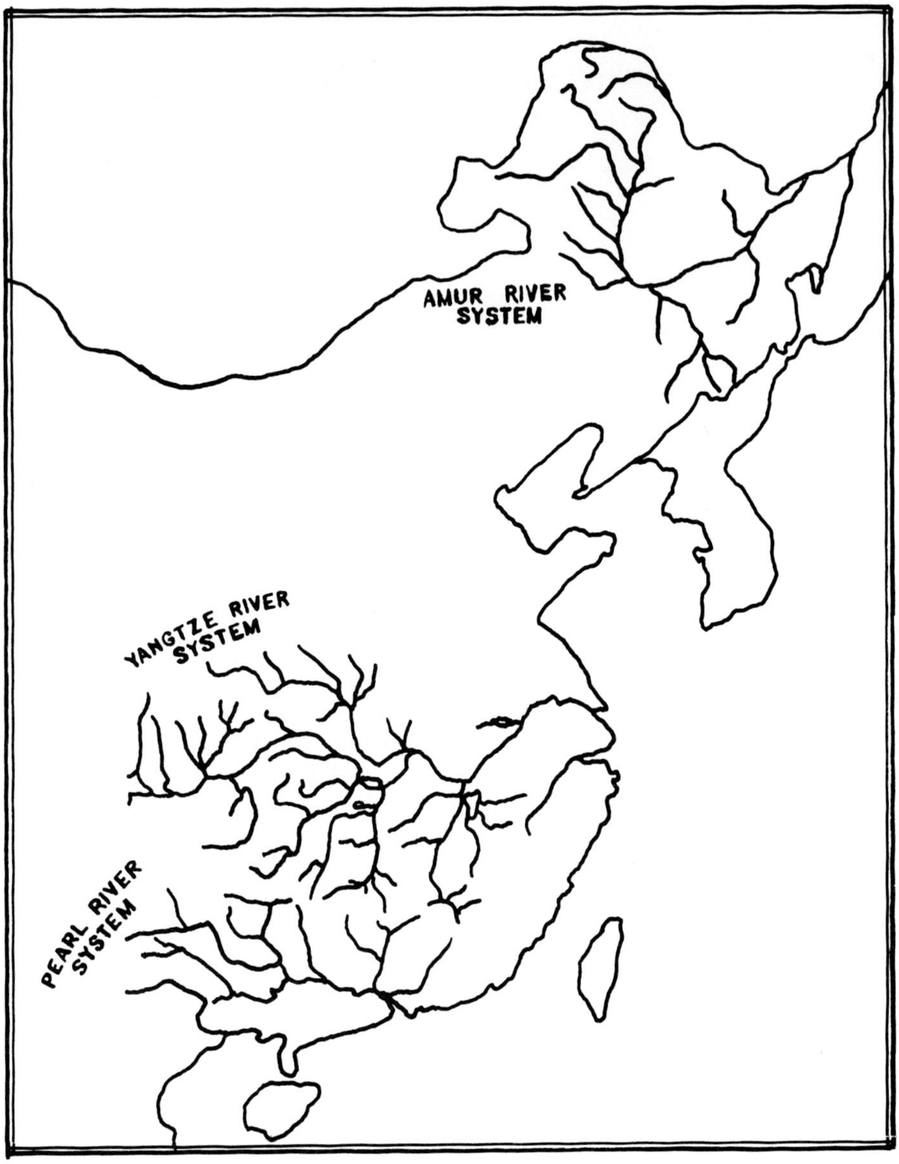


FIGURE 19.1. Fish fry producing areas in China.

*Prediction of Fry Bloom*

**Prediction of initial spawning date** — Spawning season depends a great deal upon climatic and water conditions; so does fry bloom. Therefore, it varies from year to year; the difference may be as much as 20 days. The most important factor is the temperature during the spring. The warmer the spring, the earlier the spawning and blooming. It has been estimated that if the weather stays warm after the beginning of spring, fry will appear in 80 days; if the weather is cold after that date (February 15 to 18), fry will appear in 100 days.

**Forecast of fry bloom** — Based on experience, fishermen forecast the onset of fry bloom on the basis of rain. This is because, during the late spring to early summer, a heavy rain induces fish spawning.

**Phenomena accompanying fry bloom** — The coming of fry bloom can be usually recognized by some phenomena that appear with the rise of river water. The water is usually turbid and full of debris, foam, and sometimes caddis fly larvae.

*Periods of Fry Bloom*

Fish fry are collected in five periods which are detailed in the following paragraphs.

**Mid to late April** — These are the earliest fry. They are relatively few in number and consist of a large portion of undesirable small species. They account for about 10% of the yearly total.

**Around May 6** — Fry produced in this period are larger and healthier and are more fit for transport.

**Around May 22** — During this period, water temperature rises above 20°C. Peak spawning takes place resulting in the largest production of fry. These fry are the most sought after by buyers.

**Around June 6** — This is the height of summer, and it is very hot. Fry mortality is high during transportation. Therefore, these fry are primarily used locally.

**Around June 22** — Fry production is small and is used locally.

The above periods apply to the Yangtze River. In the Pearl River, there are eight or nine blooms annually, and some of these may overlap each other.

**Selection of Collecting Site**

Most of the collecting gear used in China consists of setnets. Therefore, the collecting site is a determining factor of success or failure. Not all shoreline areas are suitable for net setting. Much depends upon the topography, current velocity, and current direction. Some of the more important points are discussed below.

**Current velocity** — Fish fry are swept into the collecting net. Current velocity, therefore, becomes the primary factor in selecting the collecting site. If the current is too rapid, it will kill the fry that are collected in the net. A nearby river bank may collapse resulting in the loss of the net. On the other hand, if the current is too slow, the net may not extend fully, and few fry will enter the net. The most suitable velocity is about 0.67 mi/hr.

**Topography** — Desirable conditions include a relatively level bottom with a gradual slope and a depth of 5 to 8 ft. The two banks should be relatively straight and parallel with the current flow. An ideal spot for net setting would be at the confluence of two opposing currents, one flowing in from midstream toward the bank and the other flowing out from shore toward midstream. A back eddy sometimes serves as a good place for setting the net if there is enough current flow.

**Nature of bottom** — It is important to avoid rocky, uneven bottoms. New delta area where there is a lot of new sedimentation should also be avoided.

**Collecting Gear**

The most prevalent gear used in the Yangtze River is the trap. There are two major kinds: the floating trap and the fixed trap. Both traps are similar in structure. The only difference is that the floating trap is attached to a log or a bamboo stem, which can move up and down with the rise and fall of water level, while the fixed trap is affixed to stakes and is not moveable.

*Description of Gear*

**Floating trap (Figure 19.2)** — This is the most popular gear used in Yangtze and Pearl Rivers. There are many variations, but they are generally composed of three parts:

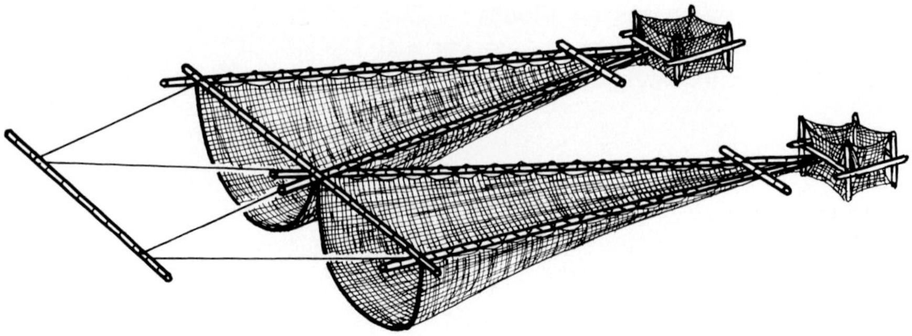

FIGURE 19.2. Floating trap which rises and falls within the water level.

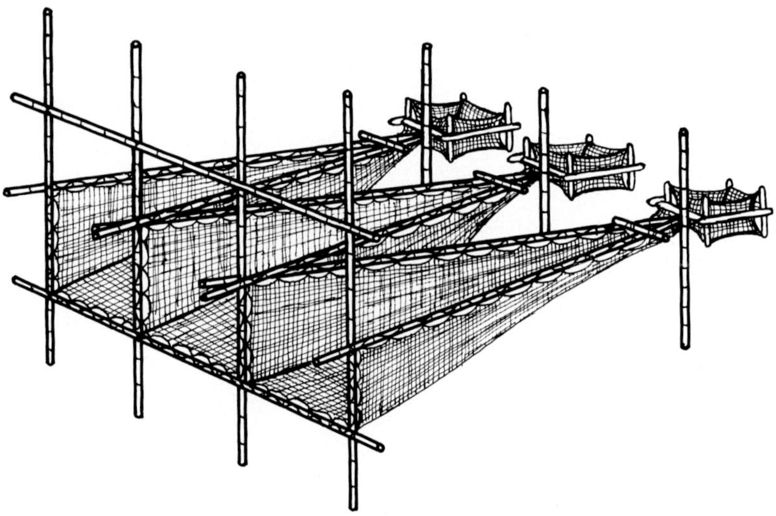

FIGURE 19.3. Fixed trap, planted with stakes in the river.

1. Body — This is shaped like a half funnel, composed of a floor, a wall, and a sleeve. The sleeve leads into the tail box. The net is about 4 m long. The diameter of the mouth is 3 to 4 m. The sleeve is 0.9 m long and has a rear diameter of 10 to 13 cm.
2. Tail box — This is the collecting box, usually 45 × 30 × 30 cm. It is mounted on a wooden support so that the net is above water surface.
3. Frame — The frame is made of bamboo or log. It is V-shaped. It is used to fasten the net as well as to provide flotation.

**Fixed trap (Figure 19.3)** — The use of the fixed trap is limited to a few localities in Hupeh, Anhei, and Kiangsu. The trap is made in the same way as the floating type. It is affixed to stakes that are planted into the river.

**Bamboo trap (Figure 19.4)** — The shape of the bamboo trap is similar to that of the floating trap and is similarly attached to a floating log or bamboo. It is made of three parts:

1. Body — This is constructed of split bamboo which is about 1 cm in width. The length of the body is 3.4 m; the width at the mouth is 1.3 m; and the diameter at

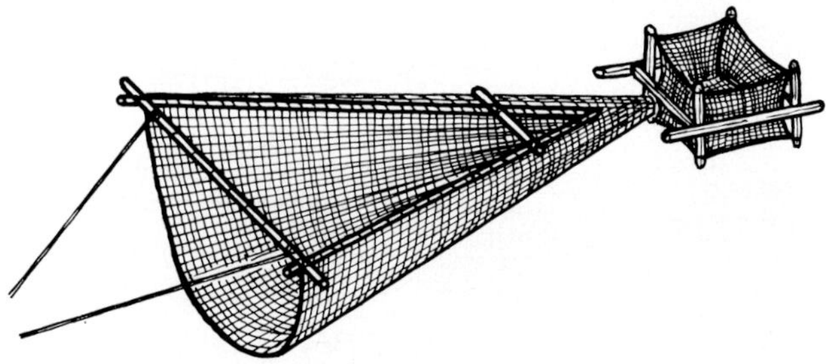

FIGURE 19.4. Floating bamboo trap.

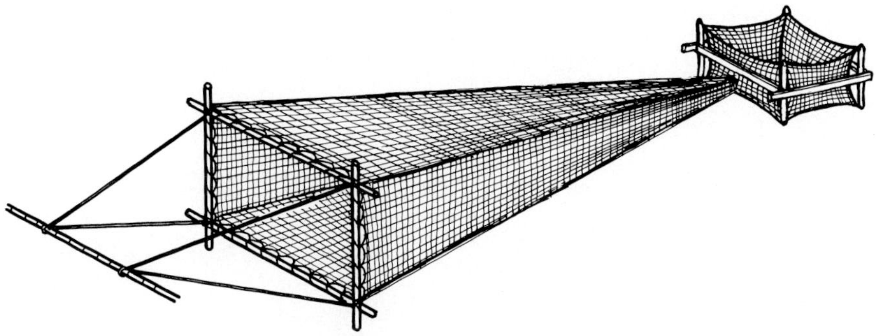

FIGURE 19.5. Cloth trap, fixed to logs planted in the river.

the tail end is 15 cm. The mesh size in front is 4 cm and the mesh size at the tail end is 1 cm.

2. Neck — This is equivalent to the sleeve of the floating trap. It is 70 cm long and 20 cm in diameter. It is made of cloth.
3. Box — This is also made of cloth or silk material. It is rectangular, 105 × 36 × 36 cm.

**Cloth trap (Figure 19.5)** — This trap is used mostly in the Pearl River area. It is four-sided and tapers off toward the tail end. The trap is made of cloth. The mouth of the trap is a rigid rectangle, 2 m long and 1 m high. A bamboo is attached to the top of the net. The sleeve is cylindrical, having a diameter of 25 cm at the mouth. The tail box is similar to that of the bamboo trap. The cloth trap is affixed to logs in the same way as the fixed trap.

*Advantages and Disadvantages of Various Gear*

**Floating trap** — This trap is low in cost and easy to set. It can rise and fall with the water level and can fish continuously. One disadvantage is that it is easily affected by winds and waves. Furthermore, it must be tended and fished from a boat.

**Fixed trap** — In contrast to the floating trap, the fixed trap is resistant to moderate winds and waves and can operate from the net frame. However, it is more costly to install, it cannot move up and down, and it is extremely difficult to move from one place to another.

**Bamboo trap** — This is the least expensive trap to make. It is easy to operate and move, does not deteriorate easily, and is easy to clean. Additionally, this trap can rise and fall with the water level and adapts better in more rapid water. Unfortunately,

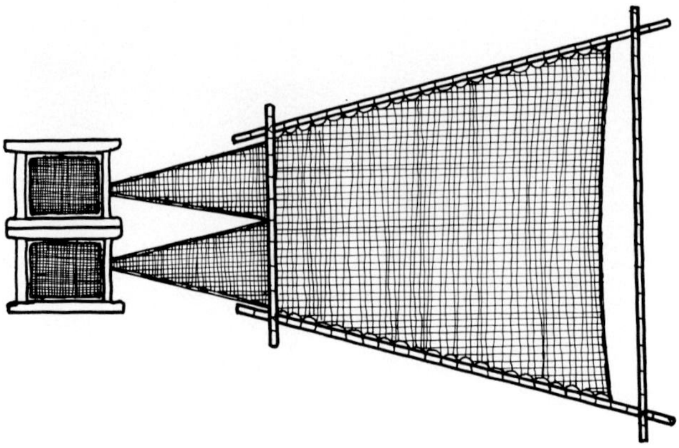

FIGURE 19.6. Large-mouth trap.

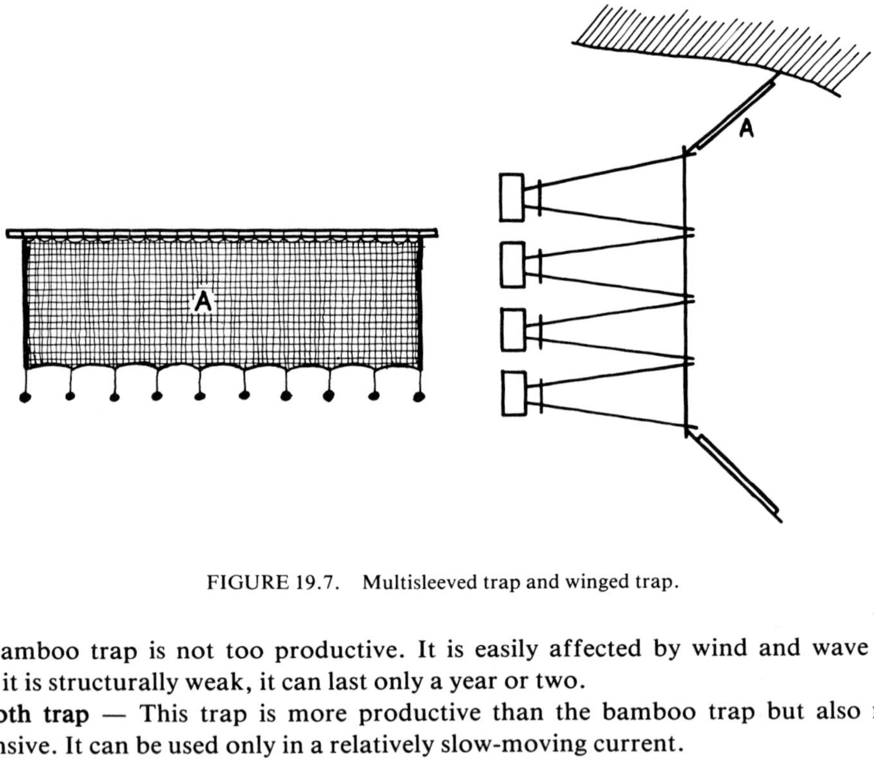


FIGURE 19.7. Multisleeved trap and winged trap.

the bamboo trap is not too productive. It is easily affected by wind and wave and, since it is structurally weak, it can last only a year or two.

**Cloth trap** — This trap is more productive than the bamboo trap but also more expensive. It can be used only in a relatively slow-moving current.

*Improvement of Traps and Fishing Methods*

In recent years many innovations have been added to the fishing traps described above. Some of the more important ones are as follows:

1. Large-mouth trap (Figure 19.6) — The opening of the trap is increased so as to filter more water and do it more efficiently.
2. Multisleeved trap (Figure 19.7) — Increase the number of sleeves to five to seven.
3. Winged trap (Figure 19.7) — Install two wings anterior and lateral to the mouth of the trap.

4. Large-meshed floor — By using large-meshed netting on the floor of the trap, the trap can be used in swifter waters and can be handled more easily. It does not, however, catch more fry.
5. Increase the depth of net opening to 1.6 to 2.0 m. This can be accomplished by merely turning the trap 90°.

In addition to the above improvements, there are also various devices which help to stabilize the trap so that it can fish during windy storms. These devices include connecting the tail boxes together, affixing the tail boxes to posts, and increasing the weight of sinkers.

Fish fry in the river are stratified in various depths. Domestic species of fry are present from 1 to 4 m. In one experiment conducted in Nanking, 43.5% of fry were caught in the upper layer, 27.6% in the middle layer, and 28.9% in the lower layer. According to the results by the Hainan Fisheries Institute, the upper layer caught 27.73%, middle layer caught 32.35%, and lower layer caught 39.92%.

### Fry Trapping

Based on prediction of fry bloom, fishermen usually put out a test trap first. When large amounts of fry are discovered, full-scale fishing is started. Test fishing is undertaken every time before a bloom.

During fishing, the catch in the tail box must be emptied at definite intervals into large containers. A large amount of debris and aquatic insects and larger pest fish species are often collected in the box. These must be removed by the use of a sieve.

In order to increase the production of fry, fishermen follow five "frequents":

1. Frequent testing — Change the location as necessary.
2. Frequent emptying — Fry in the box should be removed every 30 min or at least once every 1 or 2 hr when fishing in slack.
3. Frequent cleaning of the net — Debris in the net is injurious to fry.
4. Frequent moving of net — Fry occurrence depends upon water level, velocity, wind direction, light intensity, etc.; therefore, the net must be moved accordingly.
5. Frequent study — Study methods of improving the gear and adjust as necessary.

## NAMES OF FISH FRY AND THEIR IDENTIFICATION

### Names of Fish Fry

Fish fry are called various different names by people from different localities. Generally there are two main kinds of fry: culturing fry and pest fry. The former include grass carp, roach, bigheads, carp, goldfish, bream, etc. that are used in commercial protection, and are cultured to some extent.

### Identification of Fry

Many fish species spawn more or less at the same time in the river. The trap catch usually contains a mixture of species. Recognition of the species is a very important skill. However, because fry are so small, identification is a difficult task. Fishermen can usually tell species by naked eye, a skill gained through long experience. Fry are recognized by three methods:

1. Examined in the water — Identification is based on the movement of the fry and the morphology of the back.
2. Examined out of water — Identification is based on characteristics viewed from the side.

3. Combined method.

Characters that are used to identify fish fry are (1) body shape and size, (2) size and position of the eyes, (3) shape, size, and position of air bladder, (4) shape of caudal finfold, and (5) body color and distribution of pigments.

In production work, field identification with a naked eye is very useful, but it requires long experience. More reliable, of course, is the use of a magnifying lens.

**Grass carp fry** — Grass carp fry which are caught in a trap usually measure 7.0 to 8.5 cm long. In fry, the dorsal side of the head is light orange in color and has relatively large, star-shaped black pigments (not as dense as those in bigheads but denser than those in roach). There is a series of relatively large, star-shaped pigments on the dorsal side extending from the head to the tail. There is also a band of pigments along both the dorsal and ventral side of the notochord; near the caudal end of the ventral band there is a network of red blood vessels, giving rise to the name of "red tail". The ventral finfold is broader than that of roach. The anal finfold is short and narrow. Both finfolds are devoid of pigments. The caudal finfold is fan-shaped, with a network of pigments on the lower lobe. The air bladder is oval; its longer axis is twice as long as its shorter axis and 1.5 times the diameter of the eye. There are 28 to 32 myomeres in the trunk and 12 to 16 myomeres postanal.

**Black Chinese roach fry** — These fry are 0.37 m in length. The head is light yellow with some light green shade. Pigments are prominent and large, but scattered. The row of pigments along the back is scattered. The ventral finfold is narrower than that of grass carp and bigheads. The anal finfold is narrow and long. Both finfolds are devoid of pigments. The dorsal lobe of the caudal finfold has a pigment mass, and the ventral lobe has a network of pigments. The air bladder is oval and has a blunt front end and a pointed rear end. Its longer axis is 1 to 1.5 times the shorter axis and 2 times the eye diameter. The trunk has 22 to 24 myomeres, and there are 15 postanal myomeres.

**Fry of white bighead** — The length of these fry averages 0.29 to 0.37 m. In fry 0.37 m long, the head is light yellow and heavily dotted with granula and star-shaped pigment cells. Posterior to the eye and below the ossicle there is a row of pigments which extends along the air bladder and intestine to the tail. There are also three or four rows of irregular star-shaped pigments dorsal to the notochord. The ventral finfold is broad and has star-shaped pigments. The anal finfold is broader than that of grass carp and roach, but narrower than that of striped bighead. Both dorsal and ventral lobes of the caudal finfold have a network of pigments. The air bladder is oval; its longer axis is 1.5 times the shorter axis and 2 times the eye diameter. There are 24 to 26 trunk myomeres and 14 to 17 postanal myomeres.

**Fry of striped bighead** — In fry, the head is light yellow with scattered, star-shaped pigment cells. The body is also light yellow. In addition to the row of pigments along the intestine, there are also five rows of pigments on the dorsal and lateral sides of the body. The ventral finfold is broad with pigments; the anal finfold has a row of pigments. The dorsal lobe of the caudal finfold has some indistinct pigments, but the ventral lobe has a network of pigments. The caudal end of the notochord is curved upwards. The air bladder is oval; its longer axis is 1.5 to 2.0 times the shorter axis and 2 to 3 times the eye diameter. There are 22 to 23 trunk myomeres and 16 to 17 postanal myomeres.

**Carp fry** — The head is compressed and triangular in shape. The length is usually 7 to 10 mm. In fry, there are many irregularly laced, round pigments. There are also two rows of round pigments along the back, extending from the head to tail. The air bladder is oval, slightly more pointed toward the posterior end. The dorsal and anal finfolds have a network of many blood vessels. When the fry is above 10 mm long, the air bladder is partitioned into two lobes, pigments are greatly increased, and the body becomes dull green. There are 17 to 20 trunk myomeres and 11 to 14 postanal myomeres.

**Goldfish fry** — The eyes in fry of 6 to 7 mm in length are orange red with a light green shade. The head is short and squarish. The air bladder is oval and near the head. Irregular, small pigments are distributed on dorsal and lateral sides of body. The caudal finfold is rounded, with irregular pigments on the lower lobe. The dorsal finfold is broad, with netlike blood vessels in the posterior half. There are 22 trunk myomeres and 14 postanal myomeres.

**Bream fry** — In fry 0.28 mm long, the head is elongated, and the eyes are extremely black and dotted with some fine golden yellow spots. The area around the pupil is pinkish. The distance between the eyes is larger than the eye diameter. There are two black pigments on the snout. There is a large flowerlike pigment on the ossicle. An even larger flowerlike pigment is present on the heart. The air bladder is somewhat oval and longer than it is high. The ventral finfold is broader than the dorsal finfold. The caudal finfold is truncated. The body is covered with sensory hair. Myomeres are 25 in the trunk and 14 postanal.

**Fry of *Cirrhina molitorella*** — In fry 7 mm long, the body is pinkish, with many small, round, black dots. The air bladder is conical, with the pointed end toward the posterior end. There are a few star-shaped pigments on the dorsal side of bladder. The caudal fin is rounded. Myomeres are 24 in the trunk region and 9 in the postanal region. In fry 10 mm or longer, a network of star-shaped pigments appears on the basal part of the caudal finfold.

**Fry of *Squaliobarbus curriculus*** — In fry 6 mm long, the head is oval as viewed from the top. The front rim of the eyes is golden, showing a few black spots. The bladder is large and oblong with its anterior end close to the base of the pectoral fin. From the lateral view, there is a row of black pigments extending from below the ossicle to the caudal end. The ventral finfold is broad. Myomeres are 25 in the trunk and 12 in the tail.

**Fry of *Siniperca chautsi*** — In fry 6 mm long, fin rays are not developed on pectoral fins; other fins are undifferentiated. When the fry is 11 mm long, the tip of the notochord is curved upward slightly, and the lower lobe of the caudal fin shows bony fin rays. The intestine is short. When long, the body is compressed; the head is large and more than one fourth the body length. The snout is pointed and the mouth large. Mandibles bear fine, sharp teeth. The preopercle bears four to five sharp spines along the posterior edge; the opercle has two. The abdominal cavity is triangular in shape and is covered with several extremely large, star-shaped pigment cells. The anus is situated in midlength of body. A row of star-shaped pigments occurs postanal on the ventral finfold. The bladder is small, indistinct, and constricted in the middle. The tail is rounded. Body myomeres are 9; tail myomeres, 19.

**Fry of *Elopichthys bambusa*** — In fry 7 to 8 mm long, from the lateral view there is an oral slit which extends to mideye. The rim of the eyes is pinkish except the lower rim, which is slightly greenish. Ossicles are larger than the eyes. The fins are undifferentiated. The posterior edge of the caudal fin is truncated. When viewed from the dorsal side, the snout is rounded and short. The distance between the eyes is twice the eye diameter. The body is light yellow. When the fry is 10 mm long, the yolk sac is completely absorbed, and the bladder is oval and twice as long as high. There are 38 to 41 body myomeres and 15 to 19 tail myomeres.

**Fry of *Parasilurus asotus*** — The body is light brown; both head and body are densely covered with round or patched black pigments. The intestine is short and straight. The anus is located in the anterior half of body. In fry 8.6 mm long, the anal finfold and the caudal finfold are contiguous. When the fry is 10 mm long, these two finfolds become separated, and finrays appear also. The pectoral fin is featherlike; the caudal fin is truncated. Three pairs of barbels are present. There are 12 body myomeres and 25 tail myomeres.

**Fry of *Hemiculter leucisculus*** — In fry 0.24 mm long, the eyes are large and protruded; the ossicles are distinct; the gill arch is present; the snout is blunt; and the anus is situated posteriorly. The intestine is fine and straight. The anal finfold is short and narrow; the dorsal finfold is narrow and long. In fry 8 mm long, the bladder is present and similar to the eyes in size. In fry 0.36 mm long, pigments extend the whole body length and are light yellow. There are 32 body myomeres and 11 tail myomeres.

## DETERMINATION OF SPECIES COMPOSITION AND QUALITY OF FRY

To determine species composition of fry in a catch, a sample is placed in a white tray, and the composition is estimated by visual examination. The composition is either in terms of percentage of culture species of the entire catch, including undesirable species, or in terms of percentage of each culture species.

There are several ways of assessing the quality of fry. One method is to place fry in a white container, blow at the surface, and watch for the reaction of the fry. If fry swim against the blow, they are strong; if they are blown to the side of container, they are weak. Another way is to drain the container: if fry struggle hard and curl up head to tail, they are strong. If they stay flat against the bottom and struggle little, they are weak. A third method is to stir the water to form a whirlpool. If fry swim against the current along the periphery of the whirlpool, they are strong; if they drift passively, they are weak.

## QUANTITATIVE MEASUREMENT OF FRY

There are many factors that make the counting of fry a very difficult task. At best, it is an estimate. Some factors that affect quantitative measurement are that fry are minute; they often occur in great quantities; species are mixed; fry are not uniform in size; and they are fragile and die readily once out of water. Chinese fishermen use the methods discussed below in estimating the quantity of fry.

**Estimation** — This method depends entirely on experience. First, place an appropriate amount of water in the fish basket. Stir thoroughly to distribute fry uniformly. Take a sample out on a white dish, and estimate the species number and composition. Repeat this process several times. The total quantity of fry is estimated, based on the composition of species and volume of water.

**Counting by aliquot** — Use the smallest practical container as an aliquot unit. Count fry in this aliquot, and calculate the total number of fry by the proper factor. Several aliquots of samples are counted to arrive at a mean.

**Volumetric method** — Before measuring, the seller and buyer must be agreed on the species composition. Concentrate the fry in the center of a net by lifting the net bottom close to the surface, then measure the fry by a standard sized container, usually a bowl. Generally, each bowl is regarded as containing 10,000 fry. In Kiangsi, where a tin cup is used, calculation is based on 80,000 fry per cup.

**Partition method** — This method is adopted in Kwangtung and Kwangsi. Fry are gathered into a cloth container. They are divided into two portions, then four, then eight, and so on, until a small enough unit is obtained when fry can be counted. The one unit is picked by lot for counting, and the number is multiplied by the proper

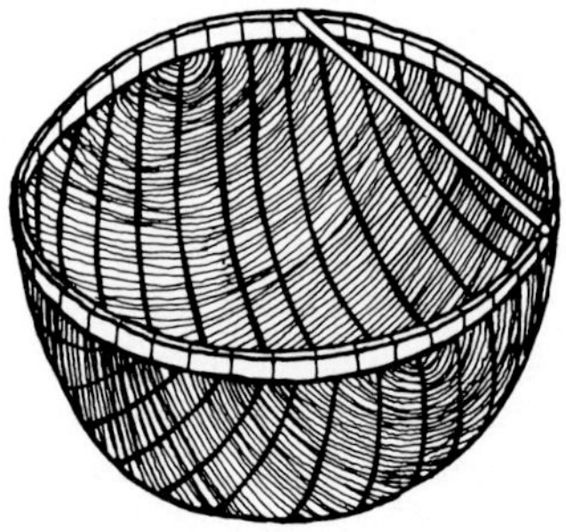

FIGURE 19.8. Fish sieve made of finely split bamboo.

factor to give the total. Of the methods discussed, the partition method is most accurate.

## PEST CONTROL

Fish fry that are caught in the river contain not only the culture species, but also many other nonculture species. Some of the latter are predators, such as catfish and Chinese perch; others are competitors, competing for food, oxygen, and space. Control of these nonculture species is a very important undertaking in fish culture business. There are many methods of pest control, the best of which are the sieve method and the squeeze method.

### The Sieve Method

This method includes sieving, selection, and separation.

*Sieving*

Fry of the same age which are caught during the same bloom are of three basic sizes. One is the smallest; it is long and transparent, with only two eyes and the bladder dark. This group consists primarily of pest fry. The second group is slightly longer, and the body is light yellow in color. This group generally includes culture species, plus some pest fish. The third group consists of the largest fish which are mostly pest species. By using different mesh-sized sieves, pest fish are mechanically separated out.

Fish sieves are made of finely split bamboo (see Figure 19.8). They are half spherical, with a diameter of 23.6 m and a depth of 18.8 m. Meshes are long and narrow. There are usually 24 sieves in a complete series. Each sieve is designated by a number to indicate the width of the mesh.

The procedure of sieving consists of the following steps. First, prepare a cloth container, and place it in a water area where work can be easily accomplished. Second, get rid of all debris and dead fry, and leave only live fry in the container. Third, use a No. 2.5 to No. 3 sieve (mesh 1.2 to 1.4 mm) to separate out the larger, pest species.

*Selection*

Oxygen consumption rate differs in different species of fry. Therefore, the tolerance toward oxygen depletion also differs. Selection of culture species is based on this principle. In a thick concentration of fry in a container, fish will surface for oxygen. The timing of surfacing differs according to species. Therefore, different species of fish become stratified in the container, which is the basis for separation.

The selection procedure consists of two steps. First, place the fish fry that passed through the No. 2.5 to No. 3 sieve in a selection basket, and fill the basket with water to the rim. Fry will generally stratify themselves in three layers. The uppermost layer generally consists of white bighead, striped bighead, black Chinese roach, and some pest species. The middle layer consists predominantly of grass carp fry. The bottom layer consists mainly of mud carp. In order to effect a higher percentage separation by species, it usually requires secondary separation, and much depends upon skill and experience.

The second step involves sieving out pest fish. Fish fry that are separated out from each layer always contain a certain portion of pest species. To get rid of pest species as much as possible, use appropriate sieves. By so doing, 95% purity can be achieved.

*Separation*

This is the procedure by which white bighead and striped bighead are separated. The remaining pest species are also removed. This is a complicated and difficult technique, and a lot of effort is required for preparatory work.

The white bighead fry occur on the top layer and striped bighead on the lower layer. Also, because of the density of fry in the separation container, oxygen becomes very low and any pest fish that are mixed in it will die off rather rapidly.

The pond selected for separation work should be conveniently located. It should be rectangular — about ½ acre in size and 0.8 m deep — and have a flat bottom. During the first week of April, the pond should be drained and sterilized with lime. Leave it exposed for 7 to 10 days, then fill it with water up to 0.6 m deep.

Several days before fry are introduced, put wilted grass in the pond. After 4 or 5 days, when large amounts of plankton are present and when the water turns from clear to dull red, introduce about 4000 long striped bighead to the acre. If these fish surface every morning, then fertility is just right for species separation. If the fish surface too much, the water is overfertilized and must be diluted. If the fish do not surface, then more grass should be added. The second purpose is to eliminate larger sized zooplankton (such as copepods and cyclops) which are too large to be used as fry food but which will compete with fry for food and oxygen.

Before introducing fry into the pond, the pond should be partitioned into two series of compartmental ponds of unequal sizes with a communicating channel 5 to 6 m wide between the two series of ponds. Each compartmental pond is about 5 m wide by 15 m long and may contain 2 million to 2.4 million fry.

First, transfer all the striped bigheads into the communicating channel and shut all the doors. Then place the selected fry into the compartmental pond. This is done at noon when the weather is clear. Wilted grass is added to the pond at this time. Then at 3:00 p.m. all the grass is removed; fry separation can be started at 4:00 p.m. A few hours after introduction of fry, pest species will largely die off due to oxygen depletion.

The separation depth is generally at 6 to 9 cm, but it should be determined by sampling. Fry taken from the upper layer usually consit of 90% of white bighead during the first separation and about 50% during the fourth separation. After separation, white bighead and striped bighead are reared in separate ponds. One final separation should be made the next day. If there are still some pest fry, they can be removed by the use of sieves.

## The Squeeze Method

The squeeze method is based entirely on the principle that tolerance of low oxygen level is much lower in pest species than in culture species. The procedure is described below.

Half fill a fish basket with clean water. Add to it a few hundred thousand fry. Create a whirlpool by stirring so that dead fish and debris will gather at the center of the

TABLE 19.1

**Water Contents of DO and $CO_2$ Before and After Squeeze**

| Number of fry (bowls) | Air temperature (°C) | Water temperature (°C) | DO (mg/ℓ) | | $CO_2$ (mg/ℓ) | |
|---|---|---|---|---|---|---|
| | | | Before | After | Before | After |
| 48 | 26.5 | 22.8 | 4.22 | 0.75 | 5.4 | 7.8 |
| 46 | 30.5 | 26.7 | 7.85 | 1.75 | — | — |
| 50 | — | 21.5 | 6.67 | 0.73 | 8.2 | 11.74 |
| 25 | — | 21.5 | 6.54 | 0.75 | — | — |
| 54 | — | 20.8 | 8.05 | 0.70 | 8.1 | 11.24 |
| 30 | 22.5 | 21.8 | 7.93 | 0.86 | 5.0 | 10.20 |

bottom. Remove dead fish and debris with a pipette. At first, fry will swim in schools along the wall of the basket. After a while, due to depletion of oxygen, they will scatter and surface. Pest species will be the first to faint and rapidly die off. Culture species will soon also become much disturbed and finally will ball up. At this time, fresh water must be added immediately. Stir the water to create a whirlpool and add mud paste into the basket so that mud will stick to dead fry and make them sink. Transfer live fry into a new basket and let them rest for a while. If pest species are still present, the process can be repeated once more.

The death of pest species is due to suffocation from the lack of oxygen. It has nothing to do with the increase of carbon dioxide. This is proven by the result of suffocation experiments conducted with cultured carps. When carbon dioxide reaches 100 mg/ℓ in water, fry of four domestic carps will gradually lose their balance and start to die. If it is over 150 mg/ℓ, then most, if not all, of the fry will die. However, in the squeeze method, carbon dioxide content only rises to about 12 mg/ℓ (Table 19.1), far below the lethal concentration. Furthermore, experiments with the four domestic carps show that the critical dissolved oxygen (DO) content is below 1 mg/ℓ. In the squeezing process, DO approaches or even goes below 1 mg/ℓ. Among the culture species, the white bighead, which has the highest oxygen consumption rate, starts to be knocked out. At the same time, pest species, which have an even higher consumption rate, have already died out. It is clear, therefore, that the squeeze method kills pest fry by the depletion of oxygen and that use of the white bighead as a signal to stop squeezing is a viable method.

**Advantages and Disadvantages of Squeeze and Sieve Methods**

The advantages of the sieve method are (1) tools are of fine quality and are least injurious to fry during processing; (2) the technique is simple; (3) most of the pest fry are returned to the river to maintain fishery resource; and (4) the four domestic species are separated for separate rearing. The disadvantage is that pest species are only partly removed.

The advantages of the squeeze method are (1) the success rate is very high; (2) survival rate is high during transportation; and (3) tools are simple. The disadvantage is that all the pest fry and some fragile culture species are killed.

Combining the merits of sieve and squeeze methods, a composite method should be practiced. The steps of the composite method are as follows:

1. Clean the pond.
2. Get rid of pest species by sieving.
3. Separate species by swimming behavior. Gather fry at one end of the pond and

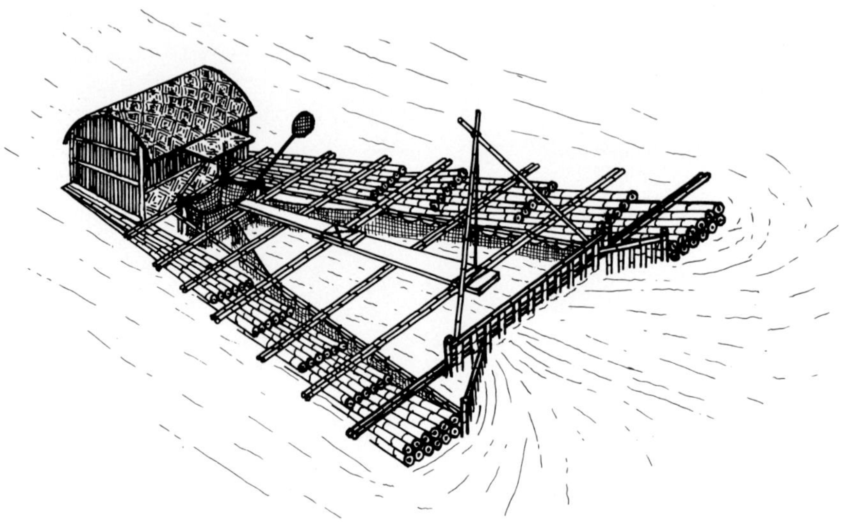

FIGURE 19.9. Fixed trap for collecting fish eggs.

release water from the other end, creating a current flow. Fry will swim against the current. Fry of the bigheads, being strong swimmers, will be in front, while the black roach and grass carp, which are weak swimmers, will lag behind. These two groups of fry can, therefore, be separated.

4. Separate species by differential oxygen tolerances. Further separation of species can be achieved by subjecting fry to oxygen depletion.
5. Get rid of pest species.

Fry collected from the river usually comprise some 50 species, most of which are of economic importance. In numbers, the culture species constitute about one fourth of the total. It is clear, therefore, that the great majority of collected fry are yet to be utilized.

## COLLECTION OF FISH EGGS

### Collection of Eggs of Grass Carp, Black Roach, White Bighead, and Striped Bighead

Although there are many rivers in which spawning of culture carps takes place, only in the Hsiang River, Human Province, is the taking of spawn a well-known practice. Among the species of eggs collected, the majority of eggs belong to grass carp and white bighead.

**Prediction of egg bloom** — In the Hsiang River, spawn are usually taken from the latter third of April to the latter third of June. Prediction of egg bloom is based on weather and water conditions.

**Selection of egg collecting sites** — A good collecting site is essential to a good harvest of eggs. Water depth, topography, and water velocity are factors which must be taken into consideration in the selection of a suitable site.

**Collecting gear** — A fixed trap (Figure 19.9) is used as the gear for collecting eggs. The trap can rise and fall with the water. A typical trap consists of two supports made of bamboo rafts, which are set up in the shape of a V. At the end is built a little hut to be used by the watchman. The trap itself consists of the body, funnel, and cod-end. The body is wide at the mouth, long, and has a mesh size of 3 mm. The diameter of the funnel is 4.4 cm. The cod-end is 1.5 m long, 80 cm wide, and 30 cm high. During fishing the net is held open by the current. The dip net has a diameter of 44 cm and is 3 cm high; the meshes are 3 mm. The grate is divided into the front and side parts. Its chief function is to prevent debris from entering the trap. A pair of fork posts is planted at the mouth of the trap to keep the mouth open and to adjust the depth of the net. Aside from the trap, there must also be a hatching box to hatch the eggs. The hatching box is rectangular in shape and is made of hemp or cotton webbing. It is $1.5 \times 0.8 \times 0.3$ m.

**Method of collecting eggs** — Prior to egg bloom, test collecting should be conducted. Once the bloom is discovered to have started, the trap is then put in operation immediately. The eggs are collected in the cod-end. They are dipped out, placed in a container, and transferred to a hatching box to hatch. When the eggs are in large quantity, they can be dipped 30 to 40 times a day; when they are in low abundance, they can be dipped 8 to 10 times a day.

**Identification of fish eggs** — Fish eggs that are collected in the river are nonadhesive. Because of their minute size, they are difficult to identify. Generally, identification is based on size, embryo size, shape, and color. The eggs of culture species are generally larger than those of pest species. In the culture species, the eggs have a slightly yellowish vegetative pole and a light-bluish animal pole. The color of pest species eggs is varied — yellow, red, black, or white — and they are nonuniform in size.

**Determination of quality** — To judge the quality of eggs, one must observe the embryo. When it has reached a stage of movement within the egg shell, the embryo is considered strong if it wriggles back and forth frequently. On the other hand, if the embryo merely moves occasionally, after hatching the fry is likely to be weak, slow growing, and have a high mortality rate. The hatching boxes should be located in small bays with clean water and good circulation to prevent accumulation of mud. After hatching, it takes 2 to 3 days before absorption of yolk. At that time, the fry should be fed well-cooked egg yolk. In a couple of days, they can be transferred into the pond for rearing.

### Collection of Carp Eggs

The common carp is a tenacious species. It can withstand both cold and hot temperatures. It is prolific and requires an environment far different than that of grass carp, black roach, and bigheads. It can spawn not only in shallow areas among the grass in slow-flowing rivers, but also in still waters, such as reservoirs, lakes, and ponds.

To collect carp eggs, it is necessary to select the spawning ground, then place artificial nests. Carp eggs will adhere to the nests, which are then collected and placed in ponds for hatching.

**Time for collecting eggs** — The common carp spawns when water temperature is between 17 and 21°C. The date varies according to geographical locality. In the Yangtze River, spawning starts in March or April; in the Amur River, June; in the Pearl River, December. Spawning usually takes place immediately before or immediately after a thunderstorm or when the weather turns clear after a long rainy spell. Spawning is most active between midnight and dawn.

**Spawning location** — Locations that are conducive to carp spawning are (1) shallow bays in rivers where the flow is slow and grass abounds; (2) near shore areas in lakes where there is grass but which is not too thick to allow free movement; and (3) slow-flowing bays in lakes where water is clean and grass grows.

**Method of collecting eggs** — Carp eggs are adhesive. A proper number of nests should be placed in spawning areas. Nests should be examined frequently between midnight and dawn. They should be removed as soon as they are full of eggs and can be replaced by new nests.

## COLLECTION OF MULLET FRY

The mullets (*Mugil cephalus* and *M. soiuy*) are marine species. They can live not only in salt water or fresh water along the sea coast, but also in fresh-water bodies inland. Therefore, they have become the favorable objects of fresh-water culture species.

Mullets are quite abundant along the Chinese coast. *M. soiuy* predominates in the north, and *M. cephalus* predominates in the south. The culture of mullets along the sea coasts in China has had a long historical record. However, the mullets do not reproduce in fresh water. Therefore, much effort must be expended each year in collecting natural spawns on the coast.

### Spawning Season

Mullets grow in littoral sea. Their fry, when 10 to 15 mm long, migrate to estuaries to feed and grow. This is where they can be collected. The appearance of mullet fry in estuaries varies according to geographical locations. In general, it is earlier in the south than in the north, and *M. cephalus* is earlier than *M. soiuy*. In southern China, from early February to April is the best time for the collection of *M. cephalus*. In the mid-Eastern China Sea, *M. cephalus* appears as early as mid-February and persists until May. In northern China, the collecting time is from late May to July. For *M. soiuy*, the collecting time is usually later.

### Habitat and Fishing Grounds of Fry

**Fry habitat** — When 10 to 15 mm long, the mullets migrate upriver and congregate in inshore areas and river mouths where the salinity is relatively low. They prefer warm waters. In spring and summer, they migrate into shallow shores to feed. In winter when water temperature drops below 5°C, the mullets migrate into the deeper parts of the ocean to winter. These fish are very active, often leap out of water, and therefore require a lot of oxygen. The young love light and frequent sand bottoms rich in organic detritus. Their food consists primarily of plankton. During the ebb tide, they school together, but on the flood tide, they scatter around in depths of 30 to 60 mm. Fry are more plentiful during calm, clear days with a big tide than on rainy, small-tide days.

**Fishing grounds** — Mullets are often gathered in (1) estuaries near the mouth of rivers; (2) deep holes in bays where water is retained after the tides goes out; and (3) in harbors where there are a lot of shelters.

### Fishing Gear and Methods

Basically there are two kinds of fishing gears and methods: portable and fixed.

#### *Portable Fishing Gear and Methods*

**Tow net** — This net is made of two wings and a bag. Shells, feathers, etc. are tied on the tow line in order to make noise to scare the fish into the net. The net is set around a school of fry, and two men tow the net toward the shore. The fry are collected in the bag. They suffer some injury, and the survival rate is 70 to 80%.

**Seine** — This is adapted to waters not deeper than 2 m. Generally, the net is rectangular and is made of cotton thread. In one tide (around 3 hr) more than 10,000 fry can be collected.

**Push net (Figure 19.10)** — This is also made of cotton thread. The net is triangular in shape and is supported by two bamboo poles on the sides. The bottom end of each pole is affixed to a wooden shoe to prevent the net from sinking into the mud. The bottom line of the net is lined with a row of eight to ten leads. This net is more suitable for smaller creeks.

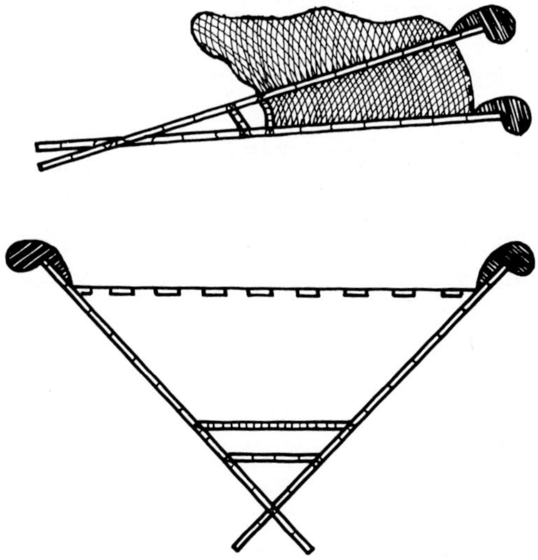

FIGURE 19.10. Push net made of cotton thread.

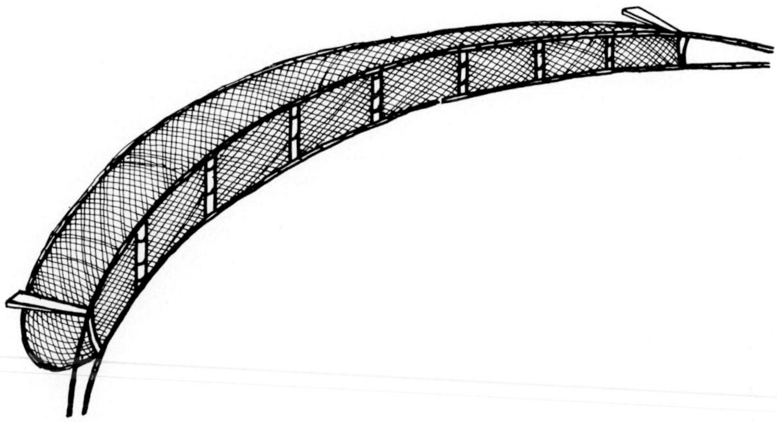

FIGURE 19.11. Bag net made of silk or linen thread.

**Bag net (Figure 19.11)** — This net is made of silk or linen thread of fine meshes. It can collect smaller fry and is operated by two men.

*Fixed Fishing Gear and Methods*

These devices save labor. They utilize light and food to attract fry so that they are very effective.

**Balloon net** — This net is used in the tidal river mouth. It is made of silk or cotton thread. The total length is 2 m, the width at opening is 5.6 m, and the height is 1.8 m. The net tapers toward the tail end, where it is 14 to 20 cm in diameter. The tail end is connected with an open box 50 × 50 × 40 cm which is supported by a wooden frame.

**Collection by bait** — During flood tide, a sieve is used as the collecting gear. Bits and pieces of fresh crab meat are placed in the sieve and the sieve is lowered 8 to 10 cm below the surface. When fry gather to feed on the crab, the sieve is lifted to collect the fry. This method is not very productive.

**Light attraction method** — This method is for night fishing. A series of 6- to 12-W lightbulbs is placed under water at intervals of 3 m. At first, all lights are turned on. After 90 min, fry will gather near the lights. Then the lights are turned off, one after the other, as fry follow the lights until the last light is left lighted. At this time, a preset seine is hauled up to collect the fry. This method is best used during the ebb tide.

### Identification of Mullet Species

*Mugil cephalus,* when 10 to 15 mm long, is laterally compressed. The head is rounded and large; the mouth is rounded and not protruded; and there is a square red pigment on the head. The back is greenish-blue with a yellowish tinge; there is a small, insignificant red spot near the basal part of the dorsal fin. The body is silvery white. The large pectoral fins are situated high, with black pigments on the sides. They are strong swimmers, often schooled, and form uninterrupted waves on the surface. On warm days, they prefer shallow waters.

## REFERENCES

1. **Roxby, P. M. and Freeman, T. W.,** *China Proper; Physical Geography, History and Peoples,* Division of Naval Intelligence, Great Britain, London, England, 1944.
2. **Whitaker, D. P.,** Area Handbook for the Peoples Republic of China, DA Pam 556-60, U.S. Government Printing Office, Washington, D.C., 1972.
3. **Gangstad, E. O., Raynes, J. J., and Burress, R. M.,** Herbivorous Fish for Aquatic Plant Control, Aquatic Plant Control Program Tech. Rep. 4, U.S. Army Engineers Waterways Experimental Station, Vicksburg, Miss., 1973, 148.
4. **Yuh-Forn Chang,** *Culture of Freshwater Fish in China,* Science Publishers, Peking, China, 1966.

Chapter 20

# TRANSPORTATION OF FISH EGGS, FRY, AND YOUNG*

## INTRODUCTION

The proper transportation and handling of fish eggs and fry is a vital part of any fish production system. Careful planning is important. The first thing to decide is the type of container to be used to hold the fish. There are buckets and barrels of many sizes and designs, each with different characteristics. A second consideration is the density to which the fry should be packed in the container. It must be as dense as possible to save space and money, but not so dense as to kill the fish. The two most important factors in determining the maximum allowable density are water temperature and dissolved oxygen levels.

Fish fry and eggs are shipped by three major means: by land, by water, and by air. This involves either trucks, trains, boats, or airplanes. Each method has advantages for different situations. During any form of transportation, water maintenance is necessary. If some sort of method for replenishing dissolved oxygen is used, replacement of water can be put off, but still not avoided. The source of fresh water must be inspected carefully before it is used. Clean pond water is the best. Any temperature difference between old and new water should be minimized. In more recent years, cultivation of mullet species has become prominent, moving in with the more traditional species, the carps and the bigheads.[1,2]

## PRETRANSPORTATION, PREPARATION, AND TRANSPORTING UTENSILS

### Preparation Work

Before fish fry are transported, proper planning is essential to success. Utensils must be selected and inspected. Agreements must be entered with transportation agencies beforehand. A survey must be made of the water source and water quality on the transporting route, and stations must be established where water change will take place.

### Transporting Utensils

**Fish basket** — This is made of bamboo splints. The inside wall is coated with tissue paper and persimmon oil so that it is water proof. It has a round mouth and a square bottom. The size varies somewhat from place to place. Although the basket is cheap to make, it is not durable, generally lasting only one season (Figure 20.1).

**Canvas bucket** — Cylindrical in shape, the bucket measures 1 m in diameter and 1.6 m in height. Inside the mouth, there is a 13 cm wide collar to stop the water from sloshing out. The canvas is supported by a metal frame. The bucket can be dissembled and is therefore convenient to use (Figure 20.2).

**Wooden barrel** — The barrel is 100 cm high, has a bottom diameter of 90 cm, and an opening diameter of 70 cm. A lid is provided. In the center of the lid is an opening with a diameter of 35 cm. This opening is for the stirring rod. There is a hole with a cork stopper 13 cm from the bottom. The advantage of the wooden barrel lies in its sturdiness and its good insulating qualities (Figure 20.3).

* From *Culture of Freshwater Fish in China*, Science Publishers, Peking, 1966, translated by Koo, T.Y., Dept. of Fisheries, U.S.D.I., Solomons, Md., and Walker, C., U.S.D.I., Washington, D.C.

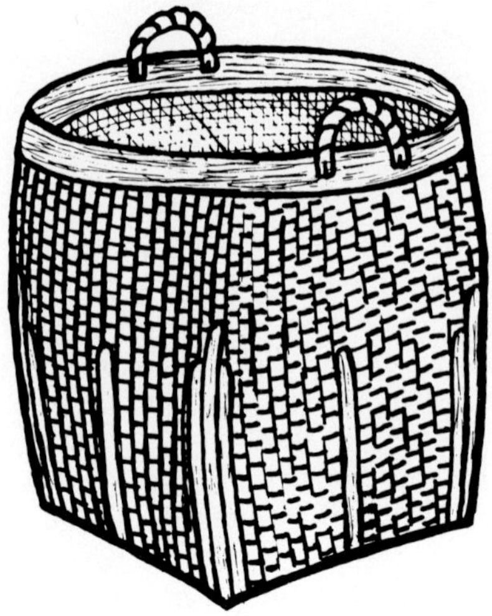

FIGURE 20.1. Fish basket for transport of fry.

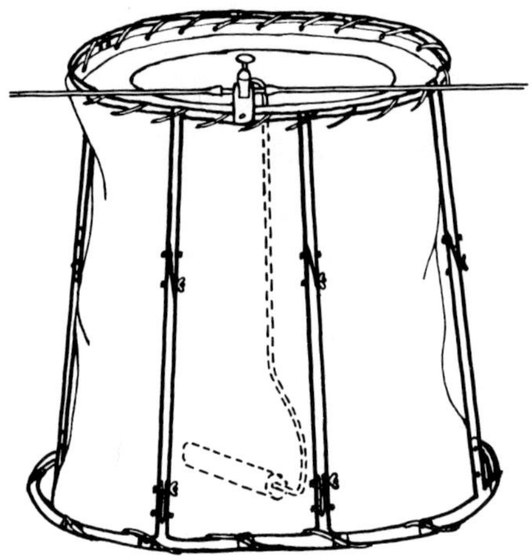

FIGURE 20.2. Canvas bucket for transport of fry.

**Carrying basket** — This basket, used for transporting fry, is carried with a pole over the shoulder. It usually has a round mouth and a square bottom (Figure 20.4).

**Suction tube** — This is used to suck out dead fish and debris from the bottom. It is a cylinder, 1 m high and 0.3 m in diameter (Figure 20.4).

**Volume measure** — A standard volume measuring cylinder.

**Water bailer** — This is made of split bamboo and is wrapped on the outside with oil skin.

In addition to the above utensils, other items must also be prepared, such as small basins, spotlight, hand light, thermometers, etc.

## TRANSPORTING DENSITY

Fry density during transportation often determines success or failure. The proper density varies according to species, fry size, condition of the fish, water temperature,

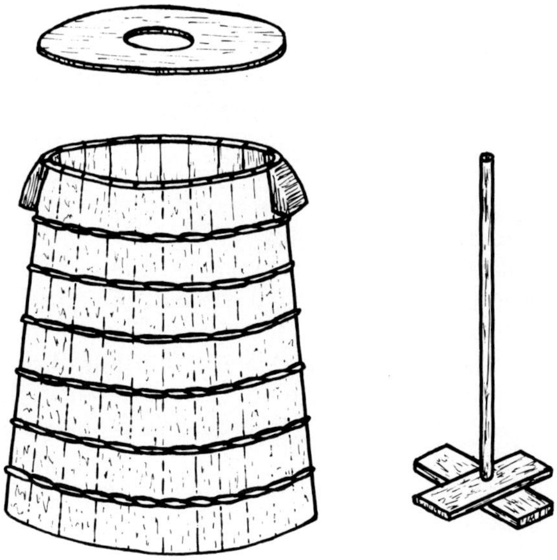

FIGURE 20.3. Wooden barrel for transport of fry.

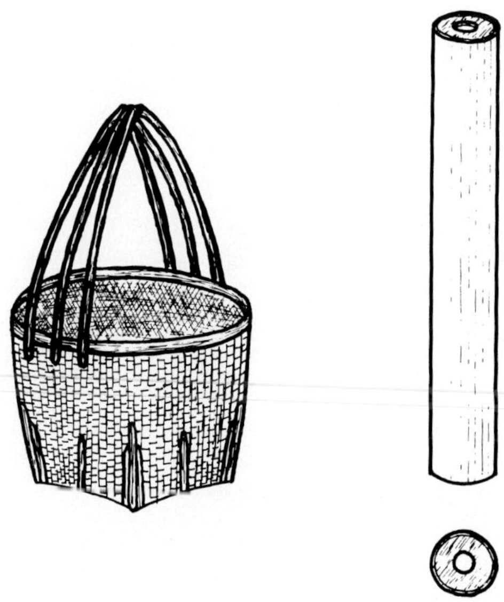

FIGURE 20.4. Carrying basket for transport of fry and suction tube to remove dead fish and debris.

distance to be traveled, and whether there would be any transference from one transporting means to another.

**Transporting Density of Fry**

When transporting is via ship, generally from 200,000 to 300,000 fry can be placed in a basket that contains approximately 200 ℓ of water (water temperature 15 to 25°C). In the case of the black roach or relatively older fry, the number should be less.

When transporting by hand carrying, the number of fry will have to depend upon the distance to be traveled, the weather condition, and the fish condition. The skill of the carrier and the condition of the road also have some bearing. Generally, if the carrying takes 1 to 2 days, 50,000 to 60,000 fry less than 1.3 cm long can be placed in each 20- to 25-pt container of water. If the fry are 1.5 to 2.0 cm long, 8000 to 12,000 should be allowed; if 2.5 cm long, only 5000 to 6000 fry.

In truck transport (3- to 4-ton truck), six baskets, each two thirds filled with water, can be loaded at a time. In the spring when the weather is not hot, each load can take 3.6 million fry if the destination is within the travel distance of 1 day; 2.4 million fry if the destination is within the travel distance of 2 days. When the weather gets hotter, the number of fry should be reduced to between 300,000 and 200,000. The two rear buckets will be subjected to the heaviest shaking and, therefore, a lesser number of fry should be placed in them.

A 15-ton rail car can take 22 to 24 baskets and transport 4 to 5 million fry. Past experience shows a great deal of variability in survival rate of fry in rail transport; it varied from 20 to 98%. Much of this variation was probably due to fry density, weather, water quality, and other factors.

A medium sized airplane can take 5 baskets; a large plane can take 10 baskets. Each basket can take 400,000 to 500,000 fry.

### Transporting Density of Summer "Fry"

When fry are reared to summer seed stage, the weight is increased many times — as much as more than 100 times. The transporting density of summer fry must, therefore, be greatly reduced.

When shipping via boat with constant change of water, at water temperature of 20 to 24°C, 30,000 to 38,000 fish may be put in each ton of water for fish of 2 cm in length. For larger fish, the number should be reduced.

### Transporting Density of "Spring Seed"

Spring fry are usually shipped during the period from December to February when water temperature is not more than 10°C. For shipping via boat, about 10,000 fish of 4 cm in length (bigheads) may be placed in each ton of water; 8000 fish for length of 6 cm; and 5000 to 6000 fish for length of 5 cm. Maintain the same density for the black roach. The density for grass carp may be increased 20 to 25%, since oxygen consumption in grass carp is lower.

The amount of dissolved oxygen (DO) in water is an important consideration of fish density in transportation. The maximum number of fish that can be transported under certain temperature and oxygen conditions can be calculated according to the following formula:

$$o - b\,\frac{o - o'}{a} = o''$$

or

$$b = a \cdot \frac{o - o''}{o - o'}$$

where o = DO before fish are put in; a = number of fish in container; o′ = DO after fish are put in (usually ½ to 1 hr); o″ = minimum requirement of DO for fish survival; and b = maximum possible number of fish to be shipped.

The minimum DO requirement for the four domestic carps is 2.0 to 2.5 mg/ℓ for the fry stage and 1.5 to 2.0 mg/ℓ for the seed stage.

## MEANS OF TRANSPORTATION

Fish fry and seed are shipped by one of three major means: water, land, and air.

**Water transportation** — This involves various sizes of ships and boats. Some take the advantage of regular passenger boats; others use specially constructed fish-transporting boats. One of the special boats is the "live water" boat, in which water intake openings are provided on the bottom or sides of the boat so that fresh water constantly enters into the tanks as the boat is traveling. Excess water is passed through a screen and is pumped out.

**Land transportation** — This involves hand carrying or the use of truck or train. Carrying by human labor is the most primitive method of transporting fish fry, but it is useful in certain areas that are not accessible by other means. However, fish mortality rate is high, sometimes 100%. Transporting fry by truck has made some improvements in the past few years, but specially built trucks for fish carrying are yet to be developed in China. Rail shipping is good for shipments carrying large quantities and going long distances. The cost is also the lowest. In recent years, due to artificial aeration and water change on the way, the survival rate of fish fry has been vastly improved.

**Air transport** — The advantage of transporting fish fry via airplane lies in speed of reaching destination. The density of fry can be higher than in other means of transportation, and the survival rate is high. In 1959, fish fry were sent to the U.S.S.R. in sealed nylon bags via airplane with good success. The survival rate was greater than 90%.

## MAINTENANCE DURING TRANSPORTATION

### Replenishment of DO in Water

*Water Change*

Regardless of transportation means, change of water is important. Generally, the time for water change is when the fish start to surface and gulp for air. In May 1959 when the Fisheries Research Institute of Academia Sinica undertook the transporting experiment, water quality data were obtained. The results are seen in Table 20.1.

It can be seen that pH varied between 7.2 and 7.6, and it did not have any adverse effect on the fish. Water temperature was 19 to 21°C; each basket contained 360,000 to 400,000 fry. In freshly changed water, DO was 5.92 to 7.36 mg/ℓ. After 4 to 5 hr, it dropped to 3.04 to 4.72 mg/ℓ, a decrease of 2.56 to 3.20 mg/ℓ. At this time, many fish start to surface and show discomfort.

The source of water must be carefully inspected before the water is used for changing. Water coming from a mine or from an overly fertile pond or salty water are all to be avoided. Tap water contains chlorine which is harmful to fish. It has been proven that when water contains chlorides to 0.2 to 0.6%, all cyprinid fry will die. Wherever possible, it is best not to use tap water entirely. A mixture of tap water and pond water has yielded good results. Temperature differential between water in the basket and fresh water must be minimized; a sudden change of water temperature may kill the fish. In any change of water, no more than two thirds of water is replaced.

*Stirring*

Stirring water to increase DO is a crude method and is very labor consuming. The effect is also minimal. In manipulating the stirring action, the stirrer should be kept under water and should be operated rather gently.

TABLE 20.1

**Change of Dissolved Oxygen (DO) Before and After Change of Water During Fry Transportation**

| Basket | Density | Date | Time | Air (°C) | Water (°C) | pH | DO in fresh water (mg/ℓ) | DO when fish surface (mg/ℓ) | Drop in DO (mg/ℓ) |
|---|---|---|---|---|---|---|---|---|---|
| 1 | 360,000 | May 22—23 | 1430 | 21.0 | 20 | 7.5 | 5.92 | — | — |
| | in 400 | | 1800 | 21.5 | 20 | 7.3 | — | 2.88 | 3.04 |
| | ℓ water | | 1930 | 21.5 | 20 | 7.6 | 6.24 | — | — |
| | | | 2345 | 19.0 | 20 | 7.4 | — | 2.62 | 3.62 |
| | | | 0012 | 20.0 | 21 | 7.6 | 6.72 | — | — |
| | | | 0500 | 20.0 | 19 | 7.4 | — | 3.04 | 3.68 |
| | | | 0600 | 19.0 | 20.5 | 7.6 | 6.24 | — | — |
| 2 | 380,000 | May 22—23 | 1430 | 21.0 | 20 | 7.5 | 6.08 | — | — |
| | in 400 | | 1800 | 21.5 | 20 | 7.3 | — | 3.04 | 3.04 |
| | ℓ water | | 1930 | 21.5 | 20 | 7.6 | 6.40 | — | — |
| | | | 2345 | 19.0 | 20 | 7.4 | — | 2.56 | 3.84 |
| | | | 0012 | 20.0 | 21 | 7.6 | 6.24 | — | — |
| | | | 0500 | 20.0 | 19 | 7.4 | — | 3.20 | 3.04 |
| | | | 0600 | 19.0 | 20.5 | 7.6 | 6.24 | — | — |
| 3 | 440,000 | May 22—23 | 1430 | 21.0 | 20 | 7.5 | 6.08 | — | — |
| | in 600 | | 1800 | 21.5 | 20 | 7.3 | — | 2.88 | 3.20 |
| | ℓ water | | 1930 | 21.5 | 20 | 7.6 | 5.92 | — | — |
| | | | 2345 | 19.0 | 20 | 7.2 | — | 2.88 | 3.04 |
| | | | 0012 | 20.0 | 21 | 7.6 | 7.36 | — | — |
| | | | 0500 | 20.0 | 19 | 7.4 | — | 2.62 | 4.72 |
| | | | 0600 | 19.0 | 20.5 | 7.6 | 7.04 | — | — |

*Aeration*

This is done either by an electric pump or by a hand pump. Rubber tubing is used to send air into fish baskets; a sand stone is connected at the end of the tubing. The stones should be placed near the center of the bottom of the baskets. Air pressure is regulated by valves. Even when electric pumps are used, some spare hand pumps should be provided in case the electric pumps fail to function.

*Running Water Spray and Aeration*

A system of running water spray and another system of forced air are installed in a railway boxcar. In each boxcar, 16 fish tanks, 0.8 $m^3$ each, are installed. A water storage tank of the same size is also installed and is connected to the fish tanks with a screen partition between them. Water is drawn off the bottom of the fish tanks and is recirculated to the overhanging water tanks which release water back to the fish tanks through overhead sprayers. Air is supplied to the fish tanks through rubber tubing and sand stones by air-compressing pumps.

*Oxygenation in a Sealed Container*

This method has been used in recent years. The container is three fourths filled with water. It is then filled with highly compressed oxygen and sealed tight. In a series of experiments conducted in 1958, the survival rate was found to be 96.6% after 5 hr in the air in a density of 2000 fry per liter of water. After 6 hr, the survival rate was 94%.

In another large-scale experiment, a cylindrical metal container with a diameter of 20 cm and a height of 80 cm was used. When two thirds full, it contains 50 ℓ of water and may take 197,000 fry. With an air temperature of 36°C and water temperature of 28°C, and after 6 hr of transportation, the survival rate was greater than 90%. Compared with transporting by truck, this method can transport 5 times the volume at 2 to 3 times the survival rate. The only drawback is that it requires pure oxygen, which imposes a definite limitation to widespread adoption of the method.

Most recently, nylon sacs have been used in the place of metal containers, and they have yielded good results.

**Water Temperature**

Fish species are adapted to special temperature regimen beyond which they will suffer mortality. Fry and seed are more delicate than adult fish, and their temperature tolerance is much lower. In transporting fish fry, it is therefore important that sudden temperature changes be avoided. Within the temperature regime, it is best to keep the temperature as low as possible. Sudden change of water temperature usually occurs when water is changed. It is a good practice to let the fresh water sit for a while before replacing the old water. Fish fry can adapt better from high to low temperatures than from low to high temperatures. The optimum temperature range for fishes is between 15 to 25°C. Fish activity fluctuates according to temperature. Under high temperature, the metabolic rate is high, breathing rate is fast, and oxygen consumption is large. In transporting fish during the summer, it is best to cool the water by using ice.

**Water Quality**

During long transport, fish fry should be fed; but feeding should be slight, or water quality will deteriorate. Egg yolk should be ground into fine granules so that it remains suspended in the water for a longer time, thereby giving the fish more opportunity to feed on it. In transporting summer fry, there is no need to feed if the destination can be reached in 1 or 2 days. In transporting spring fry, feeding should not be done regardless of the length of time in transport.

When feeding is necessary in long distance transport, it is better not to feed in the basket, but instead use the method of "pond lodging". This is to transfer fry into a live cage in a preselected pond and let the fish rest for a period of time (not too long). This has proved to be quite effective in increasing the survival rate of the fry. Dead fish and excrement in the basket should be removed promptly. In warm water, these materials decay easily and produce toxic material.

Before feeding and cleaning it is necessary to change water. Water should be changed again 5 or 6 hr after feeding. Another occasion that calls for water change is at dawn when fry usually surface to gulp for air. The frequency of water change depends upon water quality. If water change is accompanied by aeration, then once a day is usually sufficient.

Chemicals can be added to water to prevent bacterial activity and, thereby, to lessen decomposition of decaying matter. When antibiotics were used in experiments, the effect was especially pronounced where changing water was not feasible. Many chemicals have been used. When copper sulphate (0.7 ppm) was used, the mortality rate was only 5.4% after 5 days. When 0.5% salt solution was used, mortality rate was 5.8% after 4 days. A new development is in the use of tranquilizers, such as sodium amytal, to reduce the metabolic rate and decrease the consumption of oxygen by the fish.

## TRANSPORTATION OF MULLET FRY

The mullets have long been cultivated in China, but rearing these fish in fresh water has not progressed rapidly in the past due to difficulties in transportation. In more recent years, however, there has been some success in transporting these fish, and fresh-water culture of mullets has had some outstanding results.

### Elimination of Other Species

When mullets are collected in the field, there are usually some other species mixed in, including large numbers of labrids. These are especially harmful to the mullets, and therefore must be removed first. They can be removed by (1) sieves, since they are smaller; (2) selective scooping, since the labrids are in the lower layer and mullets are in the upper layer; and (3) hand picking.

### Acclimation

In the vicinity of mullet fry collecting sites, just above high tide mark, several small ponds about 3 ft deep are dug and provided with an intake and an outlet. From 200,000 to 300,000 mullet fry can be placed in each pond to acclimate. Acclimation takes four steps: first, 75% sea water and 25% fresh water; second, half and half; third, 25% sea water and 75% fresh water; fourth, all fresh water. Fish should be left in each step for 2 to 3 days. Because the density in the acclimation pond is high and natural food becomes insufficient, fish must be fed. At the end of the acclimation period, which usually takes around 10 days, the fish can be transported.

### Conditioning

Mullets are easily excited. Before transporting, they should be kept in dense schools for several days to clear them of excrements. The best time to transport is in the early morning or evening when temperature is relatively low. Transport should be avoided during days when atmospheric pressure is low.

At the present, the transportation of mullets is mostly confined to short distances — the longest not over 2 days. Generally, 1 $\ell$ of water can contain 140 fry 1 cm long, 50 fry 2.5 cm long, 20 fry 4 cm long, 10 fry 5 cm long, or 3 to 4 fry 6.5 cm long. Water should be changed every hour. If necessary, the entire carrying basket can be immersed in water in a pond to rescue fry that start to surface.

In long distance transport, survival rate seldom surpasses 70% in the summer. In one experiment conducted by Nankai University when mullet fry were sent from Mukden to Inner Mongolia, the survival rate reached 97% in 3 days of transport and when the temperature was 14 to 22.8°C. Each basket had 8550 to 9450 fry.

Based on many experiments conducted in Hopei Province, it is certain that the most important single factor affecting the survival rate of mullet fry during transportation is that the temperature be kept under 16°C; mullets can tolerate DO content down to 0.9 ppm. Although survival rate of mullet fingerlings is higher, it is still more practical to transport fry, because the number of fingerlings transported must necessarily be low and the per unit cost high.

## TRANSPORTATION OF FISH EGGS

Egg transportation in fish culture does not play an important role. This is because the hatching time of most culture species (grass carp, black roach, the bigheads) is too short to be of any practical use for transportation. The major species for which egg transportation applies is the common carp. Even here, however, the application is quite limited because carp can breed in various kinds of still water.

The important considerations in carp egg transportation are to maintain proper temperature, humidity, and air circulation, so that normal development is not hampered but, at the same time, development is retarded enough to prevent hatching en route.

### Preparation Work Prior to Transporting

A proper container should be selected. The best container is a wooden box. This can be of any shape, but small holes should be provided on the bottom as well as on the sides. The box is multilayered, each layer being 10 to 15 cm deep. Items at hand for transportation should include water sprayer, water storage vessel, thermometer, and a thermos jug with ice.

### Methods of Egg Transportation

#### *Dry Method*

The advantages of the dry method are that *Saprolegnia* does not grow, there is a high hatching rate, and handling is easy.

Generally, it is better to transport eggs before the eye stage. At this time, oxygen demand of the eggs is low. If transportation does not take more than 2 days, it can begin 4 to 10 hr after fertilization.

Dead (white, opaque) eggs should be removed and only good eggs (transparent and slightly yellowish) selected for transportation. Two to three million eggs can be placed in a cubic meter tank. Egg nests should be placed in a crisscross fashion and not too thick. The important point is to provide good ventilation. Direct sunshine and wind should be avoided.

It is important to maintain proper humidity in the egg container. Generally, additional water should be added by spraying every 30 to 60 min. En route, the temperature must be strictly controlled; sudden changes must be avoided. In hot weather, ice can be added. For best result, water temperature should be maintained at 13 to 14°C.

It is also important that mechanical damage be avoided. This can be done by cushioning the containers.

#### *Wet Method*

When eyed eggs are to be transported, the wet method should be used. This is simply to immerse eggs in a water-filled container. Water should be changed in the morning and in the evening. The wet method is relatively safe in that water temperature is not subject to sudden changes. However, it occupies a lot of space and is difficult to handle. Therefore, it is not used as often.

When fish eggs reach destination, dead eggs should be removed, and the egg nests lifted gently and transferred to a hatching pond. To increase the hatching rate, the eggs can be immersed in a 0.2 to 0.4% saline solution for 5 to 10 min. If, by this time, eggs still are not eyed, they should be kept in a dark room to continue development until the eyed stage is reached.

## SUMMARY AND CONCLUSIONS

The proper transportation and handling of fish eggs and fry are vital in any fish production system. Careful planning is important. The first thing to decide is the type of container to be used to hold the fish. There are buckets and barrels of many sizes and designs, each with different characteristics. A second consideration is the density to which the fry should be packed in the container. This must be as dense as possible to save space and money, but not so dense as to kill the fish. The two most important factors in determining the maximum allowable density are water temperature and DO levels.

Fish fry and eggs are shipped by three major means: land, water, and air. This involves either trucks, trains, boats, airplanes, or human labor. Each method has advantages for different situations. During any form of transportation, water maintenance is necessary. If some sort of method for replenishing DO is used, replacement of water can be put off, but still not avoided. The source of fresh water must be inspected carefully before it is used. Clean pond water appears to be best. Any temperature difference between old and new water should be minimized. In more recent years, cultivation of mullet species has become prominent, moving in with the more traditional species, the carps and the bigheads. Most of the techniques used for the transportation of all species, including mullet, apply to their eggs.

## REFERENCES

1. **Chen, C. S. and Lin, S. Y.**, The Fish Industry in China, *Bull. Chekiang Prov. Fish Exp. Sta.*, 1, 38, 1935.
2. **Atkinson, C. E.**, Peoples Republic of China, in *World Fish Farming,* Brown, E. E., Ed., AVI Publishing, Westport, Conn., 1977, 324.

Chapter 21

# THE REARING OF FISH FRY AND FINGERLINGS*

## INTRODUCTION

The supply of fish fry and fingerlings is one of the key questions facing fish culture industry. In order to solve the supply problem, it is necessary to increase the survival rate of fish fry to fingerlings by improving rearing techniques.[1,2]

## THE IMPORTANCE OF POND CARE

The growth and survival of fish in a fish pond depends upon the environmental conditions in the pond. It is therefore very important that the conditions of the pond be improved periodically. After the culture of fish for a year, food residue, fish feces, and material washed from the land accumulate on the bottom of the pond. Bacteria and parasites propagate; pest fish species remain. All these conditions are negative factors.

A thorough cleaning of a fish pond once a year has the following advantages:

1. The pond will become fertilized. After the draining, the pond bottom will be subjected to freezing and drying, thereby making the surface layer more porous and better ventilated. This hastens the mineralization of decaying matter and converts organic matter into nutrients. At the same time, some microbes are killed by ultraviolet radiation.
2. The rearing capacity of the pond can be increased. When bottom mud is removed, pond volume is increased, which increases the rearing capacity. Experience shows that the removal of 1 ft of bottom mud from a 5- to 6-ft pond increases rearing capacity by 30%.
3. Fish diseases are reduced. Drying and cleaning the pond usually kills off bacterial pathogens, parasites, and aquatic insects and their larvae. In one district of Chekiang Province, such cleaning of fish ponds resulted in lowering the mortality rate of grass carp and black roach to only 3%, lower than in neighboring areas where no cleaning was done. The production was also increased to 200 kg/ha.
4. Leaking ponds can be repaired.
5. Pest species can be eliminated.
6. Harmful organisms, such as filamentous algae, aquatic plants, aquatic insects, and tadpoles, can be killed.
7. The bottom mud can be utilized to fertilize farm land.
8. The volume of the pond can be increased by building up the rim of the pond. This is an easier way to increase the rearing capacity than deepening the pond.

## METHODS OF POND CARE

### Mud Removal

At 20 to 30 days prior to fry planting, bottom mud should be removed and the water drained. Settled mud along the banks should be spread over the banks, allowed to dry for 2 to 3 days, then pat down flat and smooth. If sedimentation is slight, draining

* From *Culture of Freshwater Fish in China,* Science Publishers, Peking, 1966 translated by Koo, T.Y., Dept. of Fisheries, U.S.D.I., Solomons, Md., and Walker, C., U.S.D.I., Washington, D.C.

should be done in the winter. If the pond is new and in good condition, it is not necessary to drain it every year.

### Chemical Applications and Their Effects

*Lime (CaO)*

**Application in drained ponds** — After mud removal, retain some water (2 to 3 in. deep) in the pond. The amount of lime to be used is about 1000 kg/ha. The amount should be reduced if sedimentation is apparent. Lime will absorb moisture easily and change into calcium hydroxide; therefore, if not used promptly, it should be kept in a dry place.

**Method** — Dig a few small pools scattered around the pond. Then place some lime in each pool, and while the water is hot, distribute it all over the pond. In the place of pools, the same results can be obtained by dissolving lime in buckets. The next day, stir the bottom with a long-handled stirrer so that the lime solution is thoroughly mixed with bottom mud.

**Application in nondrained ponds** — In ponds which are not close to any rivers or lakes so that it is not convenient to drain and refill the ponds, the ponds can be treated with water in them. The amount of lime to be applied is 1200 kg/ha for every meter of depth.

**Method** — One way is to dig some small pools along the side of the pond. Dissolve the lime in the pools, and while the water is still hot, disperse the solution into the pond. Since this is not practical in a large pond, there is a second method to be used. This involves putting lime in a sack or basket, hanging the sack on the side of a boat, then running the boat back and forth in the pond slowly so that the lime is distributed into the pond more or less evenly. The next morning, the bottom mud should be stirred up to enhance the effect of pest control. Application in a nondrained pond has proved to be more effective and more labor-saving than that in a drained pond.

**Effects of lime application** — Lime kills residual pest fishes, frog eggs, tadpoles, aquatic insects, snails, crabs, shrimps, filamentous algae and shallow-rooted plants, disease-causing parasites and bacteria, etc. It clarifies pond water and precipitates colloidal organic matters. It fertilizes the pond by liberating nitrogen, phosphorus, potassium, etc. from the mud. In water, lime becomes calcium hydroxide. By absorbing carbon dioxide, it becomes calcium carbonate, which causes the mud to be more porous and, therefore, more ventilated which hastens the decomposition of organic matter by bacteria. Also, together with dissolved carbon dioxide and carbonic acid, it serves as a buffer to maintain pH value, keeping it slightly basic, which is beneficial to fish. Calcium is an indispensable ingredient for green plants and animals.

*Camellia Seed Residue*

This is the residue from pressing the fruits of *Camellia sasangua* or *C. semiserrata* for oil. The residue comes in the shape of a large cake; it contains saponin and acts as a blood-dissolving toxin.

*Camellia* residue is commonly used by fishermen in Kwangtung and Kwangsi provinces. Approximately 500 kg/ha for every meter of depth is used. The cake is broken into small pieces and soaked in water for 24 hr (water temperature approximately 25°C). It must be greatly diluted before adding it to the pond.

*Camellia* residue can kill pest fish, frog eggs, tadpoles, snails, and some aquatic insects, but it is not effective against bacteria. On the other hand, it helps the growth of green algae, which is not easily digested by fish. *Camellia* residue, therefore, is not as good an agent as lime.

*Croton tiglium*

The fruits of this plant contains crotin, a toxic albumen, which can kill fish. It is also toxic to man. Some people who are more sensitive to this toxin may suffer a swollen face and swollen extremities upon contact or from breathing the fumes.

The effect of crotin is poor, It can kill only pest fish, but not parasites, bacteria, frog eggs, tadpoles, or aquatic insects. The increase of plankton in the first 7 to 8 days is even greater than in ponds that are treated with lime or *Camellia,* but plankton rapidly declines thereafter. Therefore, crotin does not improve the water quality.

*Bleaching Powder (Calcium Hypochlorite)*

Upon hydrolysis, the bleaching powder becomes hypochlorous acid,a strong bactericidal agent. Bleaching powder has been used in the U.S.S.R. to clean fish ponds for a long time. It has been adopted in China only in recent years. The bleaching powder is dissolved in water, then the solution is immediately distributed into the pond. Generally, 5 days should elapse before fish planting.

Important things to remember in connection with using bleaching powder are as follows:

1. Bleaching powder must be stored in an airtight porcelain container in a cool, dry place.
2. Use porcelain or wooden vessels to handle bleaching powder. Do not use metal containers, since metals will be oxidized by the powder.
3. Workers handling bleaching powder should use a mouth mask to avoid breathing the fume. Also avoid contacting clothes with the powder.

Bleaching powder can kill fish, frog eggs, tadpoles, snails, some mussels, aquatic insects, parasites, and bacteria. In this respect, bleaching powder is equal to lime. The amount required is small, and pond recovery is fast. The disinfectant effect is closely related to fertility of the pond. The more fertile the pond, the smaller the effect.

*Rotenone*

Rotenone is contained in legumes, *Derris leguminosus* Rozle and *D. elliptica* Benth. It is extracted from the roots of these plants. It is yellowish and can be dissolved in organic solvents. It can kill fish and aquatic insects. The effective concentration is 2 ppm. Use 20 kg/ha for every meter of depth. In use, dilute to 10 to 15 times, then spray into the pond. Rotenone has no effect on plankton, parasites, or bacteria.

*DDT*

In recent years, a few places have used DDT to clean fish ponds. The method of application is similar to that of bleaching powder. DDT kills fish and aquatic insects. It has no effect in controlling diseases or improving bottom mud conditions. Its residue inhibits the reproduction of planktonic animals and adversely affects the growth of fishes.

*Combined Use of Lime and Camellia*

For each acre-foot of water, use 50 kg/ha of lime and 270 kg/ha of *Camellia.* First soak *Camellia* in water and smash it, then add the whole mixture into lime. When lime is completely dissolved, disperse the solution into the pond. In a week, test the water with ten bighead. If the fish survive after 8 hr, the pond is ready for fry planting.

Theoretically, the combined use of lime and *Camellia* should be effective. Nevertheless, much study is needed to show whether it is economical or whether the ingredients of the two substances would counteract each other and cancel their effects.

*Combined Use of Lime and Bleaching Powder*

For each meter of water depth and hectare of pond, use 30 kg of bleaching powder and 70 kg of lime. Planting can be done 10 to 12 days after application. The effect seems to be better than either of the two substances used alone.

The effects of the various chemicals are summarized in the following paragraphs.

**The effect of eliminating trash fish and disease control** — The best chemical to eliminate trash fish is lime; *Camellia* and bleaching powder are next; rotenone, *Croton,* and DDT are least effective. As far as killing parasites and bacteria, bleaching powder is the most effective; lime is next. *Croton* is useless and *Camellia* can promote the growth of bacteria.

**The effect on productivity** — Lime can increase the production of fish because it can improve the bottom mud, accelerate the decomposition of organic matter, and, at the same time, serve as a source of calcium.

**Effect on macroorganisms** — Again, the most effective is lime, next is bleaching powder. Rotenone, *Croton,* and DDT are noneffective. In addition to killing most aquatic animals, lime can also eliminate algae and some tender aquatic vascular plants.

**Effect on plankton** — Initially, lime, bleaching powder, and DDT will kill plankton. However, 4 days after lime application, and 2 days after bleaching powder application, plankton is on the rise; after 8 and 6 days, respectively, plankton peak is reached. In the case of DDT, plankton starts to grow again 8 days after application. Plankton increases also after *Camellia* application, but the increase is rather slight. *Croton* has an opposite effect: it enhances plankton growth at first, but after 2 or 3 days, plankton decreases. After lime application, plankton life is consistently kept at a high level and for a long period. In this regard, it is followed by bleaching powder and *Camellia. Croton,* rotenone, and DDT are poor.

**Effect on pH** — When lime is first applied, pH rises above 12, but within 24 hr, it rapidly drops which is then followed by gradual decrease. When the pond reaches approximately pH 9.4, zooplankton reproduce rapidly. Bleaching powder and *Croton* increase pH somewhat, but the effect on plankton growth is not determined. No change of pH has been detected after *Camellia* application.

## THE REARING OF FRY

Because fish fry are too small (6 to 9.5 mm), they are not capable of active movement, have narrow feeding habits, and are not well adapted to environmental hazards, they must be artificially reared before being planted in a pond. In general, a 15- to 25-day rearing period is needed until they attain a length of about 1 in. At this stage, they are called "summer seed". Proper rearing of fry requires skill and technique and is essential to the success of fish culture.

## MAIN FEATURES OF LIVING HABIT OF FRY

### Food Habits

Newly hatched fry rely upon their egg yolk as nutrient. When the yolk becomes absorbed, fry start to feed on planktonic organisms. Observations show that, at the beginning, they feed on zooplankton; later, a greater variety of foods is eaten.

Two points should be explained. First, in discussing food habits, we must take into consideration the environment in which the fish live. The species and abundance of plankton in a fish pond are reflected in the food contents of fish fry. Second, the natural foods of fry fall into two categories: preferred food and obligatory food. Fish select their preferred food when there is ample supply of food; they feed on obligatory food when there is scanty supply of food.

The question of nutrition through osmosis by fish fry should not be neglected. Inorganic salts can permeate the body of carp through the gills and oral membrane.

### Growth Rate

Under normal conditions, the increase in body length and weight in fry is fairly rapid. In 19 days, the grass carp increase in weight by 65.7 times, black roach by 42 times, white bighead by 162 times, and striped bighead by 82.6 times. However, when food is deficient, fish fry will rapidly become emaciated and eventually die. Realizing this, fishermen usually inspect their fish ponds two to three times a day to make sure that food is not lacking.

### External Characteristics of Four Domestic Carps

In fry that are 1.2 to 2.2 mm in length, major differences among the different species are as follows. The grass carp and black roach can be distinguished from the two bigheads by the absence of black pigments on the ventral finfold. The grass carp is distinguished from black roach by the shape of a network of pigments on the lower corner of the base of caudal fin. In the former, this network is in the form of a round spot and is lighter in color, whereas in the latter, it is elongated and is darker. The white bighead differs from striped bighead in having distinctly arranged pigments on the ventral finfold. In the striped bighead, these pigments are not as dense and are scattered and less distinct. At the summer seed stage (about 1 in.), the grass carp is golden-yellow in color, with distinct scales and a more blunt snout. The black roach is greenish-yellow, with indistinct scales and a more pointed snout. The white bighead is silvery white; the ventral ridge extends from the anus to the pectoral region; pectoral fins extend to the base of the ventral fins. The striped bighead is golden-yellow in color; the ventral ridge is present only between anus and ventral fins; pectoral fins are long, extending over the basal part of ventral fins.

### Distribution of Four Domestic Carps in a Fish Pond

When fry are first placed in the pond, they are more or less evenly distributed. However, when they grow to about 1.5 mm in length, their distribution covers different food habitat. Grass carp and black roach are gathered near shore areas where larger planktonic animals and benthic animals are more abundant. The two bigheads live in the upper layers of the middle of the pond.

## THE SELECTION OF A FRY POND

In selecting a pond in which to raise fish fry, the environmental conditions must satisfy the biological characteristics of the fry and the feeding and harvesting requirements. Selecting conditions are discussed below.

**Ample water source** — During the course of fry culture, new fresh water must be added from time to time in order to maintain the water quality. Therefore, the fry pond should be located near the source of water so that it is convenient to add water.

**Appropriate size and depth of pond** — Generally, a fry-raising pond is from 0.152 to 0.758 acres in size and never exceeds 1.515 acres. If the pond is too large, it is difficult to manage, to add water, etc. Also, it is easily affected by wind, and the waves will injure fry. On the other hand, when the pond is too small, then environmentally induced changes will be too big. In pond depth, requirements vary according to localities. In Kiangsu, Chekiang, and Kwangtung, 3- to 6-ft deep ponds are more suitable. In Hupeh, Kiangsi, and some other provinces, ponds are generally 1.5 to 2.5 ft deep. When ponds are too shallow, water temperature and water quality can fluctuate violently due to environmental influence. When ponds are too deep, they are too expensive to fertilize, and water temperature is usually too low. Generally speaking, 3- to 6-ft-deep ponds are more appropriate.

**Regular shaped ponds** — In a regular shaped pond, it is easy to harvest by seining; it is also easy to feed and manage.

**Nonleaking** — If there are cracks and leaks, fry will tend to gather there and swim against the flow, thereby affecting feeding activity. It is also more difficult to maintain the proper volume of water.

**Level bottom** — The pond bottom should be relatively level and covered with only a small amount of mud. Too much mud will cause turbidity during seining. Mud also contains organic matter and causes oxygen depletion. It will produce hydrogen sulfide and marsh gas which are injurious to fish. A certain amount of mud, say 2 to 3 in., is beneficial because it provides the necessary adjustment of fertility for the pond.

**Plenty of sunlight** — The main food of fry is plankton, which depends upon the sunlight for growth. Strong sunshine also keeps water temperature up, which is beneficial to the growth of fish.

## REARING OF CYPRINID FRY

According to Chinese classical literature, during the Sung Dynasty the Chinese people began catching fish fry from the rivers and rearing them in ponds. Because of the poor communication, there was little exchange of experience among fish culturists in various localities, and as a result, culturing methods vary widely from place to place. The most typical methods are those used in Kwangtung and Kwangsi provinces where green grass is used as fry food, and those used in Kiangsu and Chekiang provinces where soybean milk is used. Today, fish culturists from these two areas are distributed over the entire country, exerting a great influence with their traditional methods. The following paragraphs discuss in brief the essentials of the fish fry culturing methods that are practiced in these two areas.

### The Green Grass Feeding Method of Kwangtung and Kwangsi

By this method, green grass and cow manure are mixed and then placed in fish pond. As they decay, they fertilize the water and promote the growth of plankton, which are eaten by fish fry. Where grass carp and *Cirrhina molitorella* are raised, peanut cakes are added after the third day.

#### *Preparatory Work*

**Pond cleaning** — Generally, tea cake residue is used for pond cleaning. In order to exterminate Diphylobothrium tapeworm, the pond should be cleaned 50 days before fry are introduced. A second treatment should be made 10 days before fry are introduced. Then, the day before fry introduction, some 6-6-6 powder is added to kill off detrimental insects. In areas where no tapeworm disease is present, only one treatment (10 days before fry introduction) is sufficient.

**Fertilization** — From 5 to 10 days before fry introduction, the pond can be fertilized with green grass. In Kwangtung, this is usually several species of compositae. Actually, some legumes or any other nontoxic, tender-leaved plants that decay easily can be used just as well. The grass is gathered in bundles which are placed in the lee side of the pond in shallow waters about 1 or 2 ft away from the shore. In due time, the grass will decay. The grass bundles should be turned over in a day or two. When the leaves and young stems are deteriorated, the residual stems and roots can be removed. If cow manure is used, the manure is diluted with water, and the mixture is evenly spread over the surface of the pond. The amount of fertilizer to be used depends upon the fertility of the pond, the fish species to be cultured, and the quality of grass used. In general, 1.76 to 3.51 kg of grass is used for each hectare of pond. If the grass is of inferior quality (difficult to decay), then the amount should be increased. After the introduction of fish fry, depending upon the water quality, the pond should be continually fertilized.

**Water quality testing fish** — To test whether the fertility of water in the pond is suitable, 200 to 300 young (approximately 4 in. long) striped bighead are placed in the pond several days after green grass has been added to the pond. The purpose is twofold. One purpose is to test the fertility of water. If the test fish surface for a short time every morning, it is an indication of suitable water quality. If they surface for an unduly long time, it is a sign of overfertility. If they do not surface at all or seldom surface, then the water is not fertile enough and more grass should be added. If the pond is used to raise grass carp, then the surfacing of the test striped bighead should last no longer than 2 or 3 days. The other purpose of using the test striped bighead is to let them feed on the larger zooplankton (such as the Daphnia). Larger planktonic animals cannot be used by the small fish fry, and they contribute to the depletion of dissolved oxygen.

In addition to using the striped bighead, fishermen in Kwangtung and Kwangsi also use grass carp as the test fish. The grass carp can remove filamentous algae. A recent study shows that 2-year-old grass carp can be a carrier of tapeworm and can transmit the disease; therefore, it is no longer recommended as a test fish.

Before fry planting, all test fish must be removed. If after 10 days or so of testing fry planting still has not been carried out, then 1 t of green grass should be added to each hectare of pond.

*Fry Planting*

In Kwangtung and Kwangsi, the fish fry that are collected from the river are sorted out by species through sieving, separation, etc. and reared separately.

The rearing density varies according to species reared. In general, for the white and striped bigheads, 450,000 to 600,000 fry are planted in each hectare of pond; for grass carp, 300,000 to 450,000; for the common carp, 450,000 to 800,000; for *Cirrhina molitorella*, 500,000 to one million. The density can be varied somewhat according to the condition of fry and the season in which fry are planted. In early season, the density can be greater; in late season, less.

*Rearing and Management*

***Rearing Methods***

The methods differ according to species reared, as discussed in the following paragraphs.

**The rearing of the white and striped bigheads** — Two to three days before planting, 1.5 to 2.0 t of green grass is placed in each hectare of pond. The day after planting, fry will surface — about 2 to 3 hr/day. If they do not surface, more grass should be added. Thereafter, grass should be added every 2 to 3 days. The purpose of having the fry surface is to eliminate pest species that may be mixed in the collection. The pest species are less tolerant to low oxygen, and they will die of asphyxiation under conditions where the bigheads surface. In 6 to 7 days, fry will grow to 1.6 to 1.9 cm in length. At this time, they should be netted out, and the two species separated and reared in separate ponds in which large amounts of plankton have already been cultured. They can be reared here until they are 3 cm long.

The white and striped bigheads can be separated by sieving if they are of different sizes. However, when they are similar in size and cannot be separated by sieving, they can be managed in the following manner. Place the fry in a boxed net for 4 or 5 hr. Then place some food near the shore, and release the fry from the box. The white bighead, being more active and stronger, will eat most of the food and, consequently, will grow faster. In a day or two, the sizes of these two species will become different, and they can be separated by sieving. If not, the process can be repeated once or twice.

**Rearing fry of grass carp** — From 1.5 to 2 t of greens per hectare should be added to the pond every 3 days. Also, starting from the third day, 500 to 700 kg of cow manure should be added; 100 to 200 kg of peanut residue cake should be added to each hectare every day. In 10 to 15 days, grass carp fry can grow to 2.5 to 3.0 cm.

**Rearing of fry of *Cirrhina molitorella*** — Fertilize the pond as in a grass carp pond. Generally, it requires one full month of growth before the fish can be taken out for culturing in an adult pond.

**Rearing of fry of common carp** — The common carp are reared from the eggs. At the time eggs are placed in the pond, greens can be added. Other fertilizers can be added after the eggs hatch. In 15 to 20 days, carp fry can grow to 2.5 to 3.0 cm.

#### *Pond Inspection*

Fry pond should be inspected once every morning and once every afternoon. Water coloration and behavior of fry (whether surfacing or not) must be carefully observed in order to decide whether fertilization is needed.

#### *Insect and Disease Control*

In using green plants as a fertilizer, various kinds of insects are introduced which are injurious to fish fry. Other diseases such as gill fungus, rotifers, tapeworms, etc. also can occur and must be controlled at an early date.

#### *Hardening and Transfer*

After being reared for 15 to 20 days, fish fry grow into 2.5- to 3.0-cm summer seeds. At this stage, they feed actively and demand more space. They must be removed and reared in separate new ponds. Before doing so, they should be hardened as follows.

**Method** — Select a clear day when the fry do not surface. At about 9:00 to 10:00 a.m., seine the fry into a cloth net. Care must be taken to avoid getting debris into the net. In the cloth net, fry are concentrated in a dense school. At about 2:00 p.m., they are usually hardened enough to be transferred; sometimes, however, a second hardening is called for if the fry are not hard enough. Also, if the seeds are to be transported to a distant place, then not only a second hardening is required, but also a further lean-water treatment is recommended. By lean-water treatment, fish are removed from the rearing pond and placed in cloth containers in an unfertilized pond overnight. By the next morning, they are ready to be transported.

**The purposes of hardening** — First, when disturbed, fish secrete a large amount of mucus, which is harmful. After being hardened once or twice, fish become adapted to high density, and will reduce mucus secretion during transportation. Second, in 15 to 20 days, fry grow rapidly, increasing their body weight by up to more than 100 times. Body tissue contains a high percentage of water. By concentrating the fry in a cloth container during hardening, the water content of body tissue is greatly reduced, and the muscles are hardened, making the fish much more tolerant to transferring and transporting. Third, when fry are confined in dense schools, all their feces are deposited, which is desirable for transportation purpose.

### The Soybean Milk Feeding Method of Kiangsu and Chekiang

#### *Preplanting Water Change*

After the rearing pond has been treated, it should be drained about 3 days before fry are introduced. Debris, harmful organisms (water scorpions, tadpoles, etc.), and large plankton animals are removed from the bottom. Then, fresh water is added through a gunny sack filter. This is to prevent injurious fish and eggs from entering the pond. Water should be filled to about 2.5 ft.

*Fry Planting*

As soon as fry arrive, they should be immediately placed in the rearing pond, regardless of the time of day. To each hectare of pond, 700,000 to 900,000 fry can be planted. A bamboo sieve is used to sieve out large-sized harmful fishes.

Water temperature differential between fry water and pond water should not be more than 4°C. Feeding can be started soon after introduction of fry: when fry are planted in the morning or afternoon, soybean milk can be added the same afternoon; when fry are introduced in the evening, feeding can be done the next morning.

*Feeding Method*

**Preparation of soybean milk** — First, soak the beans. The time of soaking depends upon water temperatures: 10 to 12 hr below 18°C; 6 to 7 hr below 25 to 30°C. Milk production is poor if the soaking is overdone or underdone.

When producing milk by grinding, beans and water must be fed into the grinding stone simultaneously. If water is added after grinding, precipitation will take place. After the milk is prepared, precipitation will take place if it is not used within ½ hr. Generally, 2.5 to 3.0 lb of soybean can yield 50 lb of milk.

**Feeding frequency and method** — Formerly, feeding was done twice daily. In recent years, the daily quota is divided evenly into three portions and fed thrice daily. In the morning, the soybean milk is fed into the pond first along the periphery of the pond and then over the entire pond. At noon, feeding is only along the periphery. In the afternoon about 2:00 to 3:00 p.m., the morning feeding process is repeated. The main reason for this type of feeding procedure is to assure plenty of food for shore-living fry of grass carp and black roach. Feeding should be temporarily suspended when it rains or when fry surface. If the pond is very fertile, only one feeding is enough for a day. The soybean milk should be filtered; use the filtrate to feed fry. The milk should be spread evenly over the entire pond. In small ponds, this can be done from the shore; in large ponds, the milk can be dispensed from a boat. Each day, use a rope to skim off the surface foam.

**Quantity of feed** — The quantity of milk to be used depends, not on the number of fry present, but on the size of pond. When fry are first introduced, use 50 to 75 kg/ha to each pond. Thereafter, the amount varies according to the color of water and the growth of fry. In a continuing rain, when water temperature drops, a part of the milk will accumulate on the bottom, and plankton growth is slow. Under this condition, more milk should be added. Care should be taken, however, not to use too much milk, since a sudden change of weather may start a plankton bloom and cause fish mortality.

**Additional feeding of soybean paste** — After the application of soybean milk for more than 10 days, water becomes loaded with phytoplankton, but plankton animals decrease in number. Grass carp and black roach cannot find enough large-sized planktonic animals on which to feed. These fish then gather in schools near the shore and swim wildly. To prevent this kind of condition from happening, soybean paste can be used as an additional feed. This is usually done by placing the paste in small grass bundles which are tied to small bamboo poles in shallow water. If this treatment does not work, then feed the fish with grain residue from wine making. The residue should be placed in shallow water (about ½ ft deep) along the shore. Because the wine residue has a strong flavor, it seldom fails to attract fish.

*Pond Inspection*

The pond should be inspected twice a day, once in the morning and once in the afternoon. Remove undesirable weeds along the shore area, and get rid of frog eggs and tadpoles.

*Intermittent Pond Filling*

This is an important step, as learned from experience in Kiangsu and Chekiang. During the entire raising period, water should be filled in three to four installments, each time adding 3 to 5 in. deep. Water is filtered through a gunny sack cloth to prevent pest fish from getting into the pond. Installment filling of the pond is advantageous because when fish fry are first introduced into the pond, shallow water is beneficial to fish growth since natural food organisms breed faster due to rapidly rising water temperature. However, after a few days when the fish become much larger, water quality will deteriorate fast, and the level of dissolved oxygen drops considerably, especially during the night. The addition of fresh water increases oxygen content.

*Fry Hardening and Transfer*

After 15 to 20 days of rearing, fry grow to about 3 cm long. At this time, they are ready to be transferred to new ponds for rearing. To avoid mortality during the transferring process, fry must go through the hardening procedure.

To test whether fry of white bighead are ready for hardening training, one can scare the fish by simply stomping on the shore or splashing on the water surface to see if the fry leap out of water in groups. If they do, it means that they are ready in 2 or 3 days.

**First hardening** — Spread a seine across the entire width of the pond. Drag the seine from one end of the pond to the other. When nearing the halfway point, feed some cooked soybean milk at the far end of the pond to attract the fish away from rubbing against the net. When seining is complete, gather the fish in the bagged net and leave them in the crowded condition for 10 sec or more before releasing them back into the pond. Then 2 hr later, feed some raw soybean milk. The next day, feed five or six times with cooked soybean milk.

**Second hardening** — If the fish act normally after the first hardening and if the weather is good, the fish can go through the second hardening on the third day. Collect the fry in a seine net in the same manner as in the first hardening. However, instead of releasing them almost immediately back into the pond, transfer the fry to a box net, where they are held for ½ hr before releasing. The fish should be fed with raw soybean milk the afternoon they are hardened and fed 10 to 12 times with cooked milk the next day.

**Third hardening and transferring** — After the first and second hardening and if the weather is good, the fish are usually ready to be transferred. If the weather is bad, then after a couple of days a third hardening is desirable before transferring. In transferring the fry from the fry pond into new rearing ponds, the fish are collected into a box net. Then they are segregated into four boxes by species. Generally, the two bigheads grow faster, and they make up the major portion of the first transfer. The grass carp and black roach are smaller and usually form the second transfer. As the fish are segregated, pest species are removed. Also, the number of each species is estimated by the aliquot method.

**Rearing with Mixed Fertilizers**

Experimental studies have been made in rearing fish fry with mixed fertilizers. This method has been widely adopted throughout China.

*Kinds of Fertilizers and Mixing Formulas*

One fertilizer mixture is composed of four parts green grass, two parts sheep manure, one part human night soil, and 1% lime. Another mixture consists of one part each green grass and cow manure, and 1% lime.

*Preparation*

Dig nondraining fermentation pits along the shore. Place green grass and cow manure in alternate layers in the pits; add lime solution (1 part of lime to 100 parts of total grass and manure) on top of each grass layer, and finally add enough lime solution to cover up all the fertilizer. Then seal the pits with mud and let fermentation take place. When the pits are opened for using, they should not be opened for any length of time, or nitrogen will be lost.

*Application*

From 3 to 5 days prior to the planting of fry, apply fertilizer once or twice at 3 m of water volume in the pond. From the day fry are planted, the pond should be fertilized twice a day at the rate of 120 to 160 g of fertilizer to each cubic meter of water. Place the fertilizer in a sieve, wash the fertilizer into the water, then sprinkle the water evenly over the pond. Place the residue in the corners of the pond so that further fermentation and deterioration can take place.

In 1955, the Fisheries Research Institute of Academia Sinica conducted a fertilization experiment in Tang Pond in Chekiang Province. The pond is 0.273 acres in area and 4.265 ft deep. On May 20, the pond was sterilized with 150 kg of lime. On June 8, 198,700 fry were introduced. The pond was fertilized twice a day, using the 4:2:1 proportion of green grass to sheep manure to human night soil. After 23 days of rearing, the fish were taken out on July 1. There were 189,410 fry, a survival rate of 95.3%. The pH of the pond was 7.3 to 9.2; water temperature ranged between 23.1 and 34.3°C, with an average of 27.1°C.

In 1957, another experiment was conducted in Hupeh Province. This time, different sizes of fish were segregated during the rearing process. When the fish were reared for 9 days, they were segregated into two groups by size by the use of sieving. The large ones, which were primarily the bigheads, were raised in one pond; the smaller ones, which were mainly grass carp and roach, were reared in another pond. Both ponds were sterilized with lime and fertilized. The ingredients and proportion of fertilizers were one part each green grass and cow manure with 1% lime. A total of 520,000 fry were reared. After 19 days, 480,142 summer seed were harvested, giving a survival rate of 92.33%.

Under such management, the environmental factors were favorable. The pH was 7.6 to 8.7; average water temperature, 29.5°C; dissolved oxygen (DO), 2.0 to 9.6 ppm; biochemical oxygen demand (BOD), 8.4 to 14.2 ppm. The ponds maintained a high plankton life throughout. In one pond, the plankton volume was 2.7 to 38.9 g/$m^3$; in another pond, 2.2 to 36.9 g/$m^3$. Also, because the large and small fish were separated, each size group grew satisfactorily.

**Rearing with a Mixing Feed of Organic Fertilizer and Soybean Milk**

In essence, this is a combination of the methods used in Kwangtung-Kwangsi and Chekiang-Kiangsu. Three to four days prior to fry introduction, fertilize the pond with cow manure-green grass. After fry introduction, feed with soybean milk daily. The results of one experiment[3] using the above method are as follows:

Pond, 0.175 ha in area; water depth, 0.8 m

| | |
|---|---|
| April 25 | Lime 235 kg, to cleanse the pond |
| May 11 | Green grass, 750 kg |
| May 12 | Cow manure, 500 kg |
| May 18 | Green grass, 135 kg |
| May 23 | Introduced fry, 415,800 |

Daily feed, soybean milk and soybean, 3-9 kg with one more addition of some cow manure
Produced 353,050 summer seed
Survival rate, 84.91%

In the entire process, the total amount of soybean used was 91 kg; green grass, 885 kg; and cow manure, 1644.5 kg. The pH was 7.6 to 8.4, and the average water temperature was 27.5°C.

**Growth rates of various fishes** — Black roach attained an increase rate of 29.24% before June 4; after that date, it dropped to 0.99%. In grass carp, the daily rate was 40.45% before June 31, after that, 10.49%; after June 7, it further dropped to 0.31%. In white and striped bigheads, the daily increase rate was 35.65 and 33.52%, respectively, during the early period (May 23 to June 3). After June 3, the rates dropped to 23.51 and 18.16%, respectively.

### Combination Use of Organic and Inorganic Fertilizers

This method is still in the experimental stage. The experiment conducted by Chekiang Freshwater Fisheries Institute in 1957 is briefly described below.

The experimental pond is 0.123 ha; the water depth, 1.8 m. The pond was drained April 22 to 24 and treated with 150 kg of lime on April 25. It was fertilized on May 20 with 106 kg of mixed manure (two parts green grass, four parts pig manure, one part each sheep manure and human night soil, and lime 5%), 64 kg of sheep manure, 5 kg ammonium sulfate, and 6 kg calcium phosphate. On May 22, 116,784 fry were introduced. They were fertilized once daily with the above fertilizers. In 18 days, 101,246 summer fry were produced, giving a survival rate of 86.65%.

During the entire rearing period, the following amounts of fertilizers were used: mixed fertilizer, 1433 kg; calcium phosphate, 35.42 kg; ammonium sulfate, 20.21 kg. In addition, 0.5 kg of fish meal and 5 kg of bean cake were added. Other conditions of the experiment were pH 8.4; DO maintained at 5.3 ppm; BOD, 11.11 ppm; phosphates, 0.03 to 1.7 mg/ℓ; nitrates, 0.0336 to 0.25 mg/ℓ; and nitrites, 0.005 to 0.0371 mg/ℓ. Zooplankton counts were 20,000 to 588,000/ℓ. Experiments in which only inorganic fertilizers were used yielded poor results.

### Utilization of Sewage Water

This method not only saves fertilizers, thereby lowering the cost, but also helps to solve the problem of sewage disposal in the city. In 1959, the Fisheries Research Institute of Academia Sinica used sewage for fish rearing. The results of this program are described below. The pond which used organic fertilizers is given as control.

Both experimental and control ponds were 200 $m^2$ in surface and 0.8 m deep and were treated with lime 10 days prior to fish introduction. The main chemical ingredients of sewage are as follows (in mg/ℓ):

pH, 7.2—7.4
BOD, 122—247
Inorganic salts, 6.30—17.86
Total phosphorus, 15—21.4
Ammonium *N*, 15.3—55.0
Nitrite *N*, trace
Nitrate *N*, 0.072—0.179
Organic *N*, 10.0—16.0

The sewage water was added into the pond every morning at the rate of 3 tons/100 m of water surface. The density of fry added was based on 100,000 fry per 0.805 ha. The quality of fry was rather low, with roach accounting for 58%.

In 10 days (June 30 to July 10), 48 tons of sewage was added; 1300 kg of organic fertilizers was used in the control pond.

Survival rate in the control pond was 40.72%; in the experimental pond, 43.42%. There were also fewer instances of parasitism in the experimental pond than in the control pond.

## PROBLEMS OF FRY REARING

### Density of Fry Planting

Because the density of fry controls the quantity of zooplankton on which fry depend as food, growth and survival are closely related to density. Today the density of fry planting varies tremendously in various parts of the country, varying from several hundred thousand to more than one million per 0.6 ha. Experience has shown that the survival rate is inversely proportional to density: the higher the density, the lower the survival rate. In one Chekiang experiment in 1957 with fry of four domestic carps, the results were as follows: fry density of 319,000/0.5 ha resulted in 40.83% survival; density of 280,000 yielded 68.38% survival; density of 180,000 fry per 0.5 ha resulted in survival rate of 86.98%. Also, in the pond with the greatest density, fish were not uniform in size and were much smaller. A density experiment conducted by the Fisheries Research Institute in 1959 on carp fry showed similar results.

From the results obtained at various places, it seems that the most suitable density is around 990,000 fry to each 0.405 ha of pond. If the species reared is *Cirrhina molitorella,* the density can be doubled, for these fish feed on detritus and, from a very early stage, are slow growing. The density can also be varied according to species composition, fish condition, and water quality. When there is a large proportion of bigheads, the density can be higher; when there is a large proportion of grass carp and black roach, the density should be lower. Density of early fry can be higher than that of late fry. Density in an old pond can be higher than that in a new pond.

### The Question of Separate or Mixed Rearing

To have a clear understanding of this question, it is essential to know the food habits of the various species of fry. Our preliminary observations reveal that grass carp and black roach, when 7 to 9 mm long, feed on rotifers, copepods, and nauplii; when 10 to 14.5 mm long, they feed on larger Daphnia, copepods, and rotifers; when 15 to 18 mm long, they feed on large Daphnia; and when 18 to 23 mm long, they feed on large Daphnia and some benthic animals. The bigheads feed on rotifers and nauplii when they are 7 to 9 mm long; on rotifers and small Daphnia when 10 to 11.5 mm long; on larger Daphnia and a few rotifers and nauplii when 13 to 17 mm long; and on phytoplankton in addition to rotifers, Daphnia, and copepods. When these fish reach 30 mm long, they feed less and less on zooplankton and more and more on phytoplankton.

These four cyprinids, during their early stages, all feed on zooplankton. However, when they grow to about 15 mm long, their food habits begin to differ. Also, after a week or so of rearing, the quantity of zooplankton is greatly reduced, but phytoplankton rises rapidly in number. Because of the difference in feeding intensity among the various species, the growth rate of the bigheads soon surpasses that of grass carp and black roach. For instance, in a 1955 experiment by the Fisheries Research Institute of Academia Sinica it was found that, after 23 days of rearing, the total lengths of the four summer fry were as follows: grass carp, 2.42 to 2.84 cm; black roach, 2.09 to

2.53 cm; white bighead, 3.55 to 4.15 cm; striped bighead, 2.99 to 3.36 cm. In length, the former two are one fifth to one fourth shorter. In weight, the difference is far greater; the bigheads are 2 to 4 times heavier.

From the above it can be concluded that before 15 mm in length, since the food habits of all species are the same, it makes no difference whether the different species are reared separately or mixed. However, after 15 mm, mixed rearing is not good for grass carp and black roach, and therefore, they should be reared separately.

Each fry rearing method used in one place has its merits and its drawbacks. By adopting the good points of each method and avoiding the disadvantageous practices, we suggest the following combined procedures:

1. Pond cleaning — Stop all leakages; remove weeds along the shore. Treat the pond with lime 10 to 15 days prior to fry planting. Eliminate pest fish and other injurious animals.
2. Screen out pest species among fry stock.
3. Fertilize the pond once 3 to 5 days before planting in order to cultivate planktonic organisms.
4. The day before introduction of fry, seine the pond to get rid of aquatic insects, frog eggs, tadpoles, etc. If necessary, treat the pond with 1% 6-6-6.
5. Plant 100,000 to 150,000 fry to each 0.405 ha of pond.
6. When fry reach 15 mm in length, grass carp and black roach should be separated out from bigheads and reared separately.
7. Fertilize the pond daily. In later stages, supplement with soybean cake and milk.
8. Add water intermittently every few days. Each time add 3 to 5 in. of water to increase DO.
9. Remove all injurious animals promptly. Examine for the presence of parasites every 4 to 5 days; if found, control immediately.
10. Harden the fry before transferring.

## SUMMARY AND CONCLUSIONS

The supply of fish fry and fingerlings is of primary concern to the fish culturing industry. In order to increase the supply, it is necessary to improve rearing techniques.

The first of many parts to this program is to improve pond condition. It is recommended that rearing ponds be drained and cleaned once a year. The cleaning process involves treatment with various chemical substances and simple mud removal (dredging). This reduces the incidence of disease, eliminates pest species and other harmful organisms, and facilitates repair of the pond structure. Many different chemicals can be used, although lime and bleaching powder seem to be the most effective.

The second area to be considered is that of fry nutrition. Different species of fry have different food preferences. For this reason, they can be found in different areas of the pond. For each species of fish, slightly different pond and fertilizer conditions are considered optimal. The basic fertilizers used are grass, cow manure, soybean milk, and peanut cakes. Recent experiments using human sewage have proven very successful, with exciting prospects for the future.

A third consideration is the selection of a suitable rearing pond. Characteristics of a good pond are that it has an ample water source, is of appropriate size and depth, has a circular shape and a level bottom, is nonleaking, and receives plenty of sunlight.

Another part of the rearing process is the preparation of the fry for transportation, otherwise known as hardening. The purpose of hardening is to get the fish to excrete as much of their mucus and feces as possible before being put in the transporting container, so these harmful substances will not build up to toxic levels in the container.

# REFERENCES

1. **Anon.**, Rearing of Chinese Carp fry for stocking fattening ponds, *Fish. Bull. Singapore,* 2, 5, 1949.
2. **Atkinson, C. E.**, Peoples Republic of China, in *World Fish Farming,* Brown, E. E., Ed., AVI Publishing, Westport, Conn., 1977, 321.
3. **Chen, T. P.**, Polyculture of Chinese Carps in Freshwater Ponds, Chinese-American Joint Commission on Rural Reconstruction, 1973.

Chapter 22

# THE REARING OF SUMMER FRY*

## INTRODUCTION

Fry reach about 3 cm in length 3 weeks after hatching. If left in the original pond, they will grow rather slowly because of crowded conditions and the lack of food. The summer fry, therefore, must be transferred to new ponds for rearing. This is because summer fry are small and cannot effectively escape enemy attack. They will suffer large mortality if transferred directly into large ponds. Additionally, the density of summer fry is normally about 20,000 to 50,000/ha, a density far too great for rearing fish to eatable size.[1,2]

## FERTILIZATION OF SUMMER FRY PONDS

In Kwangtung, the pond is fertilized with green grass and cow manure before fish are introduced. This is good because when the fish enter the pond, zooplankton are already blooming, thereby providing ample food for the fish. Fishermen in Kiangsu and Chekiang wait until after fish are introduced, then they add soybean milk to feed the fish.

## TRANSFERRING OUT OF THE POND

### Selection of Healthy, Uniform-Sized Summer Fry

The advantage of selecting strong fish for transferring is obvious; weaker fish cannot stand the strain of shipping, and they are less able to withstand adverse conditions in a new pond. Uniform size is also important since small fish will not be able to compete with the large ones for food. The small fish can be selected out and reared in a special pond.

Standards for determining the quality of summer fry are tabulated below:

| Good quality | Poor quality |
|---|---|
| Examination in the pond | |
| Swimming actively; schooled | Sluggish; scattered |
| Respond immediately upon being frightened | Slow response to being frightened |
| When surfacing, wander in middle of pond | When surfacing, wander along shore |
| Remain in deep water in daytime | Remain near surface in daytime |
| First to fight for food | Last to fight for food |
| Examination in a container | |
| Active | Sluggish |
| Uniform size | Irregular in size |
| Body highly iridescent | Body lacks luster |
| Head relatively small compared with body | Head large compared with body |
| Swim under surface; like to swim against current | Swim on surface; cannot make headway against current |

* From *Culture of Freshwater Fish in China,* Science Publishers, Peking, 1966, translated by Koo, T. Y., Dept. of Fisheries, U.S.D.I. Solomons, Md., and Walker, C., U.S.D.I., Washington, D.C.

**Rearing Density and Ratio of Species in Mixed Rearing**

**Determination of rearing density** — This is an important measure in fish culturing work. Generally, 30,000 to 45,000 summer fry can be reared in each hectare of pond. The following formula can be used to calculate the proper density of summer fry to be reared in a pond:

$$\text{Density} = \frac{\text{pond area} \times \text{production per unit area (estimated)}}{(\text{estimated weight per spring fry} - \text{weight per summer fry}) \times \text{estimated survival rate,\%}}$$

The estimation of production rate is based on past performance, plus planned improvement of the present year. Survival rate is estimated on the same basis.

In actual practice, because the summer fry weigh only 0.1 to 0.5 g, the weight is regarded as 0 to simplify the calculation. Summer fry at time of planting measure 2.54 cm in length; when grown into spring fry, grass carp should measure 12 cm and white bighead, 8 cm. Survival rate in grass carp is 80%; in white bighead, 95%.

The principle in mixed rearing is in selecting those species that have different food habits and that live in different strata of the water column. Unlike rearing adult fish where seven or eight species can be reared together in the same pond, no more than two or three species should be mixed in the same pond for summer fry. This is because, even though the summer fry feed differently among different species, they still eat the same food to a certain extent. Rearing too many species in the same pond will cause serious food competition to develop. It will also be difficult for the fish to attain uniform size at spring fry stage. On the other hand, rearing a single species is not economical. If the grass carp or black roach is reared separately, plankton growth will be too thick and become harmful to the growth of the fish. If the white bighead or striped bighead is reared separately, then a small amount of benthic animals will be utilized. Mixed rearing will increase the utilization rate of the pond and increase the production of fish.

In mixed rearing, it is important to have the proper proportion among species, or the growth of some fish will be affected. In extreme cases, an entire species may be eliminated.

Under natural conditions, both species of bighead feed on plankton — the white bighead mainly on phytoplankton and the striped bighead mainly on zooplankton. In rearing adults, the general ratio is four or five white bighead to one striped bighead. In rearing summer fry, it is even more important to follow strictly the proper ratio since the major source of food is artificial food. The white bighead are much more active and are stronger feeders than the striped bighead. If equal numbers of the two species of fish are reared in the same pond, the white bighead can outweigh the striped bighead 2:1 by harvest time.

In mixed rearing, therefore, the white bighead and striped bighead are usually not used together. If they must be mixed, then the number of striped bighead should not exceed 25% of the number of white bighead. For example, if 10,000 white bighead are planted, no more than 2500 striped bighead should be mixed in the group. In this way, even if the striped bighead fail to compete for artificial food, they can still rely on natural food for normal growth. A reverse ratio of these two species will not be appropriate, for the faster growth rate of the white bighead will give the striped bighead too much competition for food.

A mixed rearing experiment conducted by the Chekiang Fishery Experimental Station in 1956 resulted in a preliminary conclusion that the mixing ratio between white and striped bigheads should be 8:2, or at most, 7:3.

The relation between black roach and grass carp is similar to that between the two bigheads. The grass carp is a more active fish than is the black roach. If in rearing black roach 20% of grass carp are added, then grass carp will grow rapidly but black roach will grow poorly. On the other hand, in a grass carp pond, the addition of a small number of black roach will not cause any harm. This is because black roach can utilize some of the benthic animals that are not utilized by grass carp.

In actual production work, any one of the four carps (grass carp, black roach, common carp, or *Cirrhina molitorella* can be combined with either of the bigheads in mixed rearing. The grass carp or the black roach is also reared as a single species.

In raising grass carp and black roach in the same pond, black roach should be planted 7 to 15 days before grass carp so that they can get a head start in growth and, thereby, become more capable of competing for food.

Two points that should be taken into consideration are (1) the number of fry which can be reared in a unit area can be increased as the total surface area of the pond increases, and (2) the proportion of the number of fish for the various species in mixed rearing can be altered according to the conditions of the pond.

## REARING AND MANAGEMENT

Even though the density of summer seeds is usually smaller than that of fry, the young fish require a lot of food for rapid growth. Therefore, artificial food is necessary to satisfy this need. Artificial foods that are commonly used are as follows:

1. *Wolffia arrhiza,* a favorable food of grass carp, black roach, etc. of 1 to 10 cm in length
2. *Lemna minor,* main food of grass carp of sizes above 3 cm in length
3. *Spirodela polyrhiza,* main food of grass carp above 8 cm in length
4. *Vallisneria spiralis,* main food of grass carp above 10 cm long
5. *Hydrilla verticillata,* main food of grass carp above 12 cm long

Other foods for the various species of carps are bean cake, peanut cake, wine residue, rice bran, silkworm pupae, wheat bran, etc. Snails, mussels, clams, etc., when broken into small pieces, can be used to feed the carps, but usually are not used for grass carp because of the high cost. Various species of water plants and young, tender graminae can be used to feed grass carp that are above 8 cm in length. Additionally, grass and animal manure can be used to grow plankton, which is food for summer fry.

## FEEDING OF SUMMER FRY

Feeding success depends much upon feeding technique; it is also essential to good production. Long experience by fishermen has resulted in the principles of standardized time and location.

Summer fry of grass carp are usually reared together with the bigheads. Therefore, when they first enter the pond, grass carp can get a part of the bean cake that is fed the fish. Starting with the second day, *Wolffia* must be fed daily at the rate of 15 to 30 kg per 10,000 fish. After 20 days or so when grass carp grow to 2 in. in length, the feed can be changed to *L. minor*. For each 10,000 fish, the daily amount of *Lemna* needed is 80 kg. After 20 days of feeding with *Lemna,* grass carp may attain a length of 8 cm. At that time, the feed can be changed to aquatic plants.

It must be stressed that in daily feeding, plants must be fed first so that grass carp will be well fed by the time bean cakes are placed into the pond for the bigheads. Otherwise, the bigheads cannot compete with grass carp for food.

**Feeding of Black Roach**

Before the planting of summer fry of black roach, a large amount of Daphnia should be cultured through fertilization. Therefore, in the first 2 to 3 days after planting, no feed is necessary. Then, place 2 to 4 kg of bean cake on the feeding round (hard bottom devoid of mud in water about 1 m deep) to attract black roach to feed. If the fish are seen to feed actively, then bean cake can be fed twice a day: morning and noon, each time 10 to 15 kg. Also, a daily addition of 100 kg of *Wolffia* is a part of the feeding program. When black roach grow to 7.5 cm in length, change the diet to ground dry bean cake at a daily dose of 50 to 100 kg per 10,000 fish, fed in the morning and noon.

When black roach grow to 8 cm, cracked snails can be added to the diet at a rate of 30 kg per 10,000 fish at first, then increased to 70 to 90 kg/day. When the fish grow to above 12 cm long, small snails that are sieved out can be added to the diet.

**Feeding of White Bighead**

Under natural conditions, the summer fry of white bighead feed mainly upon phytoplankton. Before planting, therefore, the pond must be fertilized to cultivate plankton growth. However, natural food alone is not enough; it must be supplemented by artificial foods. At first, 1 to 2 kg of soybean milk residue can be added to each 10,000 fish per day. Later, wine residue (or mixed with bean cake) can be added until mid-October. The food can be placed on a feeding platform 0.5 m under the water surface.

**Feeding of Striped Bighead**

Basically, the feeding of striped bighead is the same as for white bighead, except the amount of food should be doubled.

The optimum temperature for the carps is between 25 and 35°C. When water temperature is within this range, the need for food is at a maximum. However, the amount of food must be decreased when the climate changes rapidly, either from cold to hot or from hot to cold, or when atmospheric pressure is abnormally low, and prior to and after thunderstorms.

Fish diseases must be carefully observed. If disease is present, feeding should be reduced or stopped altogether. Fish are more liable to get sick when the weather becomes abnormal. That is why feeding should be reduced during abrupt changes in weather conditions.

**Fertilization of Ponds**

In the rearing of summer fry, one can depend on fertilization alone to provide for the growth of plankton as fish food or on feeding with artificial food. The former method can be applied to light-density rearing and the latter to high-density rearing. In some cases, both methods can be applied as a combination.

**Daily Management**

**Inspect the pond once every morning** — Observe the water color and fish activity so that any abnormal conditions can be remedied immediately. Noon and evening inspections can be taken at the time of feeding and cleaning.

**Prevent fish escape** — Before the onset of rainy season, the pond and draining canal should be kept in good shape. After a heavy rain, inspect the outlet so that no fish can escape.

**Keep the pond clean** — Routinely rid the pond of weeds and decaying matter along the banks. The feeding platform should be cleaned once every 2 to 3 days and sterilized with bleaching powder twice a month. Around 3:00 p.m. each day, examine the feeding platform to see if all the food has been taken, in order to determine the amount of food needed the next day. Since large amounts of food are usually added to the pond of summer fry, water tends to become too rich. Therefore fresh water should be added from time to time to improve water quality. Generally, during the entire rearing procedure, four or five additions of water will be sufficient.

**Prevent predation by water birds** — This is especially important in ponds that are in the neighborhood of a large lake. This can be done by installing strawmen along the shores to scare the birds away or by shooting the birds with shotguns.

## CONCENTRATION AND OVER-WINTERING

After 4 to 6 months of rearing, when the weather turns cold and the water temperature drops to around 10°C, fish eat little. At this time, they should be fished out of the pond, segregated by different sizes, and placed in new, deeper ponds for over-wintering. In the majority of places in China, over-wintering of summer fry offers no problem. It is only in the Northeast, where the winter is very severe with a long ice period, that there are some peculiar problems. The latter case will be treated in a special section.

### The Purposes of Concentration

One purpose for concentration in a deeper pond is for easier management. Second, the summer seeds have already grown to 2.5 to 6 in. in length as spring seeds; thus, they are too large to be reared in the original pond. Also, since they are now sorted out according to size, they can be easily transplanted into separate adult ponds later. A third reason the original pond is emptied is to clean it for use next year. Finally, the production plan for the year can be accurately gauged.

### The Time of Concentration

Three factors must be considered in determining the best time for combining summer seeds into a common pond for over-wintering: (1) water temperature — the most suitable is around 10°C; (2) weather conditions — clear, warm day; (3) feeding — 6 to 7 days after feeding has stopped.

### Method of Concentration

**Catching** — Since concentrating is done in the winter, fish are not active; consequently, they are easy to catch, and no hardening is necessary. After two or three operations, nearly all the fish can be caught. In the case of carp and bream, a draining method can be used to catch all the fish.

**Sorting of species and sizes** — Freshly caught fish secrete a large amount of mucus. They should be rinsed in fresh water once or twice before sorting into species and different sizes by sieving. Unsorted ones can be emptied into a pan and sorted by hand.

**Counting** — Take a sample of 25 fish from each lot and measure their lengths and weight.

### Fish Quality Determination

Good quality fish are essential to good result in over-wintering and in the production into adult fish. Good quality depends upon the fat content in the body. Fish that contain little fat have to convert body protein into heat energy to maintain metabolism in the winter. This will result in emaciation and sometimes death.

### Over-Wintering

For over-wintering, ponds should be 1 to 2 ha in size and 2 m in depth. They should face south and have little mud on the bottom. Mud contains organic matter which would consume oxygen and produce toxic substance. In areas south of the Yangtze River, no special equipment is required. In areas north of Yangtze, however, some shelter should be constructed to ward off strong northeast winds. Density of fish should be appropriate, but there are not enough data to give any definite figure. Fishermen in Linghu place 50,000 to 60,000 3-in. spring fry in each 0.05-ha pond. It is important to keep the pond fertile. In most Chinese fish culture areas, winter water temperature is usually between 4 and 6°C, or higher on some days, and fish still feed to some extent during this period. Finally, care must be taken to prevent the lack of oxygen after ice forms and to prevent predation by water birds.

### Over-Wintering in Northeastern Areas

In the Northeast, the freezing period is very long — from mid-November to mid-March in Liaoning Province and from late October to mid-April in Heilungkiang Province. The ice may be more than 1 m thick. Such icy conditions cause difficult problems in fish over-wintering. Often there are mortalities in the spring: 21.3% in Liaoning in 1956, and out of 52 ponds, 21 had mortality in Heilungkiang in 1957.

There are generally three types of over-wintering ponds: (1) natural ponds, which account for the largest number at the present; (2) artificial ponds, which are relatively few; and (3) temperature-controlled ponds, which are very few in number and basically experimental in nature.

### The Effect of Physico-Chemical Conditions of the Water

During the entire over-wintering period, the physico-chemical changes of pond water are quite pronounced. Water volume decreases; water temperature drops; dissolved oxygen (DO) continually decreases due to respiratory function of fish and other aquatic animals; concentration of carbon dioxide ($CO_2$) and hydrogen sulfide ($H_2S$) increases; pH changes from slightly alkaline to slightly acidic. All these changes exert some effects on the fish.

In most ponds, DO is 11 to 13 mg/ℓ in early winter. Following the freezing, DO rapidly depletes. In poorer ponds, serious oxygen depletion develops in mid-December. In better quality ponds, there is barely enough oxygen to maintain the lives of carp and white bighead and to carry them through ice breakup.

On the other hand, $CO_2$ increases during the winter due to the decomposition of organic matter and the respiration of aquatic organisms. In poorer ponds, $CO_2$ may reach 174 mg/ℓ by late March. Under oxygen-poor conditions, $H_2S$ is produced from the bottom and rapidly depletes oxygen. Even after December when the oxidation process is slow, but because of the continuous production of $H_2S$, DO becomes badly depleted which makes it difficult for fish to live.

In most ponds, the pH changes from slightly alkaline to slightly acid, and the water changes from colorless and odorless to yellowish-brown and with a rotten egg smell.

Oxygen depletion starts at the bottom and spreads to the upper layers. This is due to respiration and decomposition of organic matters. Therefore, the distribution of oxygen in a pond during the winter is greater in upper layers and less in lower layers, more in shallow areas and less in deep areas. The movements of fish follow the oxygen condition correspondingly. During the start of freeze-up, DO is relatively ample in the entire water column, and fish are congregated primarily on the bottom. There is no upward movement. Following the decrease in oxygen in the bottom layers, fish and other aquatic organisms are forced to move into upper layers. Finally, when oxygen further depletes and it becomes difficult for the fish to breath, fish become scattered in their activities.

Bottom mud plays an important role in depleting the oxygen. For instance, in a fish pond which has been used for fish culture for 3 years, bottom mud used up 56% of oxygen. Experiments show that ponds with a strong sulfide smell consume 70% more oxygen than ponds with a sand bottom.

The ability to tolerate low oxygen varies according to species. *Percottus glehni* Dybowski can survive under extremely low oxygen conditions. Next in tolerance ability are goldfish, Ophiocephalus, and some small forms. Carp, grass carp, and roach are still next. Experiments have shown that below 2 to 5°C, some carp die when oxygen drops to 1 mg/ℓ, while others can survive to 0.2 mg/ℓ.

Under natural conditions, fish become paralyzed when water temperature drops to 1 to 0°C. Experiments conducted by the Fisheries Research Institute of Heilungkiang prove that when water temperature gradually drops to 7°C, carp can still take in some food; when temperature drops below 4°C, carp enter a hibernation state. Only the white bighead, which is a phytoplankton feeder, can still feed to a certain extent and maintain its body weight. General observations reveal that in the two winter months, if water temperature stays above 0.8°C, there is no adverse effect on over-wintering fish. However, if water temperatures drops to 0.5°C, grass carp, white bighead, and mirror carp will die.

**Mortality Factors for Over-Wintering**

Experience from various fishery stations shows that several factors are responsible for over-wintering mortality in fishes: (1) there is too much mud on the bottom; (2) the soil is too poor to retain water; (3) there is seepage of marsh water into the pond prior to freeze-up; (4) in running water ponds, the water temperature is too low; (5) fish are too weak or sick or have been subjected to injury during transportation; and (6) the density of fish seed is too high and DO is too low.

There are some basic conditions which must be met. One is a suitable water depth, which should be no less than 1.5 m in still ponds and no less than 1.0 m in running water ponds. Second, it is necessary to have a good bottom that will retain water. Sandy bottom is preferred; heavy, muddy bottoms cannot be used until improved upon. A third condition is that suitable physico-chemical water conditions be maintained. DO should be above 10 mg/ℓ before freezing it; it should be maintained at or around 5 mg/ℓ during the over-wintering period. As far as carp is concerned, DO as low as 1 mg/ℓ must be regarded as dangerous. The pH should be maintained at 7.2 to 8.4 and calcium carbonate ($CaCO_3$) above 150 mg/ℓ. The $CO_2$ content should not be more than 50 mg/ℓ and $H_2S$ not more than 1 mg/ℓ. During the over-wintering period, water temperature must be maintained above 1 to 2°C if carp and goldfish are reared or above 2 to 4°C if grass carp and white bighead are reared. Finally, the pond should be easy to drain or refill, and transportation should be convenient so that fish fry can be moved in and out rapidly.

In constructing a pond for over-wintering purposes, it is wise to make long-term observations before using. In selecting a natural pond for over-wintering purposes, some people follow the practice that, if there has been no winter kill, then the pond is suitable for use. This, however, is not always a safe measure, since different species have different tolerances. To be on the safe side, it is a good idea to use a weak species as an indicator. For instance, if pike survived, then the pond can be used for carp.

Usually no natural pond will suit all the conditions; therefore, it is necessary to improve on the conditions by human means. Regardless of the species of fish, all seeds should be reared to a large size and healthy condition before being transferred to over-wintering ponds. Grass carp, roach, white and striped bigheads, and carp should be no less than 20 g in body weight; only goldfish can be smaller. Smaller and thinner fry should be transferred to better conditioned ponds for over-wintering. In order to raise healthy fish, feeding during the last feeding period of the season is very important. After August, feed fish with high fat content food so that fish can accumulate fat in their body.

Care must be taken to avoid injury to the fish during catching and subsequent transporting. Before being placed in the new pond, fish must be inspected to detect if any diseases are present. The density of fish to be planted will depend upon the amount of DO available, fish size, species, etc. Generally speaking, in running water ponds water is completely refreshed every 10 to 30 days. Good management demands the following work:

1. Before introducing fish into the pond, repair the dams, clean the bottom, eliminate pest fish, and if necessary, sterilize the pond.
2. Test the gas content in the water, mainly oxygen, $CO_2$, and $H_2S$. Measure the water temperature and observe water color. Determine the water level, including water depth and thickness of ice, once a week. Watch for paralysis or asphyxiation of fish, and inspect dams, water gates, etc.
3. When DO drops to 3 mg/ℓ, oxygen should be supplemented; when it drops to 2 mg/ℓ, immediate measures must be taken to oxygenate the water.

In small ponds, when ice gets to be more than ½ ft thick, some hay or manure can be placed on top of the ice to reduce further freezing and to increase the water supply. Medium-sized ponds require a different treatment. Holes can be opened to supply air. As much as 1 to 5% of the total surface area can be opened as holes to be effective. Once holes are opened, water should be constantly stirred to prevent refreezing. Holes would be ineffective in large-sized ponds. Instead, water wheels or windmills can be used to increase oxygen content in water.

## THE FUTURE OF UTILIZATION OF AQUATIC AREAS

Following the rapid development of fresh-water fish-culturing ventures, much of the heretofore unutilized water area is being used for fish rearing. Consequently, the demand of fish fry becomes greater and greater, and the available rearing ponds are insufficient. At present, to enlarge the supply of fry, carp are bred artificially, and other cyprinids are caught with improved technique. Further, more species are being cultured.

The question of insufficient pond acreage can be solved by utilizing rice fields, artificial ponds, and swamps for rearing summer fry, according to the National Conference on Freshwater Culture in 1958.

In recent years, artificial ponds and swamps have been used to rear fish seeds, and some success has been obtained. If weeds are cleared away and if the density of fish is kept low (8000 to 10,000 per hectare), plus careful feeding and fertilization, there is no reason why high-quality fish fry cannot be produced.

## SURVIVAL RATE AND QUALITY OF FRY

The shortage in supply of fish fry is an unavoidable difficult problem during a rapid development of fresh-water fish culture. The problem becomes doubly acute due to the rapid increase in rearing surface and the extension of high-density rearing. The principal guidelines in solving this problem lie in the following: rear whatever is available; collect fry whenever fry are present; catch whatever is catchable; hatch whatever is hatchable; and raise survival during transportation and feeding. Let us consider, for the time being, only the means of increasing survival rate and insuring high quality of fish fry during the feeding procedure.

First, different methods of feeding the fry and seeds must be adopted at different localities. Even within the same locality, feeding methods are sometimes different. It is essential that the best method be used, considering the local environmental and economic conditions.

Generally speaking, during the stage of feeding fry, it is best to use mixed fertilizer. Since the fertilizers are easy to obtain, are cheap, do not cause deterioration of water quality, promote plankton growth, and have rather few parasites, they produce healthy and uniform-sized summer seeds.

In raising the summer fry, we should, as much as possible, use artificial feed, when and if artificial foods are plentiful. Where artificial foods are scanty, and where rearing areas are vast, the method of fertilization should be used. Artificial feeding can be used only as a supplementary means. In such instances, rearing density should be kept low.

Careful management work is also necessary in order to obtain good results. The ponds must be well cleaned. Before planting summer seeds, ponds must be fertilized, so that when fish enter into the ponds, food is already ample. The summer seeds should be uniform in size. The density should be appropriate. If artificial food is used, 30 to 45 fish per hectare is recommended; if organic and inorganic fertilizers are used, the density should be reduced to 15 to 30/ha. It is important to keep the food supply ample and/or to keep the density low, to prevent fish surfacing, diseases, and escape, and to add water from time to time. If the above conditions are met, we believe that fish quality will be good and survival rate will be high.

## SUMMARY AND CONCLUSIONS

The rearing of summer fry is an important link in the fish-culturing process. Summer fry must be transferred to a new pond, because the original pond has been outgrown. The new ponds are usually fertilized before the fry are transplanted, so there will be adequate numbers of zooplankton for food.

Different mixtures of fish species can utilize the resources of each pond the fullest, because different species feed at different levels. Care should be taken to select fish that are compatible. Appropriate foods and feeding techniques should be selected for the fish species being raised.

Daily management of the ponds is essential for maximal yields. Steps in the daily routine are to check for abnormal water conditions and fish behavior, to check the pond banks and dam for damages, and to keep the pond clear of excess food, dead fish, and other debris.

Perhaps the most important part of the fry-rearing process is over-wintering. The amount of dissolved oxygen, the water temperature, fish density, and pond construction all affect fry survival. Careful consideration of all of these factors will help yield maximum numbers of fry.

# REFERENCES

1. **Gangstad, E. O., Raynes, J. J., and Burress, R. M.,** Herbivorous Fish for Aquatic Plant Control, Aquatic Plant Control Program Tech. Rep. 4, U.S. Army Engineers Experimental Station, Vicksburg, Miss., 1973.
2. **Stanley, J. G.,** Production of Mono Sex White Amur for Aquatic Plant Control, Contract Rep. A-76-1, U.S. Army Engineers Waterways Experimental Station, Vicksburg, Miss., 1976.

Chapter 23

# NATURAL AND CULTIVATED FOOD OF FRESH-WATER FISHES*

## INTRODUCTION

The growth rate and fatness of fish depend upon the amount of food available and its nutritive value. Different fishes feed on different foods. For instance, grass carp feed on various aquatic and terrestrial plants, black roach feed on snails, white bighead feed on phytoplankton, striped bighead feed on zooplankton, and carp and goldfish are omnivorous. Generally, fish can be grouped into carnivores, herbivores, and omnivores. However, during young stages, most fish feed on plankton.

In fish culture, food is provided either by direct feeding or by enriching the water so as to promote planktonic growth. For grass carp and black Chinese roach, the first method is used; for the bigheads, the second method is used.

In the past, some researchers maintained that because cyprinids lack fiber-digesting enzymes, their ability to utilize plant material is limited or nonexistent. However, other researchers demonstrated that these fish can obtain such enzymes through animal food ingested and that they depended upon these exogenic enzymes to digest fibrous food.[1]

## PRINCIPAL COMPOSITION OF NATURAL FOOD

Natural fish food comprises the following kinds:

1. Phytoplankton — diatoms and many unicellular algae
2. Zooplankton — including protozoa, rotifers, daphnids, copepods, and some molluscan larvae
3. Benthic animals — including aquatic oligochaetes, nematodes, molluscs, aquatic insect larvae, etc.
4. Aquatic plants — mainly vascular plants, *Potamogeton, Vallisneria,* etc.
5. Dead animal bodies and attached bacteria — organic matter can serve as food for benthic species of fish

Species composition of fish food varies according to the kind of water body.

### Natural Food in Rivers

In the upper section of the river where the current is rapid and dissolved oxygen (DO) is rich, there are many kinds of algae, such as *Cladophora, Gomphonema, Navicula, Cocconeis,* etc. Also present are large numbers of insect larvae, such as Plecoptera, Ephemerida, Odonata, etc. Planktonic organisms are rare.

In the lower section of the river where the current is slow, plankton greatly increases in species and quantity. Phytoplankton include *Cyclotella, Melosira, Nitzshia, Achnanthes, Cymbella,* etc. and a variety of algae, such as *Chlamydomonas, Scenedesmus, Ankistrodesmus, Chroococcus, Chroomonas, Eurorina, Pandorina,* and *Volvox.*

Zooplankton also greatly increases in the lower section of the river. The more commonly seen species are *Difflugia, Arcella, Acanthocystis, Coleps, Tintinnidium, Strobilidium, Strombidium, Epistylis, Halteria, Stentor, Asplanchna, Polyarthra, Synchaeta, Conochilus, Cephalodella, Diurella, Lecane, Truchicerca, Colurella,*

* From *Culture of Freshwater Fish in China,* Science Publishers, Peking, 1966, translated by Koo, T. Y., Dept. of Fisheries, U.S.D.I., Solomons, Md., and Walker, C., U.S.D.I., Washington, D.C.

*Brachionus, Schizocerca, Pedalia, Daphnia, Bosmina, Alona, Sida, Chydorua, Bosminopsis, Ceriodaphnia, Simocephalus,* and many others.

Also increased in number are benthic animals. The principal ones include chironomids and oligochaetes. Others such as Ephemerida, Plecoptera, Trichoptera, Hemiptera, Coleoptera are also abundant. Mollusca include both gastropods and plecopods.

Aquatic vascular plants grow only in shallower parts of the river where current flow is slow.

**Natural Food in Lakes**

Fish food is much richer in the lakes than in rivers. There are numerous kinds of phytoplankton. Most important are many species of diatoms, green algae, and blue algae. Zooplankton species are also abundant. Many species of protozoans, rotifers, water fleas, and copepods are present.

Benthic animals consist mainly of chironomid larvae, annelids, and snails. In lakes along the lower Yangtze River, plecopods are sometimes abundant; more than ten species may be present.

There are many kinds of aquatic vascular plants. The principal ones are *Phragmitis communis, Zizania latifolia, Trapa incisa* var. *guadricaudata, Limnanthemum nymphoides, Hydrocharis asiaticus, Myriophyllum spicatum, Najas,* etc. These plants generally occur in shallow water along lake shores. In some lakes, *Chara* is very thick.

**Natural Food in Reservoirs**

Species composition of fish food in reservoirs is intermediate between that in rivers and that in lakes. Most medium- to large-sized reservoirs are built from rivers by damming. Therefore, at the beginning, the fauna in the new reservoir is the same as in the original river. However, as the ecology gradually changes to that similar to lakes, lake-type organisms also begin to appear.

**Natural Food in Ponds**

Food species in ponds are those that prefer organic nutrients. Abundantly present are phytoplankton (algae) and protozoans, many species of rotifers, and daphnia and copepods. Because physico-chemical conditions are easily altered by adding food and fertilizers, faunal make-up also changes accordingly.

## FLUCTUATIONS IN NATURAL FOOD

The number and quantity of natural food species fluctuate seasonally and vary according to the type of water body. In rivers which freeze in winter, seasonal changes of water conditions are very distinct. During the spring thaw, volume of flow increases rapidly and the temperature rises. This accelerates growth of organisms. The thaw of snow and ice brings forth a large quantity of organic material and minerals, thus enriching the river water tremendously. As a consequence, plankton and benthic animals flourish. Diatom population often reaches peak during the spring. In Heilung River, for instance, river ice breaks up in mid-May, and by mid-June phytoplankton reaches its first peak. Each liter of water contains 1,970,000 cells, of which diatoms account for 50.9%; benthic animals also reach peak at the end of June.

With the continuous outflow of water from the river, the level of nutrient material gradually decreases; plankton is also reduced, although benthic animals and aquatic plants continue to increase in quantity.

During the time the water level rises in the fall, plankton is again on the rise, and benthic animals and aquatic plants reach maximum development. Productibility lowers when freezing sets in toward the end of autumn.

In nonfreezing rivers, the main source of water supply is rain. Scouring after raining will bring mud into the river and increase turbidity, thereby reducing light penetration and retarding the growth of organisms.

In large rivers where water level changes are great, turbidity is high, and river is deep, higher aquatic plants are relatively rare. In small rivers, plants are plentiful, especially during summer to autumn.

In lakes where the water level is relatively stable and flow rather small, the growth and development of food species are quite normal and regular. Most of the lakes in China are relatively shallow and rich in nutrients. Plankton, benthic animals, and higher aquatic plants are all abundant.

Phytoplankton is at its peak during the spring, levels off in summer and autumn, and is at the minimum in the winter.

Seasonal fluctuations of zooplankton in the lakes are not as sharply defined as phytoplankton. Generally, they are more abundant in spring and summer and less abundant in autumn and winter. Protozoans increase greatly in quantity during the spring, decrease in summer, increase again in the fall, and drop off in winter. However, protozoans are seldom used directly as fish food. More important food species are rotifers, daphnia, and copepods. Rotifers are abundant in spring, decrease in summer, increase in fall, and diminish again in winter. Daphnia are more abundant from spring to autumn, but less abundant in winter. Copepods are present during all seasons, but are more abundant in fall and winter.

Chironomids constitute a major part of benthic fauna. They appear in great quantities in spring, summer, and fall. Molluscs are also abundant in lakes. Benthic animals are generally distributed along lake shores where the water is shallow, but they are distributed all over the lake in shallow lakes. In a Hupeh lake, there is an average of 36 lb of benthic animals per acre of lake surface.

When aquatic plants are thick in a lake, benthic animals are usually plentiful, but planktonic organisms are less abundant. Under these conditions, the lake is very clear and rich in dissolved oxygen. On the other hand, during a phytoplankton bloom, light penetration is limited, which retards the growth of bottom plants.

The status of zooplankton and phytoplankton in a reservoir is similar to that of an open lake. On one hand, the reservoir has the characteristics of a lake; on the other, it also has spring and autumn flood, which is similar to a river. Due to continuing sedimentation of sand and mud, both quantity and number of species are relatively limited.

## FACTORS AFFECTING NATURAL FOOD

The growth and development of organisms in a lake may be affected by physical, chemical, and biological factors. Physical factors vary seasonally and according to geographical areas. Chemical and biological factors are affected by physical factors.

Major physical factors are temperature, light, and current velocity. Major chemical factors are dissolved inorganic salts, organic matter, gas, and pH. Biological factors refer to interrelationships and competition among species.

The interrelationships among species have already been discussed. The following is a brief discussion of temperature, light, dissolved inorganic salts, and organic matter and their effects on aquatic organisms.

### Temperature

To the growth and development of life, temperature is an important factor. Under very low temperatures, vital activity diminishes and may even cause death. Gradually increasing temperatures usually increase metabolic rate. In general, for each increase of 10°C, the metabolic rate is doubled or tripled. Too high temperature, however, may also cause death.

Some species are adapted to a wide amplitude of temperature fluctuations; they are eurythermal. Others are stenothermal, i.e., those that can live within a narrow range of temperature regime. The eurythermal species can grow within wider latitudes. They can grow in tropics and can be found also in arctics. The stenothermal forms are more or less regional.

Because water is a slower heat conductor and is also slower in heat absorption and release, water temperature does not change as fast as air temperature. Many weak organisms can survive a daily temperature differential. Also, because the bodies of organisms are immersed in the water, there can be no evaporation due to heat as in land organisms.

Temperature can not only directly affect the living processes but can also influence the dissociation and chemical processes of matter, thereby indirectly affecting the absorption and utilization of materials by organisms.

### Light

Light is indispensable to plant growth. As to animals, its chief effect is related to movement and reproductive activity.

Green plants require ample light energy for photosynthesis. Light is the main source of productivity of both land and water organisms. The productivity of a water body is often determined by the condition of aquatic plants.

The main source of light is solar radiation. The duration and intensity of light is dependent upon latitude, altitude, and season.

In the northern hemisphere, the higher the latitude, the longer the sun shines in summer and the shorter in winter, and the air and water temperatures are directly proportional to the amount and intensity of sunlight during the season.

Water can absorb and radiate light, and this ability is increased when suspended or dissolved material is increased. At the same time, light penetration is reduced. Therefore, deep water is not conducive to plant growth. This is why a deeper lake is not necessarily more productive.

Different plants demand different light intensity and quality. Some plants prefer strong light; others like weak light. Some prefer light of long wavelength (red and orange); others like short wavelength (green, blue). Therefore, it is not always at low latitude or in summertime that plants grow at their best. For instance, planktonic green and blue algae like strong light and high temperature, but diatoms prefer weak light and low temperature. The distribution of aquatic plants in water is affected and limited by sunlight.

Light has an important bearing on the movement of zooplankton and benthic animals. For example, the diurnal migration of zooplankton is closely related to light. Many animals have distinctive positive or negative phototropism. Light is, therefore, an important factor that affects the distribution of aquatic animals.

### Dissolved Inorganic Salts and Organic Matter

All natural water bodies contain inorganic salts. Inland waters can be classified into four categories according to their salt content: (1) freshwater, containing 0.5‰ salt; (2) saline water, containing 0.5 to 2.5‰ salt; (3) supersaline containing 25 to 100‰ salt; and (4) hypersaline, containing 100 to 300‰ salt.

The amount of salt content in water affects osmotic pressure of living cells. Many organisms can tolerate a wide fluctuation of salinity; they can survive in both fresh water and salt water. These are called euryhaline forms. Others are adapted to a narrow amplitude of salinity fluctuation. These are called stenhaline forms. Species composition is quite different in fresh-water and salt-water lakes. Even different fresh waters contain varying amounts of calcium, which can also affect species distribution.

In fresh water, the more important salts that affect the abundance of organisms are nitrates and phosphates. Nitrogen is an indispensable ingredient for building plant protein, and phosphorus provides raw material for nuclear plasma. The amount of these salts in water is often the important factor that controls the growth of plants. Some workers classify lakes into eutrophic or oligotrophic according to contents of nitrates and phosphates. A lake is eutrophic if nitrogen is over 0.2 $m\ell/\ell$ and phosphorus is over 0.02 $m\ell/\ell$; lower than this level, the lake is considered oligotrophic. Other criteria must also be considered, for when large amounts of nitrogen and phosphorus are used in active metabolism, lake contents of these elements are drastically reduced even though the lake is eutrophic.

In addition to the above salts and elements, potassium, magnesium, iron, and manganese also have an important effect on the distribution of plants.

The organic matter that is contained in water is often an important factor affecting the breeding of zooplankton, especially protozoans, rotifers, and daphnia. Therefore, the amount of organic matter can be used as an index of productivity of the water body. Of course, if the content of organic matter becomes too high, it can deplete oxygen and seriously deteriorate the water quality (for example, by producing large amounts of hydrogen sulfide and marsh gas) and depress growth or cause mass mortality among organisms.

Phytoplankton releases a large amount of oxygen as a result of photosynthesis. This helps to split up organic matter into simpler materials or salts that can be readily available to organisms. Therefore, the introduction of sewage into fish ponds, if it can be controlled, can often fertilize the water, which increases fish food and promotes fish growth.

## CULTIVATED FOOD FOR FRESH-WATER FISHES

### Vegetative Food

In the past, it was only for rearing grass carp that fresh aquatic or terrestrial plants were used as food in China. Today five kinds of vegetative foods are used.

#### *Farm Products*

The main item used is soybean. It is used in the form of soybean milk to feed fry. Milk is prepared by grinding well-soaked soybean with water and by filtering out solids. The function of soybean is probably twofold. One is as direct food for fish, and the other is to promote plankton growth, which in turn is used by fish as food.

In South China, to rear fish fry, hay is placed in fish pond to culture plankton. It is as effective as soybean milk.

In recent years, the by-product of soybean oil has been used to produce milk. First, soybean is pressed to produce oil. Then the soybean cake is soaked and ground to produce milk. It is as good as soybean milk itself. In one rearing station, only 13 lb of soybean cake was used to rear 10,000 fry (about 1 in. long) for 20 days.

#### *By-Product and Waste Material of Food Industry*

**Bean cake** — To feed adult fish, cut the bean cake into small pieces. To feed fry, smash the cake into paste form.

**Bean residue** — This is residue that results from soybean milk preparation. It can be used to feed both fry and adult fish.

**Grain residue from wine and liquor preparation** — This is a fine food for both adult and fry. It may be made of corn, wheat, barley, or sorghum.

**Rice bran and wheat husk** — These are used mainly to feed fish fry, but can be used also to feed adult fish. To feed fry, they must be mixed with water and ground into fine particles.

**Residue from soy sauce preparation** — This residue contains salt and, therefore, must be mixed with other food in feeding fish. In mixing, it should not exceed 1%.

**Cabbage seed cake** — After oil is pressed out of cabbage seeds, the residue forms a cake. Use in the same manner as the bean cake. It is very, very good for both fry and adult fish.

**Cotton seed cake** — This is the residue of cotton seeds after oil is pressed out. It is as good as bean cake, but costs only half as much. In the U.S.S.R., it is used as the main food item for carp. In Chekiang, it is mixed with other food for feeding purpose.

*Aquatic and Terrestrial Plants*

Plants can be used either directly as food (as for grass carp) or indirectly as fertilizer which cultures plankton. Many kinds of plants can be used to feed grass carp. Among aquatic plants are *Wolffia arrhiza, Lemna minor, Spirodela polyrhiza, Vallisneria spiralis, Potamogeton malainus, P. maackianus, P. crispus, Hydrilla verticillata, Najas minor,* etc. Among terrestrial plants are *Echinochloa crus-galli, Setaria viridis, Pennisetum alopecuroides, Eleusine indica,* stems, leaves, and seeds of legumes, *Lactuca sativa* and *Taraxacum officinale* of Compositae, and waste leaves of a variety of green vegetables.

In recent years, many other plants have been introduced as fish food, such as *Alternanthera philoxeroides, Ipomoea reptans, Eichhornia crassipes, Pistia stratiotes,* etc. Even leaves of many kinds of trees are used, for example, pine, willow, tung, etc.

The culturing methods of some of the aquatic plants are decribed below:

***Wolffia arrhiza*** — In selecting a *Wolffia* pond, it is important to remember that *Wolffia* prefers still water. Therefore, the pond should be on the lee side. It should be small, not over 1/6 acre, and not more than 4 to 5 ft deep.

*Wolffia* does not have to be reseeded every year in a pond in which *Wolffia* has been grown for 2 years. Otherwise, 15 lb or so of seed *Wolffia* should be planted for every 1/6 acre. The growth rate of *Wolffia* is very closely related to temperature. Under 27°C, growth is most rapid; temperature lower than 20°C or higher than 35°C will retard growth.

Cleaning of a *Wolffia* pond should be started in late March. It is first drained, then treated with lime to kill off wild fish. Finally, it should be fertilized with lamb manure. Manure of other animals or humans can be used but not as effectively.

*Wolffia* can be harvested either in the morning or in the afternoon with a net. Each time, the take should not exceed 60% of the plant present in the pond. On the average, about 3000 lb can be taken out of an acre of pond every 2 days. It can be harvested from April to September in eastern China.

It is best to fertilize the *Wolffia* pond every day. To each acre, about 400 lb of night soil can be added. During the summer, *Wolffia* should be sprayed with water from the surface to avoid dryness. During the entire growing season, wild plants and frog eggs should be removed constantly.

*Wolffia* can be transplanted easily by transporting it in a closed container. The best time for transplantation is between March and May when air temperature is relatively low.

Iron rust, which is caused by certain algae, can be controlled in *Wolffia* by adding diluted night soil. Another disease, caused by certain green algae, can be controlled by adding 2 ppm of copper sulfate.

***Spirodela polyrhiza*** — The culture method of this plant is similar to that of *Wolffia.*

***Alternanthera philoxeroides, Pistia stratiotes, Eichornia crassipes*** — These are also cultured for the purpose of feeding fish. The method of culturing is similar to that of culturing *Wolffia.*

**By-product of farm product** — Many agricultural by-products can also be utilized to feed fish. The commonly used ones are green rice hay, squash leaves, potato stems and leaves, and peanut plants and shells. Other farm products that are used as raw materials in mixing with other food stuffs to feed fish include corn stems and leaves; stems, leaves, and shells of broad beans; cabbage; rice husks; etc.

*Animal and Human Manure*

Aside from being used as a fertilizer, manure can also be used as fish food. Manure of man, cattle, and pig can be mixed with other material and used directly to feed fish. Fecal deposits of silkworm have been used by Chekiang fishermen to feed grass carp and black Chinese roach.

*Food Substitutes*

Decaying aquatic plants and bottom mud are often used to mix with other food to feed fish.

**The Nutritional Ingredients of Plant Food**

In addition to containing protein, carbohydrate, fat, inorganic salts, and vitamins, plant food also has a large amount of cellulose, which cannot be digested very well by cyprinids. In evaluating the food quality, therefore, both the protein and cellulose contents should be measured. The higher the protein and the lower the cellulose contents, the better the quality of the food.

The chemical composition of a food material is not the only criterion that should be used for an evaluation of the quality of food. The digestibility of the various ingredients must also be considered. For instance, corn contains 6.2% of protein or 6.1% digestible protein; soy sauce residue contains 13.4% protein but only 9.8% digestible protein. Therefore, in comparing the nutritional value of corn and soy sauce residue, the ratio is not 6.2:13.4, but rather 6.1:9.8.

Vitamins and inorganic salts are also important ingredients that should be weighed in evaluating the food.

**Processing of Fish Food**

The principal purposes of processing plants for fish food are (1) to facilitate ingestion by the fish, such as grinding the bean cake into fine pieces; (2) to upgrade the nutritional value and utilization rate, such as softening and fermentation; (3) to facilitate preservation and storage; and (4) to improve the flavor. There are generally three means of processing: mechanical, chemical, and biological.

**Mechanical method** — This includes soaking, cutting, grinding, mixing, etc.

**Chemical method** — This includes heating and soaking in hot water to remove some of the soluble salts and alkali, or treating with acids or bases to hydrolyze cellulose into sugars.

**Biological method** — The principal method is via fermentation by microorganisms. This is the simplest, most effective, and cheapest method. The bacteria convert cellulose and protein into sugars and amino acids on the one hand, and supply a large amount of digestive enzymes and vitamins on the other.

There are various methods of biological processing. The two methods which are currently adopted are as follows:

**Green storage and fermentation** — Various kinds of nontoxic plants can be used. They should be cut before blooming. The green foliage and stem are cut into 4- to 8-cm pieces, compacted into a container, and tightly sealed to prevent air and water from getting in. Fermentation process can be completed in 25 days for grass and 7 to 10 days for vegetables at a temperature of 20°C.

**Rice hay powder** — Add special enzymes to pulverized hay. The hay will be converted into powder form. This is an excellent feed for hogs. Its application to fish has yet to be tried, however.

## ANIMAL FOOD

**Snails, *Viviparus (Idiopoma) quadratus* (Bensen)** — These snails abound in rivers, creeks, and lakes. They are the favorite food of black Chinese roach. They can be collected with either a square net and a bamboo scraper (a bamboo cylinder cut in half) or a push net.

**Yellow clam, *Corbicula fluminea* (Muller)** — Yellow clam is also called river clam or pea clam. It grows in river or lake mud and is usually collected together with snails. It is very good food for the fry of black Chinese roach.

**Oval mussel, *Anodonta arcaeformis* (Heude)** — This is a widely distributed species and is usually collected together with the snails. There are also a variety of other species of mussels which can be collected, but they occur in smaller quantity. When small, the mussels can be fed directly to fish; when large, they must be smashed before using.

**Mud snail, *Bullacta exarata* (Philippi)** — This is a marine species, quite abundant during low tides. As a fish food, it is not inferior to silkworm pupae.

**Fly maggots** — Fly maggots are a very good fish food. They contain 55% protein, 28% fat, 7% carbohydrate, and 10% ash. Many fly species provide larvae which are suitable for fish food. These include *Sarcophaga* spp., *Musca domestica, Lucilia sericata, Calliphora grahami,* etc.

**Grasshopper** — Collecting grasshopper nymphs as a fish food serves a dual purpose in eliminating an agricultural pest on the one hand and providing a new source of fish food on the other. The only problem is that it cannot be relied upon as a regular food.

**Earthworm** — There are over 120 species of earthworm in China. They are easy to cultivate and are good fish food.

**Daphnia** — Daphnia is often used to feed aquarium fishes. It is cultured in large scale for commercial production.

**Silkworm pupae** — After silk is spun from cocoons, the pupae can be kept for fish food. The nutritional value of the pupae is very high, containing 17.1% protein and 9.2% fat.

**Fish meal** — This is produced from scrap fish and fishery by-product and is used as a mixing agent for fish food.

## MIXED FISH FOOD

### Utilization of Mixed Fish Food in China

Only since 1958 have people in China begun to experiment with mixed food to feed fish. The advantage of mixed food is that it can convert raw material that is normally not attractive to fish into a food that is liked by fish. For instance, *Alternanthera philoxeroides* is not normally taken by fish as food, but when it is cut into small pieces, cooked and smashed, and then mixed with rice bran, salt, and soda powder to form a paste, it becomes a very good food which is readily consumed by all four domestic cyprinid species.

The Fisheries Research Institute in Chekiang Province has made available several mixing formulas which yield very good results:

| | | |
|---|---|---|
| Formula 1: | Rice bran | 50% |
| | *Wolffi aarrhiza* | 20% |
| | Silkworm pupae | 20% |
| | Snails | 10% |
| Formula 2: | Pine leaves | 5% |
| | Bean cake | 10% |
| | Silkworm pupae | 20% |
| | Rice bran | 30% |
| | *Wolffia arrhiza* | 35% |
| | Snails | 1% |
| Formula 3: | Cotton seed cake | 40% |
| | Cabbage stem | 20% |
| | Hay powder | 20% |
| | Bran | 20% |
| | Snails | 20% |

## Principles of Food Mixing

In mixing fish food, the nutritional requirements of fish must be met. A good fish food must contain protein, carbohydrate, fat, inorganic salts, and vitamins.

Protein is the principal ingredient of animal body. The growth and reproduction of animals require protein. It cannot be substituted with other food materials. A fish fed only with carbohydrate cannot maintain a protein balance; in fact, it will lose more protein than a fish that is starved.

Food requirements of fish vary according to fishes growth condition and the season in which the food is fed. Generally, young fish require more protein than adults. In summer, protein content in fish food should be increased; in winter, carbohydrate and fat should be increased.

The protein content of a food is often expressed in a protein ratio, which is the ratio between protein content and the combined content of carbohydrate and fat. For example, plankton contains 61% protein, 11% fat, and 5% carbohydrate, which gives a protein ratio of 1:0.25.

Some fishery workers in the U.S.S.R. maintain that for juvenile carp, the protein ratio should be 1:0.4 to 1:0.5; for 1-year-old carp, 1:0.7 to 1:1; for 2-year-old carp, 1:2 to 1:3; and for 3-year-old carp, 1:5. Of course, the proper protein ratio must be worked out for the particular area. In the U.S.S.R., by improving the nutrition of fish food, they are able to grow carp to 600 to 800 g in 1 year, compared with the normal growth of 30 to 40 g.

Inorganic salts in fish food also play an important role. They can enhance the utilization of carbohydrate by fish. Especially important are calcium, phosphorus, sodium, and nitrogen contents. Adding powdered bones and table salt to fish food helps to increase appetite and promotes growth of muscular and bony tissues.

Vitamins play a special part in the normal growth of fish. For instance, vitamin B can help digestion and metabolism; vitamin C can also enhance metabolic activity; and vitamin D is important to bone formation and the utilization of potassium and phosphorus.

## Merits of Mixed Food

By utilizing mixed food, a vast source of raw material is opened up. By using various ingredients, it is possible to complement each author and prepare an ideal nutritious food. Finally, mass production becomes possible by mechanization.

## REFERENCE

1. **Lin, S. Y.,** Notes on fish fry industry of China, *Proc. Indo-Pac. Fish Comm.*, 1, 65, 1949.

# *Appendix*

# APPENDIX A
# CONVERSION FACTORS FOR U.S. AND METRIC UNITS

| To convert Column 1 into Column 2, multiply by | Column 1 | Column 2 | To convert Column 2 into Column 1, multiply by |
|---|---|---|---|
| | **Length** | | |
| 0.621 | kilometer, km | mile, mi | 1.609 |
| 1.094 | meter, m | yard, yd | 0.914 |
| 0.394 | centimeter, cm | inch, in | 2.54 |
| | **Area** | | |
| 0.386 | kilometer², km² | mile², mi² | 2.590 |
| 2.469 | kilometer², km² | acre, acre | 0.00405 |
| 2.471 | hectare, ha | acre, acre | 0.405 |
| | **Volume** | | |
| 0.00973 | meter³, m³ | acre-inch | 102.8 |
| 3.532 | hectoliter, hℓ | cubic foot, ft³ | 0.2832 |
| 2.838 | hectoliter, hℓ | bushel, bu | 0.352 |
| 0.0284 | liter | bushel, bu | 35.24 |
| 1.057 | liter | quart (liquid), qt | 0.946 |
| | **Mass** | | |
| 1.102 | ton (metric) | ton (English) | 0.9072 |
| 2.205 | quintal, q | hundredweight, cwt (short) | 0.454 |
| 2.205 | kilogram, kg | pound, lb | 0.454 |
| 0.035 | gram, g | ounce (avdp), oz | 28.35 |
| | **Pressure** | | |
| 14.50 | bar | lb/inch², psi | 0.06895 |
| 0.9869 | bar | atmosphere, atm[a] | 1.013 |
| 0.9678 | kg (weight)/cm² | atmosphere, atm[a] | 1.033 |
| 14.22 | kg (weight)/cm² | lb/inch², psi | 0.07031 |
| 14.70 | atmosphere, atm[a] | lb/inch², psi | 0.06805 |
| | **Yield or Rate** | | |
| 0.446 | ton (metric)/hectare | ton (English)/acre | 2.240 |
| 0.892 | kg/ha | lb/acre | 1.12 |
| 0.892 | quintal/hectare | hundredweight/acre | 1.12 |
| 1.15 | hectoliter/ha, hℓ/ha | bu/acre | 0.87 |
| | **Temperature** | | |
| (9/5°C) + 32 | Celsius | Fahrenheit | 5/9 (°F−32) |
| | −17.8°C | 0°F | |
| | 0°C | 32°F | |
| | 20°C | 68°F | |
| | 100°C | 212°F | |
| | **Water Measurement** | | |
| 8.108 | hectare-meters, ha-m | acre-feet | 0.1233 |
| 97.29 | hectare-meters, ha-m | acre-inches | 0.01028 |
| 0.08108 | hectare-centimeters, ha-cm | acre-feet | 12.33 |
| 0.973 | hectare-centimeters, ha-cm | acre-inches | 1.028 |
| 0.00973 | meters³, m³ | acre-inches | 102.8 |
| 0.981 | hectare-centimeters/hour, ha-cm/hour | feet³/sec | 1.0194 |

# APPENDIX A (continued)
# CONVERSION FACTORS FOR U.S. AND METRIC UNITS

| To convert Column 1 into Column 2, multiply by | Column 1 | Column 2 | To convert Column 2 into Column 1, multiply by |
|---|---|---|---|
| | **Water Measurement (continued)** | | |
| 440.3 | hectare-centimeters/hour ha-cm/hour | U.S. gallons/min | 0.00227 |
| 0.00981 | meters$^3$/hour, m$^3$/hour | feet$^3$/sec | 101.94 |
| 4.403 | meters$^3$/hour, m$^3$/hour | U.S. gallons/min | 0.227 |

[a] The size of an ''atmosphere'' may be specified in either metric or U.S. units.

# GLOSSARY OF BIOMEDICAL TERMS

The definitions and explanations in this glossary apply to words as they are used in context. Most words have more comprehensive meanings.

**Acid equivalent** — The theoretical yield of parent acid from an active ingredient; it is used instead of or in addition to the active ingredient for certain herbicides

**Active ingredient** — The chemical compound in a product that is responsible for the herbicidal effects

**Adsorbed** — Held so tightly that the herbicide is rendered inactive or only slowly effective; the principal adsorbing agents of the soil are its inorganic (clay) and organic (humus) colloids

**Algal blooms** — A readily visible concentrated growth or aggregation of algae in water; commonly accepted to be a concentration of cells greater than 500/mℓ

**Algicide** — A chemical that will kill algae

**Aliphatic materials** — Chemically, those that have an open-chain molecular structure; as herbicides, they are less toxic to plants than aromatic compounds

**Angiosperms** — Taxonomic unit of plants which includes all of the potentially flowering plants in which the ovules and seeds, when formed, form *inside* a closed carpel system collectively known as fruits when mature

**Angiostrongyliasis** — Disease of the central nervous system arising from a nematode infection; various species of snails and slugs are intermediate hosts

**Anionic surfactant** — One that has a negative charge and performs best in cold water and soft water; most wetting agents and detergents and some emulsifiers are anionic

**Annual** — A plant that completes its life cycle from seed in 1 year

**Antifouling paint** — Paint applied to marine objects that contains a toxic agent that is released slowly repelling or destroying the larval form of fouling organisms

**Aquatic macrophytes** — Aquatic plants which may be seen with the naked eye, e.g., aquatic mosses, ferns, and rooted plants

**Aquatic plants** — Plants which undergo growth and development with all or part of the plant body in water or 100% water-saturated soil

**Aquatic vascular plants** — Aquatic plants which have specialized conductive tissues

**Arboviruses** — A large group of viruses mostly associated with arthropods, some causing infection in man or other animals; the manifestations of the disease are commonly encephalitic

**Aromatic oils and solvents** — Chemically, those that have unsaturated molecular structure; as herbicides, they are generally more toxic to plants than aliphatic materials

**Arthropods** — Joint-footed animals, including *Crustacea, Arachnida,* and *Insecta*

**Barnacles** — Type of crustacean that attaches to objects immersed in water, develops a shell, and becomes sessile; a major form of marine fouling

**Biennial** — A plant that completes its life cycle in 2 years; the first year it produces leaves and stores food, and the second year it produces fruits and seeds

**Biomagnification** — The accumulation and concentration of a substance in an animal tissue that is related to its position in the food web

**Biomass** — The total amount (weight) of living material in a given habitat

**Botanical plant name** — A scientific name made up of the genus and species; sometimes the variety or subspecies is included; it is more reliable and more universal than common names

**Broadcast application** — Uniform distribution of an herbicide over an entire area

**Broadleaf plants** — Botanically, those classified as dicotyledons; morphologically, those that have broad, usually compound leaves

**Bryozoa** — Form of animal fouling

**Carcinogenic** — The characteristic of a compound that causes cancer in man or other animals

**Carrier** — The liquid or solid material added to a chemical compound to facilitate its application in the field

**Cation exchange** — The exchange of positive ions (H, Ca, Mg, Na, $NH_4$) from cations; soils are able to filter out salts in much the same way a water softener removes them; some soils have a larger capacity than others for doing this; such soils can adsorb or filter out and hold large amounts of an herbicide so that it is not immediately effective

**Ceraciacide** — A chemical that will destroy or disinfect the cercarial form of the trematode parasites that afflict man and other animals

**Chemical name** — One that indicates the chemical composition of the compound and also the structure of the molecule

**Chemotherapy** — Use of drugs in the treatment of disease

**Clonorchiasis** — Disease of the bile ducts brought about by a trematode infection; snails serve as an intermediate host

**Coacervation** — Basic method of encapsulation based upon phase separation of an agent from polymer

**Common plant name** — An English name in common use; a plant may be known by several different common names, and one common name may be used for different plants in different parts of the country; many common names are local

**Compatible pesticides** — Compounds or formulations that can be mixed and applied together without undesirably altering their separate effects

**Concentration** — The amount of active ingredient or acid equivalent in a given volume of liquid or in a given weight of dry material

**Contact herbicide** — One that kills primarily by contact with plant tissue rather than as a result of translocation

**Controlled-released molluscicides** — Materials designed to destroy snails through chemical intoxication; the toxic agent is incorporated in a polymeric binder and releases at a slow, continuous, and constant rate for an extended duration

**Crustaceans** — A class of aquatic arthropoda with a chitinous exoskeleton and jointed appendages

**Detergent** — A chemical (not soap) having the ability to remove soil or grime; household detergents can be used as surfactants in herbicide sprays

**Detritus** — Fragments of dead organic matter in water

**Diffusion-dissolution** — A technique of releasing an agent from binding polymer whereby the agent, being soluble in the polymer diffuses under solution pressure to depleting surfaces, and then passes into the external media through dissolution processes

**Diluent** — Any liquid or solid material that dilutes an active ingredient in the preparation of a formulation

**Dinoflagellates** — A plant-like flagellate of the subclass phytomastigophorea, some species of which produce a potent neurotoxin that may cause severe food intoxication following ingestion of parasitized shellfish

**Diversity** — Pertaining to the variety of species within a given association of organisms; areas of high diversity are characterized by a great variety of species; areas of low diversity contain few species although large numbers of individuals may be present

**Ecosystem** — The sum of the biological, chemical, geological, and physical elements within a defined area which interact with one another

**Elastomeric** — Pertains to the type of high molecular weight polymer having interchain forces sufficient to provide an elastic nature

**Emergent plants** — Aquatic macrophytes that are rooted in the sediments of a water body, but with vegetative parts which project above the water surface; for example, Saggitaria

**Emulsifiable concentrates** — Usually liquids in which the chemical is dissolved in one or more water-insoluble solvents such as oil or benzene to which an emulsifier is added

**Emulsifier** — A surface active material that facilitates the suspension of one liquid in another

**Emulsion** — The suspension of one liquid as minute globules in another liquid; for example, oil dispersed in water

**Encapsulation** — Technique of surrounding a toxic material with a polymeric membrane

**Encephalitis** — Inflammation of the encephalon; brain

**Epiphytic** — Growing on plants

**Eutrophic** — Water high in plant nutrients

**Eutrophication** — The process of enrichment with nutrients, leading to increased organic production and possibly algal blooms

**Exfoliation** — Release of a chemical agent through actual loss of a polymeric surface containing that agent

**Extruded** — A process in which a powdered carrier mixed with the herbicide is moistened until it becomes plastic and then is forced or expelled as rods; these rods are dried, ground, and screened to the required screen mesh for a granuar formulation; this process increases internal pore volume and surface area; the product is frequently referred to as AA in contrast to A materials, which have not been subjected to an extrusion process

**Family** — Collection of species and genera of assumed close relationship; usually plant families end with the letters -aceae or -ae; see **Taxon**

**Fascioliasis** — Trematode infection of cattle and sheep usually centered in the liver; amphibious snails are hosts for this parasite

**Fasciolopsis** — Trematode infection of the small intestine; various snail species serve as the intermediate host

**Fauna** — A collective term to include all kinds of animals in an area

**Flora** — A collective term to include all kinds of plants in an area

**Fouling** — Accumulation of plant and animal organisms on the surface of an immersed object

**Genus (plural: Genera)** — See **Taxon**

**Granular products** — Formulations in which the chemical is impregnated on or in vermiculite, attaclay, or other suitable carriers and then formed into granules or pellets

**Grass** — Botanically, any plant of the Gramineae family; grasses are characterized by narrow leaves with parallel veins, by leaves composed of blade, sheath, and ligule, by jointed stems and fibrous roots, and by inconspicuous flowers usually arranged in spikelets

**Growth regulator** — An organic substance effective in minute amounts for controlling or modifying plant processes

**Habitat** — The place where a plant or animal lives, which has all of the conditions necessary to support its life and reproduction

**Hard water** — Water that contains certain minerals, usually calcium and magnesium sulfates, chlorides, or carbonates, in solution in amounts that cause a curd or precipitate instead of a lather when soap is added; generally defined as containing 322 ppm in terms of calcium carbonate; very hard water may cause precipitates in some herbicidal sprays

**Herbaceous plant** — A vascular plant that does not develop woody tissue; it dies down each year

**Herbicide** — A chemical used for killing or interrupting the normal growth of plants

**Herbarium (plural: Herbaria)** — A permanent collection of identified, dried, preserved and mounted plants for research and plant identification reference purposes; ideally, specimens each have identification data including collector, date of collection, name and taxonomic authority, site of collection, together with names of associated species and ecological data

**Herbicide** — A chemical used for killing or interrupting the normal growth of plants

**Hydrophytes** — See **Aquatic plants**

**Ichthystoxism** — A name loosely given to poisoning from eating the flesh of certain fish, and also to the poisonous principle involved

**Intermediate species** — A species whose response to an herbicide is in between the response of a susceptible and a resistant species; it is severely injured or partially controlled by higher than moderate rates

**Invert emulsion** — One in which oil is the continuous phase and water is dispersed in it

**Ionic surfactant** — One that ionizes or dissociates in water

**Isomers** — Two or more substances having the same chemical composition but different properties

**$LD_{50}$** — Lethal dose for 50% of the animals tested, often stated in milligrams per kilogram of bodyweight

**Littoral** — The shallow zone of a water body where light penetrates to the bottom; the zone of rooted aquatic vegetation

**Low-volatile ester** — Chemically, an ester with a heavy molecular weight such as the butoxy-ethanol, iso-octyl, or propylene glycol butyl ether esters; low-volatile esters do not include the methyl, ethyl, propyl, iso-propyl, butyl, amyl, and pentyl esters; biologically, an ester that is less liable than the high-volatile esters to injure plants by vapor activity

**Malaria (Malarial fever)** — Generic name for a group of diseases associated with the presence in the blood of *Polimitus malarie* and characterized by enlargement of the spleen and liver, anemia, and the presence in the blood, free or within red corpuscles, of deleterious parasites

**Marine algae** — Single cell or colonial plants found in a salt water environment

**Marine protista** — The aquatic living species of a proposed third kingdom of living things including the lowest orders of the animal and vegetable kingdoms, the Protozoa and Protophyta

**Miracidia** — Larval form of trematode parasite that enters the snail

**Molluscs** — Any member of the Mollusca, a class of soft-bodied invertebrates possessing an external and an internal shell

**Nonionic surfactant** — One that is chemically inert

**Nonselective herbicide** — A chemical that is toxic to plants generally without regard to species.

**Oligotrophic** — Water low in plant nutrients

**Onchocercosis** — Infection with *Onchocerca* (esp. *O. volvulus*) marked by nodular swellings forming a fibrous cyst enveloping the coiled parasites; blindness frequently occurs in advanced cases

**Organic matter** — Plant or animal remains that are partially decomposed; see **Detritus**

**Paragonimiasis** — Trematode infection usually of the lungs; several snail species serve as hosts

**Parasite** — An organism living in or on another organism

**Parasitic disease** — Pathogenic condition arising from the presence of parasites

**Perennial** — A plant that lives more than 2 years

**pH** — The chemist's measure of acidity and alkalinity; it is a scale in which the figure 7 indicates neutral, figures below 7 indicate acidity, and figures above 7 indicate an alkaline reaction

**Photosynthesis** — The process by which carbohydrates are manufactured by the chlorophyll-bearing cell granules (chloroplasts) from carbon dioxide and water by exposure to the energy of sunlight

**ppm** — Parts per million = milligrams per liter (mg/ℓ)

**Product** — The herbicide as it is sold commercially; it contains not only the active ingredients but also various solvents, cosolvents, surfactants, carriers, and other adjuvants that are designated as inert ingredients

**Propagule(s)** — Viable seeds and living, vegetatively (nonsexually) formed plant parts which can originate new plants under the proper conditions and which are often disseminated by various environmental and/or biological buds, free of the source plant, which can originate one complete, new plant during the following growing season, as in some *Potamogeton* species

**Proprietary mixture** — One that is commercially available

**Protozoan** — A member of the phylum Protozoa, that phylum of the animal kingdom including all unicellular forms

**psi** — Pounds per square inch

**Rate** — The amount of an herbicide applied to a unit area, usually in terms of active ingredients

**Resistant species** — One that is difficult to kill; the use of the herbicide is not recommended

**Schistosomiasis** — A debilitating parasitic disease caused by a trematode and afflicting the liver and other organs of man; various species of snails function as the host

**Selective herbicide** — A chemical that is more toxic to some plant species than to others

**Slurry** — A watery mixture or suspension of an insoluble herbicide

**Species (plural: Species)** — A morphologically, genetically, and ecologically defined biological entity to which a binomial and authority is given, e.g., *Potamogeton filiformis* Pers., the slender-leaf *Potomogeton filiformis* Pers., the slender-leaf *Potomogcton*

**Sporocyst** — Cyst within snail-containing miracidia that undergoes a sexual reproduction into cercaria

**Spot treatment** — Application of an herbicide to individual plants or small clumps of plants

**Spray drift** — The movement of airborne spray particles from the intended areas of application

**Spreader-sticker** — A surfactant closely related to wetting agents that facilitates spreading and increases sticking of an herbicide on vegetation

**Standing crop** — The quantity of living organisms present in an environment at a selected point in time

**Stem-foliage application** — An application of an herbicide to both stems and leaves of a plant

**Submergent plants** — Aquatic macrophytes that grow or are adapted to grow beneath the surface of the water; for example, *Potamogeton*

**Succession** — The orderly process of compositional change in populations and communities

**Surfactant** — A material that improves the emulsifying, dispersing, spreading, wetting, and other surface-modifying properties of herbicide formulations

**Susceptible species** — One that can be killed with moderate rates of a herbicide

**Suspension** — A system consisting of very finely divided solid particles dispersed in a liquid

**Sustained-release** — Same as controlled release

**Taxon (plural: Taxa)** — A unit of plant classification of any chosen rank; e.g., Potamogetonaceae (the Potamogeton family), *Potamogeton* (everything included in this genus), *Potamogeton foliosus* Raf. (everything included in this species), or *Potamogeton foliosus* Raf. variety *macellus* Fern. (everything included in this variety of *Potamogeton foliosus* Raf.)

**Taxonomic authority** — Biologist who described the morphological, ecological, and genetic limits of any taxon or who more recently revised the taxonomic position of a taxon

**Total alkalinity** — Total of hydroxice ($OH^-$), carbonate ($CO_3^=$), and biocarbonate ($HCO_3^-$) ions in water; the capability of water to accept protons

**Translocated herbicide** — One that is moved within the plant from the point of entry

**Trematode** — Parasitic flatworm

**Tularemia** — A disease caused by *Francisella tularensis* and transmitted to man from rodents through the bite of a deer fly, *Chrysops discalis*, and causing a swelling of the lymph nodes

**Vapor drift** — The movement of herbicidal vapors from the area of application

**Viscosity of oil** — Expressed in time (seconds) required for 60 cc of heated oil to flow through a Saybolt Universal Viscosimeter

**Volatile** — A compound is volatile when it evaporates or vaporizes (changes from a liquid to a gas) at ordinary temperatures on exposure to the air

**Water hardness** — The total concentration of calcium, magnesium, and the metallic ions that have the capacity to precipitate soap (soft water = 50 mg/ℓ ; hard water = 150 mg/ℓ)

**Water-soluble powder** — A finely ground powder which will dissolve in water

**Weed eradication** — The complete elimination of all live plants, plant parts, and seeds of a weed infestation from an area

**Wetland plants** — Terrestrial plants which preferentially grow in 100% water-saturated soil; for example, Carex

**Wettable powder** — A finely ground powder plus a wetting agent to keep the particles suspended in, but not floating on, the water to which it is added

**Wetting agent** — A compound that, when added to a spray, causes the spray to contact plant surfaces more thoroughly

# *Index*

# INDEX

## A



## B

## C

## D



## E

## F



## G



## H

## I



## J



## L



## M

## N



## O



## P

## R



## S

## X



## Y



## Z